MUSIC: A MATHEMATICAL OFFERING

Since the time of the Ancient Greeks, much has been written about the relation between mathematics and music: from harmony and number theory to musical patterns and group theory. Benson provides a wealth of information here to enable the teacher, the student or the interested amateur to understand, at varying levels of technicality, the real interplay between these two ancient disciplines.

The story is long as well as broad and involves physics, biology, psychoacoustics, the history of science and digital technology as well as, of course, mathematics and music. Fundamental to it is how we actually hear sound, so the book starts with the structure of the human ear and its relationship with Fourier analysis. Combining this with the mathematics of musical instruments leads to the ideas of consonance and dissonance, and from there to an understanding of the development of scales and temperaments. Later chapters introduce some separate but related threads involving symmetry in music and the modern introduction of digital techniques to produce and analyze music and sound. This is a must-have book if you want to know about the music of the spheres or digital music, and many things in between.

DAVE BENSON is Sixth Century Professor of Pure Mathematics at the University of Aberdeen. He has held positions in Georgia, Oxford, and at Northwestern and Yale, and visiting positions at many places throughout the world. He is a keen amateur singer and has performed in many operas.

MUSIC: A MATHEMATICAL OFFERING

DAVE BENSON

CAMBRIDGE
UNIVERSITY PRESS

CAMBRIDGE UNIVERSITY PRESS
Cambridge, New York, Melbourne, Madrid, Cape Town, Singapore, São Paulo

Cambridge University Press
The Edinburgh Building, Cambridge CB2 2RU, UK

Published in the United States of America by Cambridge University Press, New York

www.cambridge.org
Information on this title: www.cambridge.org/9780521853873

First published 2007

Printed in the United Kingdom at the University Press, Cambridge

A catalogue record for this publication is available from the British Library

ISBN-13   978-0-521-85387-3 hardback
ISBN-10   0-521-85387-7 hardback

ISBN-13   978-0-521-61999-8 paperback
ISBN-10   0-521-61999-8 paperback

To Christine Natasha

Ode to an Old Fiddle from *The Musical World* (London, 1834), quoted in Nicolas Slonimsky's *Book of Musical Anecdotes*, Schirmer, 1998.

THE POOR FIDDLER'S ODE TO HIS OLD FIDDLE

Torn
Worn
Oppressed I mourn
B a d
S a d
Three-quarters mad
Money gone
Credit none
Duns at door
Half a score
Wife in lain
Twins again
Others ailing
Nurse a railing
Billy hooping
Betsy crouping
Besides poor Joe
With fester'd toe.
Come, then, my Fiddle,
Come, my time-worn friend,
With gay and brilliant sounds
Some sweet tho' transient solace lend,
Thy polished neck in close embrace
I clasp, whilst joy illumines my face.
When o'er thy strings I draw my bow,
My drooping spirit pants to rise;
A lively strain I touch—and, lo!
I seem to mount above the skies.
There on Fancy's wing I soar
Heedless of the duns at door;
Oblivious all, I feel my woes no more;
But skip o'er the strings,
As my old Fiddle sings,
"Cheerily oh! merrily go!
"PRESTO! good master,
"You very well know
"I will find Music,
"If you will find bow,
"From E, up in alto, to G, down below."
Fatigued, I pause to change the time
For some *Adagio*, solemn and sublime.
With graceful action moves the sinuous arm;
My heart, responsive to the soothing charm,
Throbs equably; whilst every health-corroding care
Lies prostrate, vanquished by the soft mellifluous air.
More and more plaintive grown, my eyes with tears o'erflow,
And Resignation mild soon smooths my wrinkled brow.
Reedy Hautboy may squeak, wailing Flauto may squall,
The Serpent may grunt, and the Trombone may bawl;
But, by Poll,* my old Fiddle's the prince of them all.
Could e'en Dryden return, thy praise to rehearse,
His Ode to Cecilia would seem rugged verse.
Now to thy case, in flannel warm to lie,
Till call'd again to pipe thy master's eye.
*Apollo.

Contents

Preface

This book has been a long time in the making. My interest in the connections between mathematics and music started in earnest in the early nineties, when I bought a second-hand synthesizer. This beast used a simple frequency modulation model to produce its sounds, and I was fascinated at how interesting and seemingly complex the results were. Trying to understand what was going on led me on a long journey through the nature of sound and music and its relations with mathematics, a journey that soon outgrew these origins.

Eventually, I had so much material that I decided it would be fun to try to teach a course on the subject. This ran twice as an undergraduate mathematics course in 2000 and 2001, and then again in 2003 as a Freshman Seminar. The responses of the students were interesting: each seemed to latch onto certain aspects of the subject and find others less interesting; but which parts were interesting varied radically from student to student.

With this in mind, I have tried to put together this book in such a way that different sections can be read more or less independently. Nevertheless, there *is* a thread of argument running through the book; it is described in the introduction. I strongly recommend the reader not to try to read this book sequentially, but at least to read the introduction first for orientation before dipping in.

The mathematical level of different parts of the book varies tremendously. So if you find some parts too taxing, don't despair. Just skip around a bit.

I've also tried to write the book in such a way that it can be used as the text for an undergraduate course. So there are exercises of varying difficulty, and outlines of answers in an appendix in the online version.

Cambridge University Press has kindly allowed me to keep a version of this book available for free online at www.maths.abdn.ac.uk/˜bensondj/html/maths-music.html. No version of the online book will ever be identical to the printed book. Some ephemeral information is contained in the online version that would be inappropriate for the printed version; and the quality of the images in the printed version is much higher than in the online version. Moreover, the online version is likely to continue to evolve, so that *references to it will always be unstable*.

Acknowledgements

I would like to thank Manuel Op de Coul for reading an early draft of these notes, making some very helpful comments on Chapters 5 and 6, and making me aware of some fascinating articles and recordings (see Appendix G). Thanks to Paul Erlich, Xavier Gracia and Herman Jaramillo for emailing me various corrections and other helpful comments. Thanks to Robert Rich for responding to my request for information about the scales he uses in his recordings (see Section 6.1 and Appendix G). Thanks to Heinz Bohlen for taking an interest in these notes and for numerous email discussions regarding the Bohlen–Pierce scale in Section 6.7. Thanks to an anonymous referee for carefully reading an early version of the manuscript and making many suggestions for improvement. Thanks to my students, who patiently listened to my attempts at explanation of this material, and who helped me to clean up the text by understanding and pointing out improvements, where it was comprehensible, and by not understanding where it was incomprehensible. Finally, thanks as always to David Tranah of Cambridge University Press for accommodating my wishes concerning the details of publication.

This document was typeset with AMS LATEX. The musical examples were typeset using MusicTEX, the graphs were made as encapsulated postscript (eps) files using MetaPost, and these and other pictures were included in the text using the graphicx package.

Introduction

What is it about intervals such as an octave and a perfect fifth that makes them more consonant than other intervals? Is this cultural, or inherent in the nature of things? Does it have to be this way, or is it imaginable that we could find a perfect octave dissonant and an octave plus a little bit consonant?

The answers to these questions are not obvious, and the literature on the subject is littered with misconceptions. One appealing and popular, but incorrect, explanation is due to Galileo Galilei, and has to do with periodicity. The argument goes that if we draw two sine waves an exact octave apart, one has exactly twice the frequency of the other, so their sum will still have a regularly repeating pattern

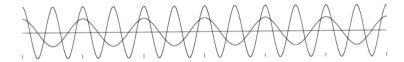

whereas a frequency ratio slightly different from this will have a constantly changing pattern, so that the ear is 'kept in perpetual torment'.

Unfortunately, it is easy to demonstrate that this explanation cannot be correct. For pure sine waves, the ear detects nothing special about a pair of signals exactly an octave apart, and a mistuned octave does not sound unpleasant. Interval recognition among trained musicians is a factor being deliberately ignored here. On the other hand, a pair of pure sine waves whose frequencies only differ slightly give rise to an unpleasant sound. Moreover, it is possible to synthesize musical sounding tones for which the exact octave sounds unpleasant, while an interval of slightly more than an octave sounds pleasant. This is done by stretching the spectrum from what would be produced by a natural instrument. These experiments are described in Chapter 4.

The origin of the consonance of the octave turns out to be the instruments we play. Stringed and wind instruments naturally produce a sound that consists of exact integer multiples of a fundamental frequency. If our instruments were different, our musical scale would no longer be appropriate. For example, in the Indonesian gamelan, the instruments are all percussive. Percussive instruments do not produce exact integer multiples of a fundamental, for reasons explained in Chapter 3. So the western scale is inappropriate, and indeed not used, for gamelan music.

We begin the first chapter with another fundamental question that needs sorting out before we can properly get as far as a discussion of consonance and dissonance. Namely, what's so special about sine waves anyway, that we consider them to be the 'pure' sound of a given frequency? Could we take some other periodically varying wave and define it to be the pure sound of this frequency?

The answer to this has to do with the way the human ear works. First, the mathematical property of a pure sine wave that's relevant is that it is the general solution to the second order differential equation for simple harmonic motion. Any object that is subject to a returning force proportional to its displacement from a given location vibrates as a sine wave. The frequency is determined by the constant of proportionality. The basilar membrane inside the cochlea in the ear is elastic, so any given point can be described by this second order differential equation, with a constant of proportionality that depends on the location along the membrane.

The result is that the ear acts as a harmonic analyzer. If an incoming sound can be represented as a sum of certain sine waves, then the corresponding points on the basilar membrane will vibrate, and that will be translated into a stimulus sent to the brain.

This focuses our attention on a second important question. To what extent can sound be broken down into sine waves? Or to put it another way, how is it that a string can vibrate with several different frequencies at once? The mathematical subject that answers this question is called Fourier analysis, and is the subject of Chapter 2. The version of the theory in which periodic sounds are decomposed as a sum of integer multiples of a given frequency is the theory of *Fourier series*. Decomposing more general, possibly nonperiodic sounds gives rise to a continuous frequency spectrum, and this leads to the more difficult theory of *Fourier integrals*. In order to accommodate discrete spectra into the theory of Fourier integrals, we need to talk about *distributions* rather than functions, so that the frequency spectrum of a sound is allowed to have a positive amount of energy concentrated at a single frequency.

Chapter 3 describes the mathematics associated with musical instruments. This is done in terms of the Fourier theory developed in Chapter 2, but it is really only necessary to have the vaguest of understanding of Fourier theory for this purpose. It is certainly not necessary to have worked through the whole of Chapter 2. For the discussion of drums and gongs, where the answer does not give integer multiples of a fundamental frequency, the discussion depends on the theory of Bessel functions, which is also developed in Chapter 2.

Chapter 4 is where the theory of consonance and dissonance is discussed. This is used as a preparation for the discussion of scales and temperaments in Chapters 5 and 6. The fundamental question here is: why does the modern western scale consist of twelve equally spaced notes to an octave? Where does the twelve come from? Has it always been this way? Are there other possibilities?

The emphasis in these chapters is on the relationship between rational numbers and musical intervals. We concentrate on the development of the standard Western scales, from the Pythagorean scale through just intonation, the meantone scale, and the irregular temperaments of the sixteenth to nineteenth centuries until finally we reach the modern equal tempered scale.

We also discuss a number of other scales such as the 31 tone equal temperament that gives a meantone scale with arbitrary modulation. There are even some scales not based on the octave, such as the Bohlen–Pierce scale based on odd harmonics only, and the scales of Wendy Carlos.

These discussions of scale lead us into the realm of continued fractions, which give good rational approximations to numbers such as $\log_2(3)$ and $\log_2(\sqrt[4]{5})$.

After our discussion of scales, we break off our main thread to consider a couple of other subjects where mathematics is involved in music. The first of these is computers and digital music. In Chapter 7 we discuss how to represent sound and music as a sequence of zeros and ones, and again we find that we are obliged to use Fourier theory to understand the result. So, for example, Nyquist's theorem tells us that a given sample rate can only represent sounds whose spectrum stops at half that frequency. We describe the closely related z-transform for representing digital sounds, and then use this to discuss signal processing, as a method both of manipulating sounds and of producing them.

This leads us into a discussion of digital synthesizers in Chapter 8, where we find that we are again confronted with the question of what it is that makes musical instruments sound the way they do. We discover that most interesting sounds do not have a static frequency spectrum, so we have to understand the evolution of spectrum with time. It turns out that for many sounds, the first small fraction of a second contains the critical clues for identifying the sound, while the steadier part of the sound is less important. We base our discussion around FM synthesis; although this is an old-fashioned way to synthesize sounds, it is simple enough to be able to understand a lot of the salient features before taking on more complex methods of synthesis.

In Chapter 9 we change the subject almost completely, and look into the role of symmetry in music. Our discussion here is at a fairly low level, and one could write many books on this subject alone. The area of mathematics concerned with symmetry is *group theory*, and we introduce the reader to some of the elementary ideas from group theory that can be applied to music.

I should close with a disclaimer. Music is not mathematics. While we're discussing mathematical aspects of music, we should not lose sight of the evocative power of music as a medium of expression for moods and emotions. About the numerous interesting questions this raises, mathematics has little to say.

Why do rhythms and melodies, which are composed of sound, resemble the feelings, while this is not the case for tastes, colors or smells? Can it be because they are motions, as actions are also motions? Energy itself belongs to feeling and creates feeling. But tastes and colours do not act in the same way.

(Aristotle, Prob. *xix. 29*)

1

Waves and harmonics

1.1 What is sound?

The medium for the transmission of music is sound. A proper understanding of
music entails at least an elementary understanding of the nature of sound and how
we perceive it.

Sound consists of vibrations of the air. To understand sound properly, we must
first have a good mental picture of what air looks like. Air is a gas, which means
that the atoms and molecules of the air are not in such close proximity to each other
as they are in a solid or a liquid. So why don't air molecules just fall down on the
ground? After all, Galileo's experiment at the leaning tower of Pisa tells us that
objects should fall to the ground with equal acceleration independently of their size
and mass.

The answer lies in the extremely rapid motion of these atoms and molecules.
The mean velocity of air molecules at room temperature under normal conditions
is around 450–500 meters per second (or somewhat over 1000 miles per hour),
which is considerably faster than an express train at full speed. We don't feel the
collisions with our skin, only because each air molecule is extremely light, but the
combined effect on our skin is the air pressure which prevents us from exploding!

The mean free path of an air molecule is 6×10^{-8} meters. This means that
on average, an air molecule travels this distance before colliding with another air
molecule. The collisions between air molecules are perfectly elastic, so this does
not slow them down.

We can now calculate how often a given air molecule is colliding. The collision
frequency is given by

$$\text{collision frequency} = \frac{\text{mean velocity}}{\text{mean free path}} \sim 10^{10} \text{ collisions per second.}$$

So now we have a very good mental picture of why the air molecules don't fall
down. They don't get very far down before being bounced back up again. The effect

of gravity is then observable just as a gradation of air pressure, so that if we go up to a high elevation, the air pressure is noticeably lower.

So air consists of a large number of molecules in close proximity, continually bouncing off each other to produce what is perceived as air pressure. When an object vibrates, it causes waves of increased and decreased pressure in the air. These waves are perceived by the ear as sound, in a manner to be investigated in the next section, but first we examine the nature of the waves themselves.

Sound travels through the air at about 340 meters per second (or 760 miles per hour). This does not mean that any particular molecule of air is moving in the direction of the wave at this speed (see above), but rather that the local distur-bance to the pressure propagates at this speed. This is similar to what is happening on the surface of the sea when a wave moves through it; no particular piece of water moves along with the wave, it is just that the disturbance in the surface is propagating.

There is one big difference between sound waves and water waves, though. In the case of the water waves, the local movements involved in the wave are up and down, which is at right angles to the direction of propagation of the wave. Such waves are called *transverse waves*. Electromagnetic waves are also transverse. In the case of sound, on the other hand, the motions involved in the wave are in the same direction as the propagation. Waves with this property are called *longitudinal waves*.

   Longitudinal waves

$\longrightarrow$ Direction of motion

Sound waves have four main attributes which affect the way they are perceived. The first is *amplitude*, which means the size of the vibration, and is perceived as loudness. The amplitude of a typical everyday sound is very minute in terms of physical displacement, usually only a small fraction of a millimeter. The second attribute is *pitch*, which should at first be thought of as corresponding to frequency of vibration. The third is *timbre*, which corresponds to the shape of the frequency spectrum of the sound (see Sections 1.7 and 2.15). The fourth is *duration*, which means the length of time for which the note sounds.

These notions need to be modified for a number of reasons. The first is that most vibrations do not consist of a single frequency, and naming a 'defining' frequency can be difficult. The second related issue is that these attributes should really be defined in terms of the perception of the sound, and not in terms of the sound itself. So, for example, the perceived pitch of a sound can represent a frequency not actually present in the waveform. This phenomenon is called the 'missing fundamental', and is part of a subject called psychoacoustics.

Attributes of sound

Physical	Perceptual
Amplitude	Loudness
Frequency	Pitch
Spectrum	Timbre
Duration	Length

Further reading

Harvey Fletcher, Loudness, pitch and the timbre of musical tones and their relation to the intensity, the frequency and the overtone structure, *J. Acoust. Soc. Amer.* **6** (2) (1934), 59–69.

1.2 The human ear

In order to get much further with understanding sound, we need to study its perception by the human ear. This is the topic of this section. I have borrowed extensively from Gray's *Anatomy* for this description.

The ear is divided into three parts, called the outer ear, the middle ear or *tympanum* and the inner ear or *labyrinth*. See Figure 1.1. The outer ear is the visible part on the outside of the head, called the *pinna* (plural *pinnæ*) or *auricle*, and is ovoid in form. The hollow middle part, or *concha*, is associated with focusing and thereby magnifying the sound, while the outer rim, or *helix*, appears to be associated with vertical spatial separation, so that we can judge the height of a source of sound.

The concha channels the sound into the auditory canal, called the *meatus auditorius externus* (or just *meatus*). This is an air filled tube, about 2.7 cm long and

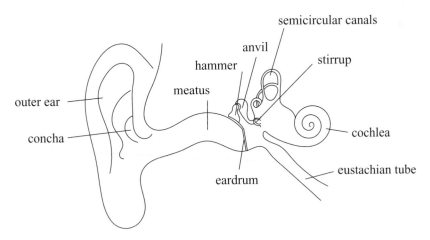

Figure 1.1 The human ear.

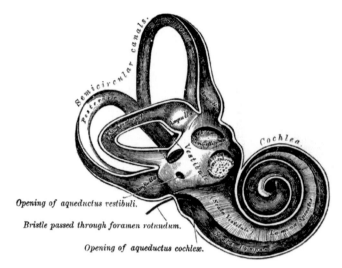

Figure 1.2 The osseous labyrinth laid open. (Enlarged.) From Gray (1901).

0.7 cm in diameter. At the inner end of the meatus is the ear drum, or *tympanic membrane*.

The ear drum divides the outer ear from the middle ear, or *tympanum*, which is also filled with air. The tympanum is connected to three very small bones (the *ossicular chain*) which transmit the movement of the ear drum to the inner ear. The three bones are the hammer, or *malleus*, the anvil, or *incus*, and the stirrup, or *stapes*. These three bones form a system of levers connecting the ear drum to a membrane covering a small opening in the inner ear. The membrane is called the *oval window*.

The inner ear, or *labyrinth*, consists of two parts, the *osseous labyrinth*,[1] see Figure 1.2, consisting of cavities hollowed out from the substance of the bone, and the *membranous labyrinth*, contained in it. The osseous labyrinth is filled with various fluids, and has three parts, the *vestibule*, the *semicircular canals* and the *cochlea*. The vestibule is the central cavity which connects the other two parts and which is situated on the inner side of the tympanum. The semicircular canals lie above and behind the vestibule, and play a role in our sense of balance. The cochlea is at the front end of the vestibule, and resembles a common snail shell in shape. See Figure 1.3. The purpose of the cochlea is to separate out sound into various frequency components (the meaning of this will be made clearer in Chapter 2) before passing it onto the nerve pathways. It is the functioning of the cochlea which is of most interest in terms of the harmonic content of a single musical note, so let us look at the cochlea in more detail.

---

[1] Illustrations taken from the 1901 edition of Henry Gray, F.R.S. *Anatomy, Descriptive and Surgical*, reprinted by Running Press, 1974.

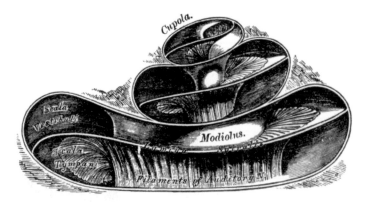

Figure 1.3 The cochlea laid open. (Enlarged.) From Gray (1901).

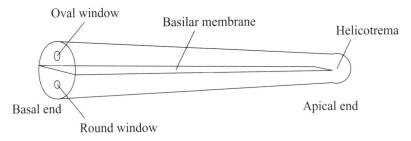

Figure 1.4 The cochlea, uncoiled.

The cochlea twists roughly two and three quarter times from the outside to the inside, around a central axis called the *modiolus* or *columnella*. If it could be unrolled, it would form a tapering conical tube roughly 30 mm (a little over an inch) in length. See Figure 1.4.

At the wide (*basal*) end where it meets the rest of the inner ear it is about 9 mm (somewhat under half an inch) in diameter, and at the narrow (*apical*) end it is about 3 mm (about a fifth of an inch) in diameter. There is a bony shelf or ledge called the *lamina spiralis ossea* projecting from the modiolus, which follows the windings to encompass the length of the cochlea. A second bony shelf called the *lamina spiralis secundaria* projects inwards from the outer wall. Attached to these shelves is a membrane called the *membrana basilaris* or *basilar membrane*. This tapers in the opposite direction to the cochlea (see Figure 1.5), and the bony shelves take up the remaining space.

The basilar membrane divides the interior of the cochlea into two parts with approximately semicircular cross-section. The upper part is called the *scala vestibuli* and the lower is called the *scala tympani*. There is a small opening called the *helicotrema* at the apical end of the basilar membrane, which enables the two parts

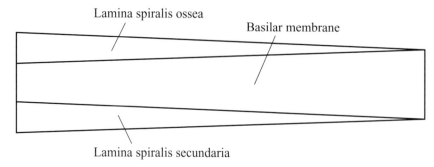

Figure 1.5 The basilar membrane.

to communicate with each other. At the basal end there are two windows allowing communication of the two parts with the vestibule. Each window is covered with a thin flexible membrane. The stapes is connected to the membrane called the *membrana tympani secundaria* covering the upper window; this window is called the *fenestra ovalis* or *oval window*, and has an area of 2.0–3.7 mm^2. The lower window is called the *fenestra rotunda* or *round window*, with an area of around 2 mm^2, and the membrane covering it is not connected to anything apart from the window. There are small hair cells along the basilar membrane which are connected with numerous nerve endings for the auditory nerves. These transmit information to the brain via a complex system of neural pathways. The hair cells come in four rows, and form the *organ of Corti* on the basilar membrane.

Now consider what happens when a sound wave reaches the ear. The sound wave is focused into the meatus, where it vibrates the ear drum. This causes the hammer, anvil and stapes to move as a system of levers, and so the stapes alternately pushes and pulls the membrana tympani secundaria in rapid succession. This causes fluid waves to flow back and forth round the length of the cochlea, in opposite directions in the scala vestibuli and the scala tympani, and causes the basilar membrane to move up and down.

Let us examine what happens when a pure sine wave is transmitted by the stapes to the fluid inside the cochlea. The speed of the wave of fluid in the cochlea at any particular point depends not only on the frequency of the vibration but also on the area of cross-section of the cochlea at that point, as well as the stiffness and density of the basilar membrane. For a given frequency, the speed of travel decreases towards the apical end, and falls to almost zero at the point where the narrowness causes a wave of that frequency to be too hard to maintain. Just to the wide side of that point, the basilar membrane will have to have a peak of amplitude of vibration in order to absorb the motion. Exactly where that peak occurs depends on the frequency. So by examining which hairs are sending the neural signals to the brain, we can ascertain the frequency of the incoming sine wave.

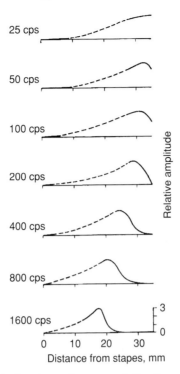

Figure 1.6 Von Békésy's drawings of patterns of vibration of the basilar membrane. The solid lines are from measurements, while the dotted lines are extrapolated. Copyright The McGraw-Hill Companies, Inc.

The statement that the ear picks out frequency components of an incoming sound is known as 'Ohm's acoustic law'. The description above of how the brain 'knows' the frequency of an incoming sine wave is due to Hermann Helmholtz, and is known as the place theory of pitch perception.

Measurements made by von Békésy in the 1950s support this theory. The drawings in Figure 1.6 are taken from his 1960 book (Fig. 11–43). They show the patterns of vibration of the basilar membrane of a cadaver for various frequencies.

The spectacular extent to which the ear can discriminate between frequencies very close to each other is not completely explained by the *passive* mechanics of the cochlea alone, as reflected by von Békésy's measurements. More recent research shows that a sort of psychophysical feedback mechanism sharpens the tuning and increases the sensitivity. In other words, there is information carried both ways by the neural paths between the cochlea and the brain, and this provides active amplification of the incoming acoustic stimulus. The outer hair cells are not just recording information, they are actively stimulating the basilar membrane. See Figure 1.7.

One result of this feedback is that if the incoming signal is loud, the gain will be turned down to compensate. If there is very little stimulus, the gain is turned up

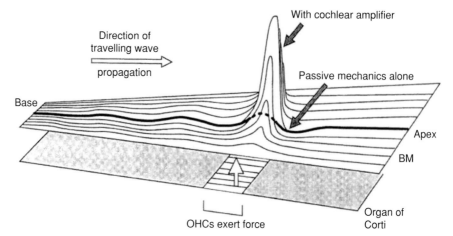

Figure 1.7 Feedback in the cochlea, picture from Jonathan Ashmore's article in Kruth and Stobart (2000). In this figure, OHC stands for 'outer hair cells' and BM stands for 'basilar membrane'.

until the stimulus is detected. An annoying side effect of this is that if mechanical damage to the ear causes deafness, then the neural feedback mechanism turns up the gain until random noise is amplified, so that singing in the ear, or *tinnitus* results. The deaf person does not even have the consolation of silence.

The phenomenon of *masking* is easily explained in terms of Helmholtz's theory. Alfred Meyer discovered in 1876 that an intense sound of a lower pitch prevents us from perceiving a weaker sound of a higher pitch, but an intense sound of a higher pitch never prevents us from perceiving a weaker sound of a lower pitch. The explanation of this is that the excitation of the basilar membrane caused by a sound of higher pitch is closer to the basal end of the cochlea than that caused by a sound of lower pitch. So to reach the place of resonance, the lower pitched sound must pass the places of resonance for all higher frequency sounds. The movement of the basilar membrane caused by this interferes with the perception of the higher frequencies.

Further reading

Anthony W. Gummer, Werner Hemmert and Hans-Peter Zenner, Resonant tectorial membrane motion in the inner ear: its crucial role in frequency tuning, *Proc. Natl. Acad. Sci. (US)* **93** (16) (1996), 8727–8732.

James Keener and James Sneyd, *Mathematical Physiology*, Springer-Verlag, 1998. Chapter 23 of this book describes some fairly sophisticated mathematical models of the cochlea.

Moore, Brian C. J. (1997), *Psychology of Hearing*.

James O. Pickles (1988), *An Introduction to the Physiology of Hearing*.

Christopher A. Shera, John J. Guinan, Jr. and Andrew J. Oxenham, Revised estimates of human cochlear tuning from otoacoustic and behavioral measurements, *Proc. Natl. Acad. Sci. (US)* **99** (5) (2002), 3318–3323.

William A. Yost (1977), *Fundamentals of Hearing. An Introduction.*
Eberhard Zwicker and H. Fastl (1999), *Psychoacoustics: Facts and Models.*

1.3 Limitations of the ear

In music, frequencies are measured in Hertz (Hz), or cycles per second. The approximate range of frequencies to which the human ear responds is usually taken to be from 20 Hz to 20 000 Hz. For frequencies outside this range, there is no resonance in the basilar membrane, although sound waves of frequency lower than 20 Hz may often be *felt* rather than heard.[2] For comparison, here is a table of hearing ranges for various animals.[3]

Species	Range (Hz)
Turtle	20–1000
Goldfish	100–2000
Frog	100–3000
Pigeon	200–10 000
Sparrow	250–12 000
Human	20–20 000
Chimpanzee	100–20 000
Rabbit	300–45 000
Dog	50–46 000
Cat	30–50 000
Guinea pig	150–50 000
Rat	1000–60 000
Mouse	1000–100 000
Bat	3000–120 000
Dolphin (*Tursiops*)	1000–130 000

[2] But see also: Tsutomi Oohashi, Emi Nishina, Norie Kawai, Yoshitaka Fuwamoto and Hiroshi Imai, *High-frequency Sound Above the Audible Range Affects brain electric activity and sound perception*, Audio Engineering Society preprint No. 3207 (91st convention, New York City). In this fascinating paper, the authors describe how they recorded gamelan music with a bandwidth going up to 60 KHz. They played back the recording through a speaker system with an extra tweeter for the frequencies above 26 KHz, driven by a separate amplifier so that it could be switched on and off. They found that the EEG (electroencephalogram) of the listeners' response, as well as the subjective rating of the recording, was affected by whether the extra tweeter was on or off, even though the listeners denied that the sound was altered by the presence of this tweeter, or that they could hear anything from the tweeter played alone. They also found that the EEG changes persisted afterwards, in the absence of the high frequency stimulation, so that long intervals were needed between sessions.

Another relevant paper is: Martin L. Lenhardt, Ruth Skellett, Peter Wang and Alex M. Clarke, *Human ultrasonic speech perception, Science*, **253**, 5 July 1991, 82–85. In this paper, they report that bone-conducted ultrasonic hearing has been found capable of supporting frequency discrimination and speech detection in normal, older hearing-impaired, and profoundly deaf human subjects. They conjecture that the mechanism may have to do with the *saccule*, which is a small spherical cavity adjoining the scala vestibuli of the cochlea.

Research of James Boyk has shown that unlike other musical instruments, for the cymbal, roughly 40% of the observable energy of vibration is at frequencies between 20 kHz and 100 kHz, and showed no signs of dropping off in intensity even at the high end of this range. This research appears in *There's life above 20 kilohertz: a survey of musical-instrument spectra up to 102.4 kHz*, published on the Caltech Music Lab web site in 2000.

[3] Taken from R. Fay, *Hearing in Vertebrates. A Psychophysics Databook*. Hill-Fay Associates, 1988.

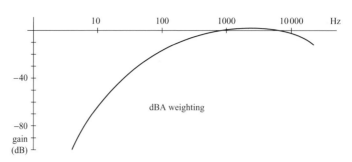

Figure 1.8 Curve A.

Sound intensity is measured in *decibels* or dB. Zero decibels represents a power intensity of 10^{-12} watts per square meter, which is somewhere in the region of the weakest sound we can hear. Adding ten decibels (one *bel*) multiplies the power intensity by a factor of ten. So multiplying the power by a factor of b adds $10\log_{10}(b)$ decibels to the level of the signal. This means that the scale is logarithmic, and n decibels represents a power density of $10^{(n/10)-12}$ watts per square meter.

Often, decibels are used as a relative measure, so that an intensity *ratio* of ten to one represents an increase of ten decibels. As a relative measure, decibels refer to ratios of powers whether or not they directly represent sound. So for example, the power gain and the signal to noise ratio of an amplifier are measured in decibels. It is worth knowing that $\log_{10}(2)$ is roughly 0.3 (to five decimal places it is 0.30103), so that a power ratio of 2:1 represents a difference of about 3 dB.

To distinguish from the relative measurement, the notation dB SPL (sound pressure level) is sometimes used to refer to the absolute measurement of sound described above. It should also be mentioned that, rather than using dB SPL, use is often made of a weighting curve, so that not all frequencies are given equal importance. There are three standard curves, called A, B and C. It is most common to use curve A, which has a peak at about 2000 Hz and drops off substantially to either side. See Figure 1.8. Curves B and C are flatter, and only drop off at the extremes. Measurements made using curve A are quoted as dBA, or dBA SPL to be pedantic.

The *threshold of hearing* is the level of the weakest sound we can hear. Its value in decibels varies from one part of the frequency spectrum to another. Our ears are most sensitive to frequencies a little above 2000 Hz, where the threshold of hearing of the average person is a little above 0 dB. At 100 Hz the threshold is about 50 dB, and at 10 000 Hz it is about 30 dB. The average whisper is about 15–20 dB, conversation usually happens at around 60–70 dB, and the threshold of pain is around 130 dB.

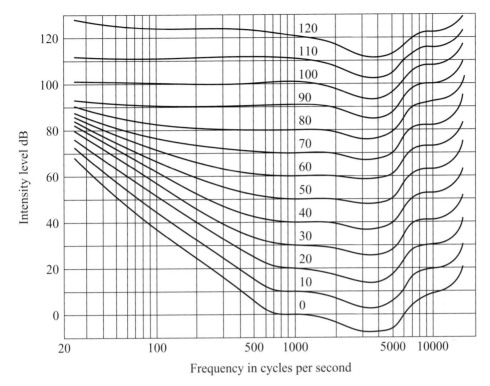

Figure 1.9 Equal loudness curves.

The relationship between sound pressure level and perception of loudness is frequency dependent. The graph in Figure 1.9, due to Fletcher and Munson[4] shows equal loudness curves for pure tones at various frequencies.

The unit of loudness is the *phon*, which is defined as follows. The listener adjusts the level of the signal until it is judged to be of equal intensity to a standard 1000 Hz signal. The phon level is defined to be the signal pressure level of the 1000 Hz signal of the same loudness. The curves in this graph are called *Fletcher–Munson curves*, or *isophons*.

The amount of power in watts involved in the production of sound is very small. The clarinet at its loudest produces about one twentieth of a watt of sound, while the trombone is capable of producing up to five or six watts of sound. The average human speaking voice produces about 0.000 02 watts, while a bass singer at his loudest produces about a thirtieth of a watt.

The *just noticeable difference* or *limen* is used both for sound intensity and frequency. This is usually taken to be the smallest difference between two successive

---

[4] H. Fletcher and W. J. Munson, Loudness, its definition, measurement and calculation, *J. Acoust. Soc. Amer.* **5** (2) (1933), 82–108.

tones for which a person can name correctly 75% of the time which is higher (or louder). It depends in both cases on both frequency and intensity. The just noticeable difference in frequency will be of more concern to us than the one for intensity, and the following table is taken from Pierce (1983). The measurements are in cents, where 1200 cents make one octave (for further details of the system of cents, see Section 5.4).

Frequency	Intensity (dB)										
(Hz)	5	10	15	20	30	40	50	60	70	80	90
31	220	150	120	97	76	70					
62	120	120	94	85	80	74	61	60			
125	100	73	57	52	46	43	48	47			
250	61	37	27	22	19	18	17	17	17	17	
550	28	19	14	12	10	9	7	6	7		
1,000	16	11	8	7	6	6	6	6	5	5	4
2,000	14	6	5	4	3	3	3	3	3	3	
4,000	10	8	7	5	5	4	4	4	4		
8,000	11	9	8	7	6	5	4	4			
11,700	12	10	7	6	6	6	5				

It is easy to see from this table that our ears are much more sensitive to small changes in frequency for higher notes than for lower ones. When referring to the above table, bear in mind that it refers to *consecutive* notes, not simultaneous ones. For simultaneous notes, the corresponding term is the *limit of discrimination*. This is the smallest difference in frequency between simultaneous notes, for which two separate pitches are heard. We shall see in Section 1.8 that simultaneous notes cause beats, which enable us to notice far smaller differences in frequency. This is very important to the theory of scales, because notes in a scale are designed for harmony, which is concerned with clusters of simultaneous notes. So scales are much more sensitive to very small changes in tuning than might be supposed.

Vos[5] studied the sensitivity of the ear to the exact tuning of the notes of the usual twelve tone scale, using two-voice settings from Michael Praetorius' *Musæ Sioniæ*, Part VI (1609). His conclusions were that scales in which the intervals were not more than 5 cents away from the 'just' versions of the intervals (see Section 5.5) were all close to equally acceptable, but then with increasing difference the acceptability decreases dramatically. In view of the fact that in the modern equal tempered twelve tone system, the major third is about 14 cents away from just, these conclusions

---

[5] J. Vos, Subjective acceptability of various regular twelve-tone tuning systems in two-part musical fragments, *J. Acoust. Soc. Amer.* **83** (6) (1988), 2383–2392.

are very interesting. We shall have much more to say about this subject in Chapter 5.

Exercises

1. Power intensity is proportional to the square of amplitude. How many decibels represent a doubling of the amplitude of a signal?
2. (Multiple choice) Two independent 70 dB sound sources are heard together. How loud is the resultant sound, to the nearest dB?
   (a) 140 dB, (b) 76 dB, (c) 73 dB, (d) 70 dB, (e) None of the above.

1.4 Why sine waves?

What is the relevance of sine waves to the discussion of perception of pitch? Could we make the same discussion using some other family of periodic waves that go up and down in a similar way?

The answer lies in the differential equation for simple harmonic motion, which we discuss in the next section. To put it briefly, the solutions to the differential equation

$$\frac{d^2 y}{dt^2} = -\kappa y$$

are the functions

$$y = A \cos \sqrt{\kappa} t + B \sin \sqrt{\kappa} t,$$

or equivalently

$$y = c \sin(\sqrt{\kappa} t + \phi)$$

(see Section 1.8 for the equivalence of these two forms of the solution). See Figure 1.10. The above differential equation represents what happens when an object is subject to a force towards an equilibrium position, the magnitude of the force being proportional to the distance from equilibrium.

In the case of the human ear, the above differential equation may be taken as a close approximation to the equation of motion of a particular point on the basilar membrane, or anywhere else along the chain of transmission between the outside air and the cochlea. Actually, this is inaccurate in several regards. The first is that we should really set up a second order partial differential equation describing the motion of the surface of the basilar membrane. This does not really affect the results of the analysis much except to explain the origins of the constant κ. The second inaccuracy is that we should really think of the motion as *forced damped harmonic motion* in which there is a damping term proportional to velocity, coming from the viscosity of the fluid and the fact that the basilar membrane is not perfectly

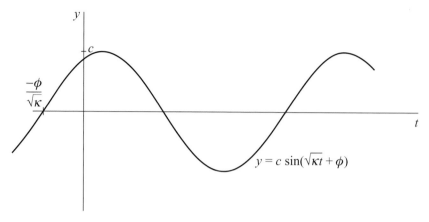

Figure 1.10.

elastic. In Sections 1.10–1.11, we shall see that forced damped harmonic motion is also sinusoidal, but contains a rapidly decaying transient component. There is a *resonant frequency* corresponding to the maximal response of the damped system to the incoming sine wave. The third inaccuracy is that for loud enough sounds the restoring force may be nonlinear. This will be seen to be the possible origin of some interesting acoustical phenomena. Finally, most musical notes do not consist of a single sine wave. For example, if a string is plucked, a periodic wave will result, but it will usually consist of a sum of sine waves with various amplitudes. So there will be various different peaks of amplitude of vibration of the basilar membrane, and a more complex signal is sent to the brain. The decomposition of a periodic wave as a sum of sine waves is called Fourier analysis, which is the subject of Chapter 2.

1.5 Harmonic motion

Consider a particle of mass m subject to a force F towards the equilibrium position, $y = 0$, and whose magnitude is proportional to the distance y from the equilibrium position,

$$F = -ky.$$

Here, k is just the constant of proportionality. Newton's laws of motion give us the equation

$$F = ma,$$

where

$$a = \frac{d^2 y}{dt^2}$$

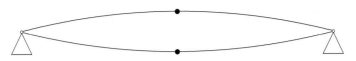

Figure 1.11 A vibrating string.

is the acceleration of the particle and t represents time. Combining these equations, we obtain the second order differential equation

$$\frac{d^2 y}{dt^2} + \frac{ky}{m} = 0. \tag{1.5.1}$$

We write $\dot{y}$ for $\dfrac{dy}{dt}$ and $\ddot{y}$ for $\dfrac{d^2 y}{dt^2}$ as usual, so that this equation takes the form

$$\ddot{y} + ky/m = 0.$$

The solutions to this equation are the functions

$$y = A\cos(\sqrt{k/m}\,t) + B\sin(\sqrt{k/m}\,t). \tag{1.5.2}$$

The fact that these are the solutions of this differential equation is the explanation of why the sine wave, and not some other periodically oscillating wave, is the basis for harmonic analysis of periodic waves. For this is the differential equation governing the movement of any particular point on the basilar membrane in the cochlea, and hence governing the human perception of sound.

Exercises

1. Show that the functions (1.5.2) satisfy the differential equation (1.5.1).
2. Show that the general solution (1.5.2) to equation (1.5.1) can also be written in the form

$$y = c\sin(\sqrt{k/m}\,t + \phi).$$

Describe c and ϕ in terms of A and B. (If you get stuck, take a look at Section 1.8).

1.6 Vibrating strings

In this section, we make a first pass at understanding vibrating strings. In Section 3.2 we return to this topic and do a better analysis using partial differential equations.

Consider a vibrating string, anchored at both ends. Suppose at first that the string has a heavy bead attached to the middle of it, so that the mass m of the bead is much greater than the mass of the string. See Figure 1.11. Then the string exerts a force F on the bead towards the equilibrium position whose magnitude, at least for small

Figure 1.12 Vibrating at twice the frequency.

Figure 1.13 Vibrating a three times the frequency.

displacements, is proportional to the distance y from the equilibrium position,

$$F = -ky.$$

According to the last section, we obtain the differential equation

$$\frac{d^2 y}{dt^2} + \frac{ky}{m} = 0,$$

whose solutions are the functions

$$y = A \cos(\sqrt{k/m}\, t) + B \sin(\sqrt{k/m}\, t),$$

where the constants A and B are determined by the initial position and velocity of the bead.

If the mass of the string is uniformly distributed, then more vibrational 'modes' are possible. For example, the midpoint of the string can remain stationary while the two halves vibrate with opposite phases. See Figure 1.12. On a guitar, this can be achieved by touching the midpoint of the string while plucking and then immediately releasing. The effect will be a sound exactly an octave above the natural pitch of the string, or exactly twice the frequency. The use of harmonics in this way is a common device among guitar players. If each half is vibrating with a pure sine wave then the motion of a point other than the midpoint will be described by the function

$$y = A \cos(2\sqrt{k/m}\, t) + B \sin(2\sqrt{k/m}\, t).$$

If a point exactly one third of the length of the string from one end is touched while plucking, the effect will be a sound an octave and a perfect fifth above the natural pitch of the string, or exactly three times the frequency. See Figure 1.13. Again, if the three parts of the string are vibrating with a pure sine wave, with the middle third in the opposite phase to the outside two thirds, then the motion of a non-stationary point on the string will be described by the function

$$y = A \cos(3\sqrt{k/m}\, t) + B \sin(3\sqrt{k/m}\, t).$$

Figure 1.14 440 Hz.

In general, a plucked string will vibrate with a mixture of all the modes described by multiples of the natural frequency, with various amplitudes. The amplitudes involved depend on the exact manner in which the string is plucked or struck. For example, a string struck by a hammer, as happens in a piano, will have a different set of amplitudes than that of a plucked string. The general equation of motion of a typical point on the string will be

$$y = \sum_{n=1}^{\infty} \left(A_n \cos(n\sqrt{k/m}\,t) + B_n \sin(n\sqrt{k/m}\,t) \right).$$

This leaves us with a problem, to which we shall return in the next two chapters. How can a string vibrate with a number of different frequencies at the same time? This forms the subject of the theory of Fourier series and the wave equation. Before we are in a position to study Fourier series, we need to understand sine waves and how they interact. This is the subject of Section 1.8. We shall return to the subject of vibrating strings in Section 3.2, where we shall develop the wave equation and its solutions.

1.7 Sine waves and frequency spectrum

Since angles in mathematics are measured in radians, and there are 2π radians in a cycle, a sine wave with *frequency* ν in Hertz, *peak amplitude* c and *phase* ϕ will correspond to a sine wave of the form

$$c\sin(2\pi\nu t + \phi). \tag{1.7.1}$$

The quantity $\omega = 2\pi\nu$ is called the *angular velocity*. The role of the angle ϕ is to tell us where the sine wave crosses the time axis (look back at the graph in Section 1.4). For example, a cosine wave is related to a sine wave by the equation $\cos x = \sin(x + \frac{\pi}{2})$, so a cosine wave is really just a sine wave with a different phase.

For example, modern concert pitch[6] places the note A above middle C (Figure 1.14) at 440 Hz, so this would be represented by a wave of the form

$$c\sin(880\pi t + \phi).$$

---

[6] Historically, this was adopted as the USA. Standard Pitch in 1925, and in May 1939 an international conference in London agreed that this should be adopted as the modern concert pitch. Before that time, a variety of standard frequencies were used. For example, in the time of Mozart, the note A had a value closer to 422 Hz, a little under a semitone flat to modern ears. Before this time, in the Baroque and earlier, there was even more variation. For example, in Tudor Britain, secular vocal pitch was much the same as modern concert pitch, while domestic keyboard pitch was about three semitones lower and church music pitch was more than two semitones higher.

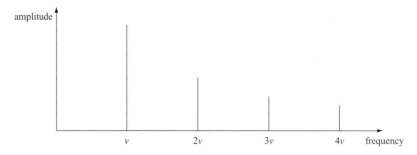

Figure 1.15 Spectrum of a vibrating string.

This can be converted to a linear combination of sines and cosines using the standard formulae for the sine and cosine of a sum:

$$\sin(A + B) = \sin A \cos B + \cos A \sin B, \tag{1.7.2}$$
$$\cos(A + B) = \cos A \cos B - \sin A \sin B. \tag{1.7.3}$$

So we have

$$c \sin(\omega t + \phi) = a \cos \omega t + b \sin \omega t,$$

where

$$a = c \sin \phi, \qquad b = c \cos \phi.$$

Conversely, given a and b, c and ϕ can be obtained via

$$c = \sqrt{a^2 + b^2}, \qquad \tan \phi = a/b.$$

We end this section by introducing the concept of *spectrum*, which plays an important role in understanding musical notes. The spectrum of a sound is a graph indicating the amplitudes of various different frequencies in the sound. We shall make this more precise in Section 2.15. But, for the moment, we leave it as an intuitive notion, and illustrate with a picture of the spectrum of a vibrating string with fundamental frequency $v = \sqrt{k/m}/2\pi$ as above, see Figure 1.15.

This graph illustrates a sound with a *discrete* frequency spectrum with frequency components at integer multiples of a fundamental frequency, and with the amplitude dropping off for higher frequencies. Some sounds, such as white noise (see Figure 1.16), have a *continuous* frequency spectrum, as in the diagram below. Making sense of what these terms might mean will involve us in Fourier theory and the theory of distributions.

It is worth noticing that some information is lost when passing to the frequency spectrum. Namely, we have lost all information about the *phase* of each frequency component.

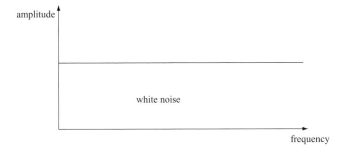

Figure 1.16 White noise.

Exercises

1. Use the equation $\cos\theta = \sin(\pi/2 + \theta)$ and equations (1.8.9)–(1.8.10) to express $\sin u + \cos v$ as a product of trigonometric functions.

1.8 Trigonometric identities and beats

What happens when two pure sine or cosine waves are played at the same time? For example, why is it that when two very close notes are played simultaneously, we hear 'beats'? Since this is the method by which strings on a piano are tuned, it is important to understand the origins of these beats.

The answer to this question also lies in the trigonometric identities (1.7.2) and (1.7.3). Since $\sin(-B) = -\sin B$ and $\cos(-B) = \cos B$, replacing B by $-B$ in equations (1.7.2) and (1.7.3) gives

$$\sin(A - B) = \sin A \cos B - \cos A \sin B, \qquad (1.8.1)$$

$$\cos(A - B) = \cos A \cos B + \sin A \sin B. \qquad (1.8.2)$$

Adding equations (1.7.2) and (1.8.1),

$$\sin(A + B) + \sin(A - B) = 2 \sin A \cos B, \qquad (1.8.3)$$

which may be rewritten as

$$\sin A \cos B = \tfrac{1}{2}(\sin(A + B) + \sin(A - B)). \qquad (1.8.4)$$

Similarly, adding and subtracting equations (1.7.3) and (1.8.2) gives

$$\cos(A + B) + \cos(A - B) = 2 \cos A \cos B, \qquad (1.8.5)$$

$$\cos(A - B) - \cos(A + B) = 2 \sin A \sin B, \qquad (1.8.6)$$

or

$$\cos A \cos B = \tfrac{1}{2}(\cos(A + B) + \cos(A - B)), \qquad (1.8.7)$$

$$\sin A \sin B = \tfrac{1}{2}(\cos(A - B) - \cos(A + B)). \qquad (1.8.8)$$

Figure 1.17 $y = \sin(12t) + \sin(10t) = 2\sin(11t)\cos(t)$.

This enables us to write any product of sines and cosines as a sum or difference of sines and cosines. So, for example, if we wanted to integrate a product of sines and cosines, this would enable us to do so.

We are actually interested in the opposite process. So we set $u = A + B$ and $v = A - B$. Solving for A and B, this gives $A = \frac{1}{2}(u + v)$ and $B = \frac{1}{2}(u - v)$. Substituting in equations (1.8.3), (1.8.5) and (1.8.6), we obtain

$$\sin u + \sin v = 2\sin \tfrac{1}{2}(u + v)\cos \tfrac{1}{2}(u - v), \qquad (1.8.9)$$

$$\cos u + \cos v = 2\cos \tfrac{1}{2}(u + v)\cos \tfrac{1}{2}(u - v), \qquad (1.8.10)$$

$$\cos v - \cos u = 2\sin \tfrac{1}{2}(u + v)\sin \tfrac{1}{2}(u - v). \qquad (1.8.11)$$

This enables us to write any sum or difference of sine waves and cosine waves as a product of sines and cosines. See Figure 1.17, for example. Exercise 1 at the end of this section explains what to do if there are mixed sines and cosines.

So, for example, suppose that a piano tuner has tuned one of the three strings corresponding to the note A above middle C to 440 Hz. The second string is still out of tune, so that it resonates at 436 Hz. The third is being damped so as not to interfere with the tuning of the second string. Ignoring phase and amplitude for a moment, the two strings together will sound as

$$\sin(880\pi t) + \sin(872\pi t).$$

Using equation (1.8.9), we may rewrite this sum as

$$2\sin(876\pi t)\cos(4\pi t).$$

This means that we perceive the combined effect as a sine wave with frequency 438 Hz, the average of the frequencies of the two strings, but with the amplitude modulated by a slow cosine wave with frequency 2 Hz, or half the difference between the frequencies of the two strings. This modulation is what we perceive as beats. The amplitude of the modulating cosine wave has two peaks per cycle, so the number of beats per second will be four, not two. So the number of beats per second is exactly the difference between the two frequencies. The piano tuner tunes the second string to the first by tuning out the beats, namely by adjusting the string so that the beats slow down to a standstill.

If we wish to include terms for phase and amplitude, we write

$$c \sin(880\pi t + \phi) + c \sin(872\pi t + \phi'),$$

where the angles ϕ and ϕ' represent the phases of the two strings. This gets rewritten as

$$2c \sin\left(876\pi t + \tfrac{1}{2}(\phi + \phi')\right) \cos\left(4\pi t + \tfrac{1}{2}(\phi - \phi')\right),$$

so this equation can be used to understand the relationship between the phase of the beats and the phases of the original sine waves.

If the amplitudes are different, then the beats will not be so pronounced because part of the louder note is 'left over'. This prevents the amplitude going to zero when the modulating cosine takes the value zero.

Exercises

1. A piano tuner comparing two of the three strings on the same note of a piano hears five beats a second. If one of the two notes is concert pitch A (440 Hz), what are the possibilities for the frequency of vibration of the other string?

2. Evaluate $\displaystyle\int_0^{\pi/2} \sin(3x) \sin(4x) \, dx.$

3. (a) Setting $A = B = \theta$ in formula (1.8.7) gives the double angle formula

$$\cos^2 \theta = \tfrac{1}{2}(1 + \cos(2\theta)). \tag{1.8.12}$$

Draw graphs of the functions $\cos^2 \theta$ and $\cos(2\theta)$. Try to understand formula (1.8.12) in terms of these graphs.

(b) Setting $A = B = \theta$ in formula (1.8.8) gives the double angle formula

$$\sin^2 \theta = \tfrac{1}{2}(1 - \cos(2\theta)). \tag{1.8.13}$$

Draw graphs of the functions $\sin^2 \theta$ and $\cos(2\theta)$. Try to understand formula (1.8.13) in terms of these graphs.

4. In the formula (1.7.1), the factor c is called the *peak amplitude*, because it determines the highest point on the waveform. In sound engineering, it is often more useful to know the *root mean square* or RMS amplitude, because this is what determines things like power consumption. The RMS amplitude is calculated by integrating the square of the

value over one cycle, dividing by the length of the cycle to obtain the mean square, and then taking the square root. For a pure sine wave given by formula (1.7.1), show that the RMS amplitude is given by

$$\sqrt{\nu \int_0^{\frac{1}{\nu}} [c \sin(2\pi \nu t + \phi)]^2 \, dt} = \frac{c}{\sqrt{2}}.$$

5. Use equation (1.8.8) to write $\sin kt \, \sin \frac{1}{2}t$ as $\frac{1}{2}(\cos(k - \frac{1}{2})t - \cos(k + \frac{1}{2})t)$. Show that

$$\sum_{k=1}^{n} \sin kt = \frac{\cos \frac{1}{2}t - \cos (n + \frac{1}{2})t}{2 \sin \frac{1}{2}t} = \frac{\sin \frac{1}{2}(n + 1)t \, \sin \frac{1}{2}nt}{\sin \frac{1}{2}t}. \qquad (1.8.14)$$

Similarly, show that

$$\sum_{k=1}^{n} \cos kt = \frac{\sin (n + \frac{1}{2})t - \sin \frac{1}{2}t}{2 \sin \frac{1}{2}t} = \frac{\cos \frac{1}{2}(n + 1)t \, \sin \frac{1}{2}nt}{\sin \frac{1}{2}t}. \qquad (1.8.15)$$

6. Two pure sine waves are sounded. One has frequency slightly greater or slightly less than twice that of the other. Would you expect to hear beats? (See also Exercise 1 in Section 8.10.)

1.9 Superposition

Superposing two sounds corresponds to adding the corresponding wave functions. This is part of the concept of *linearity*. In general, a system is linear if two conditions are satisfied. The first, *superposition*, is that the sum of two simultaneous input signals should give rise to the sum of the two outputs. The second condition, *homogeneity*, says that magnifying the input level by a constant factor should multiply the output level by the same constant factor.

Superposing harmonic motions of the same frequency works as follows. Two simple harmonic motions with the same frequency, but possibly different amplitudes and phases, always add up to give another simple harmonic motion with the same frequency. We saw some examples of this in the last section. In this section, we see that there is an easy graphical method for carrying this out in practice.

Consider a sine wave of the form $c \sin(\omega t + \phi)$, where $\omega = 2\pi \nu$. This may be regarded as the y-component of circular motion of the form

$$x = c \cos(\omega t + \phi),$$
$$y = c \sin(\omega t + \phi).$$

Since $\sin^2 \theta + \cos^2 \theta = 1$, squaring and adding these equations shows that the point (x, y) lies on the circle

$$x^2 + y^2 = c^2$$

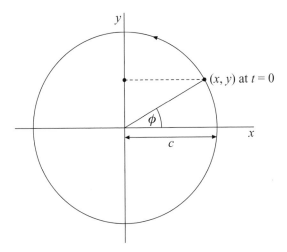

Figure 1.18 Circular motion.

with radius c, centred at the origin. As t varies, the point (x, y) travels counterclock-wise round this circle ν times in each second, so ν is really measuring the number of cycles per second around the origin, and ω is measuring the angular velocity in radians per second. The phase ϕ is the angle, measured counterclockwise from the positive x-axis, subtended by the line from $(0, 0)$ to (x, y) when $t = 0$. See Figure 1.18.

Now suppose that we are given two sine waves of the same frequency, say $c_1 \sin(\omega t + \phi_1)$ and $c_2 \sin(\omega t + \phi_2)$. The corresponding vectors at $t = 0$ are

$$(x_1, y_1) = (c_1 \cos \phi_1, c_1 \sin \phi_1),$$
$$(x_2, y_2) = (c_2 \cos \phi_2, c_2 \sin \phi_2).$$

To superpose (i.e., add) these sine waves, we simply add these vectors to give

$$(x, y) = (c_1 \cos \phi_1 + c_2 \cos \phi_2, c_1 \sin \phi_1 + c_2 \sin \phi_2)$$
$$= (c \cos \phi, c \sin \phi).$$

We draw a copy of the line segment $(0, 0)$ to (x_1, y_1) starting at (x_2, y_2), and a copy of the line segment $(0, 0)$ to (x_2, y_2) starting at (x_1, y_1), to form a parallelogram. The amplitude c is the length of the diagonal line drawn from the origin to the far corner (x, y) of the parallelogram formed this way. The angle ϕ is the angle subtended by this line, measured as usual counterclockwise from the x-axis. See Figure 1.19.

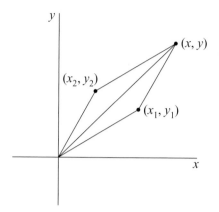

Figure 1.19 Superposition of sine waves.

Exercises

1. Write the following expressions in the form $c \sin(2\pi \nu t + \phi)$:
   (i) $\cos(2\pi t)$
   (ii) $\sin(2\pi t) + \cos(2\pi t)$
   (iii) $2 \sin(4\pi t + \pi/6) - \sin(4\pi t + \pi/2)$.
2. Use Euler's equation $e^{i\theta} = \cos\theta + i\sin\theta$ to interpret the graphical method described in this section as motion in the complex plane of the form
$$z = ce^{i(\omega t + \phi)}.$$

1.10 Damped harmonic motion

Damped harmonic motion arises when in addition to the restoring force $F = -ky$, there is a frictional force proportional to velocity,

$$F = -ky - \mu\dot{y}.$$

For positive values of μ, the extra term damps the motion, while for negative values of μ it promotes or forces the harmonic motion. In this case, the differential equation we obtain is

$$m\ddot{y} + \mu\dot{y} + ky = 0. \tag{1.10.1}$$

This is what is called a linear second order differential equation with constant coefficients. To solve such an equation, we look for solutions of the form

$$y = e^{\alpha t}.$$

Then $\dot{y} = \alpha e^{\alpha t}$ and $\ddot{y} = \alpha^2 e^{\alpha t}$. So for y to satisfy the original differential equation, α has to satisfy the *auxiliary equation*

$$mY^2 + \mu Y + k = 0. \tag{1.10.2}$$

If the quadratic equation (1.10.2) has two different solutions, $Y = \alpha$ and $Y = \beta$, then $y = e^{\alpha t}$ and $y = e^{\beta t}$ are solutions of (1.10.1). Since equation (1.10.1) is linear, this implies that any combination of the form

$$y = Ae^{\alpha t} + Be^{\beta t}$$

is also a solution. The *discriminant* of the auxiliary equation (1.10.2) is

$$\Delta = \mu^2 - 4mk.$$

If $\Delta > 0$, corresponding to large damping or forcing term, then the solutions to the auxiliary equation are

$$\alpha = (-\mu + \sqrt{\Delta})/2m,$$
$$\beta = (-\mu - \sqrt{\Delta})/2m,$$

and so the solutions to the differential equation (1.10.1) are

$$y = Ae^{(-\mu+\sqrt{\Delta})t/2m} + Be^{(-\mu-\sqrt{\Delta})t/2m}. \qquad (1.10.3)$$

In this case, the motion is so damped that no sine waves can be discerned. The system is then said to be *overdamped*, and the resulting motion is called *dead beat*.

If $\Delta < 0$, as happens when the damping or forcing term is small, then the system is said to be *underdamped*. In this case, the auxiliary equation (1.10.2) has no real solutions because Δ has no real square roots. But $-\Delta$ is positive, and so it has a square root. In this case, the solutions to the auxilary equation are

$$\alpha = (-\mu + i\sqrt{-\Delta})/2m,$$
$$\beta = (-\mu - i\sqrt{-\Delta})/2m,$$

where $i = \sqrt{-1}$. So the solutions to the original differential equation are

$$y = e^{-\mu t/2m}\left(Ae^{it\sqrt{-\Delta}/2m} + Be^{-it\sqrt{-\Delta}/2m}\right).$$

We are really interested in real solutions. To this end, we use Euler's equation $e^{i\theta} = \cos\theta + i\sin\theta$ to write this as

$$y = e^{-\mu t/2m}((A + B)\cos(t\sqrt{-\Delta}/2m) + i(A - B)\sin(t\sqrt{-\Delta}/2m)).$$

So we obtain real solutions by taking $A' = A + B$ and $B' = i(A - B)$ to be real numbers, giving

$$y = e^{-\mu t/2m}(A'\sin(t\sqrt{-\Delta}/2m) + B'\cos(t\sqrt{-\Delta}/2m)). \qquad (1.10.4)$$

The interpretation of this is harmonic motion with a damping factor of $e^{-\mu t/2m}$.

The special case $\Delta = 0$ has solutions

$$y = (At + B)e^{-\mu t/2m}. \tag{1.10.5}$$

This borderline case resembles the case $\Delta > 0$, inasmuch as harmonic motion is not apparent. Such a system is said to be *critically damped*.

Examples

1. The equation

$$\ddot{y} + 4\dot{y} + 3y = 0 \tag{1.10.6}$$

is overdamped. The auxiliary equation

$$Y^2 + 4Y + 3 = 0$$

factors as $(Y + 1)(Y + 3) = 0$, so it has roots $Y = -1$ and $Y = -3$. It follows that the solutions of (1.10.6) are given by

$$y = Ae^{-t} + Be^{-3t}.$$

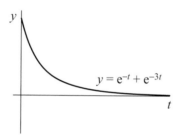

2. The equation

$$\ddot{y} + 2\dot{y} + 26y = 0 \tag{1.10.7}$$

is underdamped. The auxiliary equation is

$$Y^2 + 2Y + 26 = 0.$$

Completing the square gives $(Y + 1)^2 + 25 = 0$, so the solutions are $Y = -1 \pm 5i$. It follows that the solutions of (1.10.7) are given by

$$y = e^{-t}(Ae^{5it} + Be^{-5it}),$$

or

$$y = e^{-t}(A' \cos 5t + B' \sin 5t). \tag{1.10.8}$$

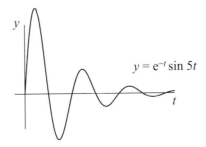

$$y = e^{-t} \sin 5t$$

3. The equation

$$\ddot{y} + 4\dot{y} + 4y = 0 \qquad (1.10.9)$$

is critically damped. The auxiliary equation

$$Y^2 + 4Y + 4 = 0$$

factors as $(Y + 2)^2 = 0$, so the only solution is $Y = -2$. It follows that the solutions of (1.10.9) are given by

$$y = (At + B)e^{-2t}.$$

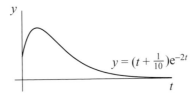

$$y = (t + \tfrac{1}{10})e^{-2t}$$

Exercises

1. Show that if $\Delta = \mu^2 - 4mk > 0$ then the functions (1.10.3) are real solutions of the differential equation (1.10.1).
2. Show that if $\Delta = \mu^2 - 4mk < 0$ then the functions (1.10.4) are real solutions of the differential equation (1.10.1).
3. Show that if $\Delta = \mu^2 - 4mk = 0$ then the auxiliary equation (1.10.2) is a perfect square, and the functions (1.10.5) satisfy the differential equation (1.10.1).

1.11 Resonance

Forced harmonic motion is where there is a forcing term $f(t)$ (often taken to be periodic) added into equation (1.10.1) to give an equation of the form

$$m\ddot{y} + \mu\dot{y} + ky = f(t). \qquad (1.11.1)$$

This represents a damped system with an external stimulus $f(t)$ applied to it. We are particularly interested in the case where $f(t)$ is a sine wave, because this represents forced harmonic motion. Forced harmonic motion is responsible for the production of sound in most musical instruments, as well as the perception of sound in the cochlea. We shall see that forced harmonic motion is what gives rise to the phenomenon of *resonance*.

There are two steps to the solution of the equation. The first is to find the general solution to equation (1.10.1) without the forcing term, as described in Section 1.10, to give the *complementary function*. The second step is to find by any method, such as guessing, a single solution to equation (1.11.1). This is called a *particular integral*. Then the general solution to equation (1.11.1) is the sum of the particular integral and the complementary function.

Examples

1. Consider the equation

$$\ddot{y} + 4\dot{y} + 5y = 10t^2 - 1. \qquad (1.11.2)$$

We look for a particular integral of the form $y = at^2 + bt + c$. Differentiating, we get $\dot{y} = 2at + b$ and $\ddot{y} = 2a$. Plugging these into (1.11.2) gives

$$2a + 4(2at + b) + 5(at^2 + bt + c) = 10t^2 + t - 3.$$

Comparing coefficients of t^2 gives $5a = 10$ or $a = 2$. Then comparing coefficients of t gives $8a + 5b = 1$, so $b = -3$. Finally, comparing constant terms gives $2a + 4b + 5c = -3$, so $c = 1$. So we get a particular integral of $y = 2t^2 - 3t + 1$. Adding the complementary function (1.10.8), we find that the general solution to (1.11.2) is given by

$$y = 2t^2 - 3t + 1 + e^{-2t}(A' \cos t + B' \sin t).$$

2. As a more interesting example, to solve

$$\ddot{y} + 4\dot{y} + 5y = \sin 2t, \qquad (1.11.3)$$

we look for a particular integral of the form

$$y = a \cos 2t + b \sin 2t.$$

Equating coefficients of $\cos 2t$ and $\sin 2t$ we get two equations:

$$-8a + b = 1,$$
$$a + 8b = 0.$$

Solving these equations, we get $a = -\frac{8}{65}$, $b = \frac{1}{65}$. So the general solution to (1.11.3) is

$$y = \frac{\sin 2t - 8 \cos 2t}{65} + e^{-2t}(A' \cos t + B' \sin t).$$

The case of forced harmonic motion of interest to us is the equation

$$m\ddot{y} + \mu\dot{y} + ky = R\cos(\omega t + \phi). \qquad (1.11.4)$$

This represents a damped harmonic motion (see Section 1.10) with forcing term of amplitude R and angular velocity ω.

We could proceed as above to look for a particular integral of the form

$$y = a\cos\omega t + b\sin\omega t$$

and proceed as in the second example above. However, we can simplify the calculation by using complex numbers. Since this differential equation is linear, and since

$$Re^{i(\omega t + \phi)} = R(\cos(\omega t + \phi) + i\sin(\omega t + \phi)),$$

it will be enough to find a particular integral for the equation

$$m\ddot{y} + \mu\dot{y} + ky = Re^{i(\omega t + \phi)}, \qquad (1.11.5)$$

which represents a complex forcing term with unit amplitude and angular velocity ω. Then we take the real part to get a solution to equation (1.11.4).

We look for solutions of equation (1.11.5) of the form $y = Ae^{i(\omega t + \phi)}$, with A to be determined. We have $\dot{y} = Ai\omega e^{i(\omega t + \phi)}$ and $\ddot{y} = -A\omega^2 e^{i(\omega t + \phi)}$. So plugging into equation (1.11.5) and dividing by $e^{i(\omega t + \phi)}$, we get

$$A(-m\omega^2 + i\mu\omega + k) = R$$

or

$$A = \frac{R}{-m\omega^2 + i\mu\omega + k}.$$

So the particular integral, which actually represents the eventual 'steady state' solution to the equation since the complementary function is decaying, is given by

$$y = \frac{Re^{i(\omega t + \phi)}}{-m\omega^2 + i\mu\omega + k}.$$

The denominator of this expression is a complex constant, and so this solution moves around a circle in the complex plane. The real part is then a sine wave with the radius of the circle as amplitude and with a phase determined by the argument of the denominator.

The amplitude of the resulting vibration, and therefore the degree of resonance (since we started with a forcing term of unit amplitude) is given by taking the absolute value of this solution,

$$|y| = \frac{R}{\sqrt{(k - m\omega^2)^2 + \mu^2\omega^2}}.$$

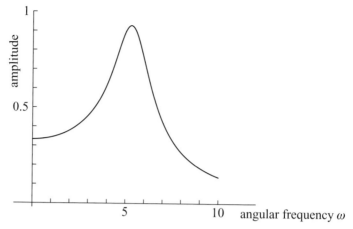

Figure 1.20 Amplitude of forced, underdamped equation.

This amplitude magnification reaches its maximum when the derivative of $(k - m\omega^2)^2 + \mu^2\omega^2$ vanishes, namely when

$$\omega = \sqrt{\frac{k}{m} - \frac{\mu^2}{2m^2}},$$

when we have amplitude $mR/(\mu\sqrt{km - \mu^2/4})$. The above value of ω is called the *resonant frequency* of the system. Note that this value of ω is slightly less than the value which one may expect from equation (1.10.4) for the complementary function:

$$\omega = \frac{\sqrt{-\Delta}}{2m} = \sqrt{\frac{k}{m} - \frac{\mu^2}{4m^2}},$$

which is already less than the value of ω for the corresponding undamped system:

$$\omega = \sqrt{\frac{k}{m}}.$$

Example

Consider the forced, underdamped equation

$$\ddot{y} + 2\dot{y} + 30y = 10\sin\omega t.$$

The above formula tells us that the amplitude of the resulting steady state sine wave solution is $10/\sqrt{900 - 56\omega^2 + \omega^4}$, which has its maximum value at $\omega = \sqrt{28}$. See Figure 1.20.

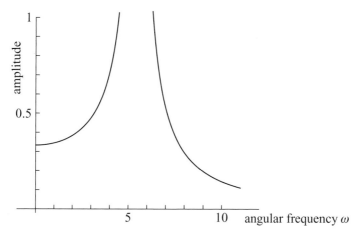

Figure 1.21 Amplitude of undamped equation.

Without the damping term, the amplitude of the steady state solution to the equation

$$\ddot{y} + 30y = 10 \sin \omega t,$$

is equal to $10/|30 - \omega^2|$. It has an 'infinitely sharp' peak at $\omega = \sqrt{30}$. See Figure 1.21.

At this stage, it seems appropriate to introduce the terms *resonant frequency* and *bandwidth* for a resonant system. The resonant frequency is the frequency for which the amplitude of the steady state solution is maximal. Bandwidth is a vague term, used to describe the width of the peak in the above graphs. So in the damped example above, we might want to describe the bandwidth as being from roughly $4\frac{1}{2}$ to $6\frac{1}{2}$, while for the undamped example it would be somewhat wider. Sometimes, the term is made precise by taking the interval between the two points either side of the peak where the amplitude is $1/\sqrt{2}$ times that of the peak. Since power is proportional to square of amplitude, this corresponds to a factor of two in the power, or a difference of $10 \log_{10}(2)$ dB, or roughly 3 dB.

2

Fourier theory

To be sung to the tune of Gilbert and Sullivan's *Modern Major General*:

I am the very model of a genius mathematical,
    For I can do mechanics, both dynamical and statical,
Or integrate a function round a contour in the complex plane,
    Yes, even if it goes off to infinity and back again;
Oh, I know when a detailed proof's required and when a guess'll do
    I know about the functions of Laguerre and those of Bessel too,
I've finished every tripos question back to 1948;
    There ain't a function you can name that I can't differentiate!
    There ain't a function you can name that he can't differentiate [Tris]

I've read the text books and I can extremely quickly tell you where
    To look to find Green's Theorem or the Principle of d'Alembert
Or I can work out Bayes' rule when the loss is not Quadratical
    In short I am the model of a genius mathematical!

For he can work out Bayes' rule when the loss is not Quadratical
    In short he is the model of a genius mathematical!

Oh, I can tell in seconds if a graph is Hamiltonian,
    And I can tell you if a proof of $4CC$'s a phoney 'un
I read up all the journals and I'm ready with the latest news,
    And very good advice about the Part II lectures you should choose.
Oh, I can do numerical analysis without a pause,
    Or comment on the far-reaching significance of Newton's laws
I know when polynomials are soluble by radicals,
    And I can reel off simple groups, especially sporadicals.

For he can reel off simple groups, especially sporadicals [Tris]

Oh, I like relativity and know about fast moving clocks
    And I know what you have to do to get round Russell's paradox
In short, I think you'll find concerning all things problematical
    I am the very model of a genius mathematical!

In short we think you'll find concerning all things problematical
He is the very model of a genius mathematical!

Oh, I know when a matrix will be diagonalisable
And I can draw Greek letters so that they are recognizable
And I can find the inverse of a non-zero quaternion
I've made a model of a rhombicosidodecahedron;
Oh, I can quote the theorem of the separating hyperplane
I've read MacLane and Birkhoff not to mention Birkhoff and MacLane
My understanding of vorticity is not a hazy 'un
And I know why you should (and why you shouldn't) be a Bayesian!

For he knows why you should (and why you shouldn't) be a Bayesian! [Tris]

I'm not deterred by residues and really I am quite at ease
When dealing with essential isolated singularities,
In fact as everyone agrees (and most are quite emphatical)
I am the very model of a genius mathematical!

In fact as everyone agrees (and most are quite emphatical)
He is the very model of a genius mathematical!
– from CUYHA songbook, Cambridge (privately distributed) 1976.

2.1 Introduction

How can a string vibrate with a number of different frequencies at the same time? This problem occupied the minds of many of the great mathematicians and musicians of the seventeenth and eighteenth century. Among the people whose work contributed to the solution of this problem are Marin Mersenne, Daniel Bernoulli, the Bach family, Jean-le-Rond d'Alembert, Leonhard Euler, and Jean Baptiste Joseph Fourier.

In this chapter, we discuss Fourier's theory of harmonic analysis. This is the decomposition of a periodic wave into a (usually infinite) sum of sines and cosines. The frequencies involved are the integer multiples of the fundamental frequency of the periodic wave, and each has an amplitude which can be determined as an integral. A superb book on Fourier series and their continuous frequency spectrum counterpart, Fourier integrals, is Tom Körner (1988). The reader should be warned, however, that the level of sophistication of Körner's book is much greater than the level of this chapter.

Mathematically, this chapter is probably more demanding than the rest of the book. It is not necessary to understand everything in this chapter before reading further, but some familiarity with the concepts of Fourier theory will certainly be useful.

Figure 2.1 Engraving of Jean Baptiste Joseph Fourier (1768–1850) by Boilly (1823) Académie des Sciences, Paris.

2.2 Fourier coefficients

Fourier (see Figure 2.1) introduced the idea that periodic functions can be analyzed by using trigonometric series as follows.[1] The functions $\cos\theta$ and $\sin\theta$ are *periodic* with period 2π, in the sense that they satisfy

$$\cos(\theta + 2\pi) = \cos\theta,$$
$$\sin(\theta + 2\pi) = \sin\theta.$$

In other words, translating by 2π along the θ axis leaves these functions unaffected. There are many other functions $f(\theta)$ which are periodic of period 2π, meaning that they satisfy the equation

$$f(\theta + 2\pi) = f(\theta).$$

[1] The basic ideas behind Fourier series were introduced in Jean Baptiste Joseph Fourier, *La Théorie Analytique de la Chaleur*, F. Didot, 1822. Fourier was born in Auxerre, France in 1768 as the son of a tailor. He was orphaned in childhood and was educated by a school run by the Benedictine order. He was politically active during the French Revolution, and was almost executed. After the revolution, he studied in the then new Ecole Normale in Paris with teachers such as Lagrange, Monge and Laplace. In 1822, with the publication of the work mentioned above, he was elected *secretaire perpetuel* of the Académie des Sciences in Paris. Following this, his role seems principally to have been to encourage younger mathematicians such as Dirichlet, Liouville and Sturm, until his death in 1830.

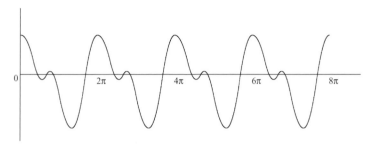

Figure 2.2 A periodic function with period 2π.

We need only specify the function f on the half-open interval $[0, 2\pi)$ in any way we please, and then the above equation determines the value at all other values of θ. See Figure 2.2.

Other examples of periodic functions with period 2π are the constant functions, and the functions $\cos(n\theta)$ and $\sin(n\theta)$ for any positive integer n. Negative values of n give us no more, since

$$\cos(-n\theta) = \cos(n\theta),$$
$$\sin(-n\theta) = -\sin(n\theta).$$

More generally, we can write down any series of the form

$$f(\theta) = \tfrac{1}{2}a_0 + \sum_{n=1}^{\infty}(a_n \cos(n\theta) + b_n \sin(n\theta)), \tag{2.2.1}$$

where the a_n and b_n are constants. So $\frac{1}{2}a_0$ is just a constant – the reason for the factor of $\frac{1}{2}$ will be explained in due course. Such a series is called a *trigonometric series*. Provided that there are no convergence problems, such a series will always define a function satisfying $f(\theta + 2\pi) = f(\theta)$.

The question which naturally arises at this stage is: to what extent can we find a trigonometric series whose sum is equal to a given periodic function? To begin to answer this question, we first ask: given a function defined by a trigonometric series, how can the coefficients a_n and b_n be recovered?

The answer lies in the formulae (for $m \geq 0$ and $n \geq 0$)

$$\int_0^{2\pi} \cos(m\theta) \sin(n\theta)\, d\theta = 0; \tag{2.2.2}$$

$$\int_0^{2\pi} \cos(m\theta) \cos(n\theta)\, d\theta = \begin{cases} 2\pi & \text{if } m = n = 0, \\ \pi & \text{if } m = n > 0, \\ 0 & \text{otherwise}; \end{cases} \tag{2.2.3}$$

$$\int_0^{2\pi} \sin(m\theta) \sin(n\theta)\, d\theta = \begin{cases} \pi & \text{if } m = n > 0, \\ 0 & \text{otherwise}. \end{cases} \tag{2.2.4}$$

These equations can be proved by using equations (1.8.4)–(1.8.8) to rewrite the product of trigonometric functions inside the integral as a sum before integrating.[2] The extra factor of 2 in (2.2.3) for $m = n = 0$ will explain the factor of $\frac{1}{2}$ in front of a_0 in (2.2.1).

This suggests that in order to find the coefficient a_m, we multiply $f(\theta)$ by $\cos(m\theta)$ and integrate. Let us see what happens when we apply this process to equation (2.2.1). Provided we can pass the integral through the infinite sum, only one term gives a nonzero contribution. So for $m > 0$ we have

$$\int_0^{2\pi} \cos(m\theta)f(\theta)\,d\theta = \int_0^{2\pi} \cos(m\theta)\left(\frac{1}{2}a_0 + \sum_{n=1}^{\infty}(a_n\cos(n\theta) + b_n\sin(n\theta))\right)d\theta$$

$$= \frac{1}{2}a_0\int_0^{2\pi}\cos(m\theta)\,d\theta$$

$$+ \sum_{n=1}^{\infty}\left(a_n\int_0^{2\pi}\cos(m\theta)\cos(n\theta)\,d\theta + b_n\int_0^{2\pi}\cos(m\theta)\sin(n\theta)\,d\theta\right)$$

$$= \pi a_m.$$

Thus we obtain, for $m > 0$,

$$a_m = \frac{1}{\pi}\int_0^{2\pi}\cos(m\theta)f(\theta)\,d\theta. \tag{2.2.5}$$

A standard theorem of analysis says that, provided the sum converges uniformly, the integral can be passed through the infinite sum in this way.[3] Under the same conditions, we obtain, for $m > 0$,

$$b_m = \frac{1}{\pi}\int_0^{2\pi}\sin(m\theta)f(\theta)\,d\theta. \tag{2.2.6}$$

The functions a_m and b_m defined by equations (2.2.5) and (2.2.6) are called the *Fourier coefficients* of the function $f(\theta)$.

We can now explain the appearance of the coefficient of $\frac{1}{2}$ in front of the a_0 in equation (2.2.1). Namely, since π is one half of 2π and $\cos(0\theta) = 1$, we have

$$a_0 = \frac{1}{\pi}\int_0^{2\pi}\cos(0\theta)f(\theta)\,d\theta, \tag{2.2.7}$$

[2] The relations (2.2.2)–(2.2.4) are sometimes called *orthogonality relations*. The idea is that the integrable periodic functions form an infinite dimensional vector space with an inner product given by $\langle f, g\rangle = \frac{1}{2\pi}\int_0^{2\pi}f(\theta)g(\theta)\,d\theta$. With respect to this inner product, the functions $\sin(m\theta)$ $(m > 0)$ and $\cos(m\theta)$ $(m \geq 0)$ are *orthogonal*, or perpendicular.

[3] A series of functions f_n on $[a, b]$ converges *uniformly* to a function f if given $\varepsilon > 0$ there exists $N > 0$ (not depending on x) such that for all $x \in [a, b]$ and all $n \geq N$, $|f_n(x) - f(x)| < \varepsilon$. See, for example, Rudin, *Principles of Mathematical Analysis*, third edn., McGraw-Hill 1976, Corollary to Theorem 7.16. We shall have more to say about this definition in Section 2.5.

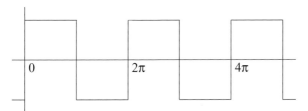

Figure 2.3 Square wave.

which means that the formula (2.2.5) for the coefficient a_m holds for all $m \geq 0$.

It would be nice to think that when we use equations (2.2.5), (2.2.6) and (2.2.7) to define a_m and b_m, the right hand side of equation (2.2.1) always converges to $f(\theta)$. This is true for nice enough functions f, but unfortunately, not for all functions f. In Section 2.4, we investigate conditions on f which ensure that this happens.

Of course, any interval of length 2π, representing one complete period, may be used instead of integrating from 0 to 2π. It is sometimes more convenient, for example, to integrate from $-\pi$ to π:

$$a_m = \frac{1}{\pi} \int_{-\pi}^{\pi} \cos(m\theta) f(\theta) \, d\theta,$$

$$b_m = \frac{1}{\pi} \int_{-\pi}^{\pi} \sin(m\theta) f(\theta) \, d\theta.$$

In practice, the variable θ will not quite correspond to time, because the period is not necessarily 2π seconds. If the period is T seconds then the fundamental frequency is given by $\nu = 1/T$ Hz (Hertz, or cycles per second). The correct substitution is $\theta = 2\pi\nu t$. Setting $F(t) = f(2\pi\nu t) = f(\theta)$ and substituting in (2.2.1) gives a Fourier series of the form

$$F(t) = \tfrac{1}{2}a_0 + \sum_{n=1}^{\infty}(a_n \cos(2n\pi\nu t) + b_n \sin(2n\pi\nu t)),$$

and the following formula for Fourier coefficients:

$$a_m = 2\nu \int_{0}^{T} \cos(2m\pi\nu t) F(t) \, dt, \tag{2.2.8}$$

$$b_m = 2\nu \int_{0}^{T} \sin(2m\pi\nu t) F(t) \, dt. \tag{2.2.9}$$

Example

The square wave sounds vaguely like the waveform produced by a clarinet, where odd harmonics dominate. It is the function $f(\theta)$ defined by $f(\theta) = 1$ for $0 \leq \theta < \pi$

and $f(\theta) = -1$ for $\pi \leq \theta < 2\pi$ (and then extend to all values of θ by making it periodic, $f(\theta + 2\pi) = f(\theta)$).

This function has Fourier coefficients

$$
\begin{aligned}
a_m &= \frac{1}{\pi} \left(\int_0^\pi \cos(m\theta) \, d\theta - \int_\pi^{2\pi} \cos(m\theta) \, d\theta \right) \\
&= \frac{1}{\pi} \left(\left[\frac{\sin(m\theta)}{m} \right]_0^\pi - \left[\frac{\sin(m\theta)}{m} \right]_\pi^{2\pi} \right) = 0 \\
b_m &= \frac{1}{\pi} \left(\int_0^\pi \sin(m\theta) \, d\theta - \int_\pi^{2\pi} \sin(m\theta) \, d\theta \right) \\
&= \frac{1}{\pi} \left(\left[-\frac{\cos(m\theta)}{m} \right]_0^\pi - \left[-\frac{\cos(m\theta)}{m} \right]_\pi^{2\pi} \right) \\
&= \frac{1}{\pi} \left(-\frac{(-1)^m}{m} + \frac{1}{m} + \frac{1}{m} - \frac{(-1)^m}{m} \right) \\
&= \begin{cases} 4/m\pi & (m \text{ odd}), \\ 0 & (m \text{ even}). \end{cases}
\end{aligned}
$$

Thus the Fourier series for this square wave is

$$
\tfrac{4}{\pi} \left(\sin \theta + \tfrac{1}{3} \sin 3\theta + \tfrac{1}{5} \sin 5\theta + \cdots \right). \tag{2.2.10}
$$

Let us examine the first few terms in this series, as shown in Figure 2.4.

Some features of this example are worth noticing. The first observation is that these graphs seem to be converging to a square wave. But they seem to be converging quite slowly, and getting more and more bumpy in the process. Next, observe what happens at the point of discontinuity of the original function. The Fourier coefficients did not depend on what value we assigned to the function at the discontinuity, so we do not expect to recover that information. Instead, the series is converging to a value which is equal to the average of the higher and the lower values of the function. This is a general phenomenon, which we shall discuss in Section 2.5.

Finally, there is a very interesting phenomenon which is happening right near the discontinuity. There is an overshoot, which never seems to get any smaller.

Does this mean that the series is not converging properly? Well, not quite. At each given value of θ, the series is converging just fine. It's just when we look at values of θ closer and closer to the discontinuity that we find problems. This is because of a lack of *uniform* convergence. This overshoot is called the Gibbs phenomenon, and we shall discuss it in more detail in Section 2.5.

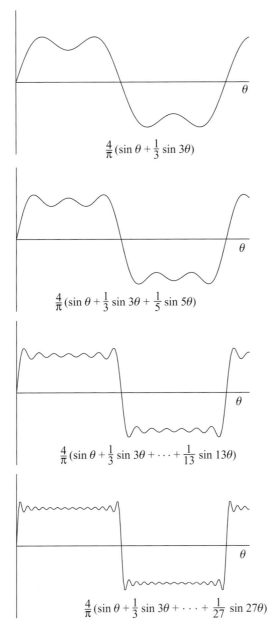

$$\tfrac{4}{\pi}(\sin\theta + \tfrac{1}{3}\sin 3\theta)$$

$$\tfrac{4}{\pi}(\sin\theta + \tfrac{1}{3}\sin 3\theta + \tfrac{1}{5}\sin 5\theta)$$

$$\tfrac{4}{\pi}(\sin\theta + \tfrac{1}{3}\sin 3\theta + \cdots + \tfrac{1}{13}\sin 13\theta)$$

$$\tfrac{4}{\pi}(\sin\theta + \tfrac{1}{3}\sin 3\theta + \cdots + \tfrac{1}{27}\sin 27\theta)$$

Figure 2.4 Fourier series for the square wave.

Exercises

1. Prove equations (2.2.2)–(2.2.4) by rewriting the products of trigonometric functions inside the integral as sums before integrating.

2. Are the following functions of θ periodic? If so, determine the smallest period, and which multiples of the fundamental frequency are present. If not, explain why not.

    (i) $\sin\theta + \sin\frac{5}{4}\theta$.
    (ii) $\sin\theta + \sin\sqrt{2}\theta$.
    (iii) $\sin^2\theta$.
    (iv) $\sin(\theta^2)$.
    (v) $\sin\theta + \sin(\theta + \frac{\pi}{3})$.

3. Draw graphs of the functions $\sin(220\pi t) + \sin(440\pi t)$ and $\sin(220\pi t) + \cos(440\pi t)$. Explain why these sound the same, even though the graphs look quite different.

2.3 Even and odd functions

A function $f(\theta)$ is said to be *even* if $f(-\theta) = f(\theta)$, and it is said to be *odd* if $f(-\theta) = -f(\theta)$. For example, $\cos\theta$ is even, while $\sin\theta$ is odd. Of course, most functions are neither even nor odd. If a function happens to be both even and odd, then it is zero, because we have $f(\theta) = f(-\theta) = -f(\theta)$.

Given any function $f(\theta)$, we can obtain an even function by taking the average of $f(\theta)$ and $f(-\theta)$, i.e., $\frac{1}{2}(f(\theta) + f(-\theta))$. Similarly, $\frac{1}{2}(f(\theta) - f(-\theta))$ is an odd function. These add up to give the original function $f(\theta)$, so we have written $f(\theta)$ as a sum of its *even part* and its *odd part*,

$$f(\theta) = \frac{f(\theta) + f(-\theta)}{2} + \frac{f(\theta) - f(-\theta)}{2}.$$

To see that this is the unique way to write the function as a sum of an even function and an odd function, let us suppose that we are given two expressions $f(\theta) = g_1(\theta) + h_1(\theta)$ and $f(\theta) = g_2(\theta) + h_2(\theta)$ with g_1 and g_2 even, and h_1 and h_2 odd. Rearranging $g_1 + h_1 = g_2 + h_2$, we get $g_1 - g_2 = h_2 - h_1$. The left side is even and the right side is odd, so their common value is both even and odd, and hence zero. This means that $g_1 = g_2$ and $h_1 = h_2$.

Multiplication of even and odd functions works like addition (and *not* multiplication) of even and odd numbers:

$\times$	even	odd
even	even	odd
odd	odd	even

Now for any odd function $f(\theta)$, and for any $a > 0$, we have

$$\int_{-a}^{0} f(\theta)\,d\theta = -\int_{0}^{a} f(\theta)\,d\theta$$

so that

$$\int_{-a}^{a} f(\theta) \, d\theta = 0.$$

So, for example, if $f(\theta)$ is even and periodic with period 2π, then $\sin(m\theta)f(\theta)$ is odd, and so the Fourier coefficients b_m are zero, since

$$b_m = \frac{1}{\pi} \int_0^{2\pi} \sin(m\theta) f(\theta) \, d\theta = \frac{1}{\pi} \int_{-\pi}^{\pi} \sin(m\theta) f(\theta) \, d\theta = 0.$$

Similarly, if $f(\theta)$ is odd and periodic with period 2π, then $\cos(m\theta)f(\theta)$ is odd, and so the Fourier coefficients a_m are zero, since

$$a_m = \frac{1}{\pi} \int_0^{2\pi} \cos(m\theta) f(\theta) \, d\theta = \frac{1}{\pi} \int_{-\pi}^{\pi} \cos(m\theta) f(\theta) \, d\theta = 0.$$

This explains, for example, why $a_m = 0$ in the example on page 41. The square wave is not quite an even function, because $f(\pi) \neq f(-\pi)$, but changing the value of a function at a finite set of points in the interval of integration never affects the value of an integral, so we just replace $f(\pi)$ and $f(-\pi)$ by zero to obtain an even function with the same Fourier coefficients.

There is a similar explanation for why $b_{2m} = 0$ in the same example, using a different symmetry. The discussion of even and odd functions depended on the symmetry $\theta \mapsto -\theta$ of order two. For periodic functions of period 2π, there is another symmetry of order two, namely $\theta \mapsto \theta + \pi$. The functions $f(\theta)$ satisfying $f(\theta + \pi) = f(\theta)$ are *half-period symmetric*, while functions satisfying $f(\theta + \pi) = -f(\theta)$ are *half-period antisymmetric*. Any function $f(\theta)$ can be decomposed into half-period symmetric and antisymmetric parts:

$$f(\theta) = \frac{f(\theta) + f(\theta + \pi)}{2} + \frac{f(\theta) - f(\theta + \pi)}{2}.$$

Multiplying half-period symmetric and antisymmetric functions works in the same way as for even and odd functions.

If $f(\theta)$ is half-period antisymmetric, then

$$\int_{\pi}^{2\pi} f(\theta) \, d\theta = - \int_0^{\pi} f(\theta) \, d\theta$$

and so

$$\int_0^{2\pi} f(\theta) \, d\theta = 0.$$

Now the functions $\sin(m\theta)$ and $\cos(m\theta)$ are both half-period symmetric if m is even, and half-period antisymmetric if m is odd. So we deduce that if $f(\theta)$ is half-period symmetric, $f(\theta + \pi) = f(\theta)$, then the Fourier coefficients with odd indices

$(a_{2m+1}$ and $b_{2m+1})$ are zero, while if $f(\theta)$ is antisymmetric, $f(\theta + \pi) = -f(\theta)$, then the Fourier coefficients with even indices a_{2m} and b_{2m} are zero (check that this holds for a_0 too!). This corresponds to the fact that half-period symmetry is really the same thing as being periodic with half the period, so that the frequency components have to be even multiples of the defining frequency; while half-period antisymmetric functions only have frequency components at odd multiples of the defining frequency.

In the example on page 41, the function is half-period antisymmetric, and so the coefficients a_{2m} and b_{2m} are zero.

Exercises

1. Evaluate $\displaystyle\int_0^{2\pi} \sin(\sin\theta) \sin(2\theta)\, d\theta$.

2. Think of $\tan\theta$ as a periodic function with period 2π (even though it could be thought of as having period π). Using the theory of even and odd functions, and the theory of half-period symmetric and antisymmetric functions, which Fourier coefficients of $\tan\theta$ have to be zero? Find the first nonzero coefficient.

3. Which Fourier coefficients vanish for a periodic function $f(\theta)$ of period 2π satisfying $f(\theta) = f(\pi - \theta)$? What about $f(\theta) = -f(\pi - \theta)$?

   (Hint: Consider the symmetry $\theta \mapsto \pi - \theta$, and compare $\displaystyle\int_{-\pi/2}^{\pi/2} f(\theta)\, d\theta$ with $\displaystyle\int_{\pi/2}^{3\pi/2} f(\theta)\, d\theta$ for antisymmetric functions with respect to this symmetry.)

2.4 Conditions for convergence

Unfortunately, it is not true that if we start with a periodic function $f(\theta)$, form the Fourier coefficients a_m and b_m according to equations (2.2.5) and (2.2.6) and then form the sum (2.2.1), then we recover the original function $f(\theta)$. The most obvious problem is that if two functions differ just at a single value of θ then the Fourier coefficients will be identical. So we cannot possibly recover the function from its Fourier coefficients without some further conditions. However, if the function is nice enough, it can be recovered in the manner indicated. The following is a consequence of the work of Dirichlet.

Theorem 2.4.1 *Suppose that $f(\theta)$ is periodic with period 2π, and that it is contin-uous and has a bounded continuous derivative except at a finite number of points in the interval $[0, 2\pi]$. If a_m and b_m are defined by equations (2.2.5) and (2.2.6) then the series defined by equation (2.2.1) converges to $f(\theta)$ at all points where $f(\theta)$ is continuous.*

Proof See Körner (1988), Theorem 1 and Chapters 15 and 16.   □

An important special case of the above theorem is the following. A C^1 function is defined to be a function which is differentiable with continuous derivative. If $f(\theta)$ is a periodic C^1 function with period 2π, then $f'(\theta)$ is continuous on the closed interval $[0, 2\pi]$, and hence bounded there. So $f(\theta)$ satisfies the conditions of the above theorem.

It is important to note that continuity, or even differentiability of $f(\theta)$ is not sufficient for the Fourier series for $f(\theta)$ to converge to $f(\theta)$. Paul DuBois-Reymond constructed an example of a continuous function for which the coefficients a_m and b_m are not bounded. The construction is by no means easy and we shall not give it here. The reader may form the impression at this stage that the only purpose of the existence of such functions is to beset theorems such as the above with conditions, and that in real life, all functions are just as differentiable as we would like them to be. This point of view is refuted by the observation that many phenomena in real life are governed by some form of Brownian motion. Functions describing these phenomena will tend to be everywhere continuous but nowhere differentiable.[4] In music, noise is an example of the same phenomenon. Many of the functions employed in musical synthesis are not even continuous. Sawtooth functions and square waves are typical examples.

However, the question of convergence of the Fourier series is not the same as the question of whether the function $f(\theta)$ can be reconstructed from its Fourier coefficients a_n and b_n. At the age of 19, Fejér proved the remarkable theorem that any continuous function $f(\theta)$ can be reconstructed from its Fourier coefficients. His idea was that if the partial sums s_m defined by

$$s_m = \tfrac{1}{2}a_0 + \sum_{n=1}^{m}(a_n \cos(n\theta) + b_n \sin(n\theta)) \tag{2.4.1}$$

converge, then their averages

$$\sigma_m = \frac{s_0 + \cdots + s_m}{m + 1}$$

converge to the same limit. But it is conceivable that the σ_m could converge without the s_m converging. This idea for smoothing out the convergence had already been around for some time when Fejér approached the problem. It had been used by Euler and extensively studied by Cesàro, and goes by the name of Cesàro summability.

---

[4] The first examples of functions which are everywhere continuous but nowhere differentiable were constructed by Weierstrass, *Abhandlungen aus der Functionenlehre*, Springer (1886), p. 97. He showed that if $0 < b < 1$, a is an odd integer, and $ab > 1 + \frac{3\pi}{2}$ then $f(t) = \sum_{n=1}^{\infty} b^n \cos a^n (2\pi v)t$ is a uniformly convergent sum, and that $f(t)$ is everywhere continuous but nowhere differentiable. G. H. Hardy, Weierstrass's non-differentiable function, *Trans. Amer. Math. Soc.* **17** (1916), 301–325, showed that the same holds if the bound on ab is replaced by $ab > 1$. Manfred Schroeder, *Fractals, Chaos and Power Laws*, W. H. Freeman and Co., 1991, p. 96, points out that functions of this form can be thought of as *fractal* waveforms. For example, if we set $a = 2^{13/12}$, then doubling the speed of this function will result in a tone which sounds similar to the original, but lowered by a semitone and softer by a factor of b. This sort of self-similarity is characteristic of fractals.

Theorem 2.4.2 (Fejér) *If $f(\theta)$ is a Riemann integrable periodic function then the Cesàro sums σ_m converge to $f(\theta)$ as m tends to infinity at every value of θ where $f(\theta)$ is continuous.*[5]

Proof We shall prove this theorem in Section 2.7. See also Körner (1988), Chapter 2.   □

We shall interpret this theorem as saying that every continuous function may be reconstructed from its Fourier coefficients. But the reader should bear in mind that if the function does not satisfy the hypotheses of Theorem 2.4.1 then the reconstruction of the function is done via Cesàro sums, and not simply as the sum of the Fourier series.

There are other senses in which we could ask for a Fourier series to converge. One of the most important ones is *mean square convergence*.

Theorem 2.4.3 *Let $f(\theta)$ be a continuous periodic function with period 2π. Then among all the functions $g(\theta)$ which are linear combinations of $\cos(n\theta)$ and $\sin(n\theta)$ with $0 \leq n \leq m$, the partial sum s_m defined in equation (2.4.1) minimizes the mean square error of $g(\theta)$ as an approximation to $f(\theta)$,*

$$\frac{1}{2\pi} \int_0^{2\pi} |f(\theta) - g(\theta)|^2 \, d\theta.$$

Furthermore, in the limit as m tends to infinity, the mean square error of s_m as an approximation to $f(\theta)$ tends to zero.

Proof See Körner (1988), Chapters 32–34.   □

Exercises

1. Show that the function $f(x) = x^2 \sin(1/x^2)$ is differentiable for all values of x, but its derivative is unbounded around $x = 0$.

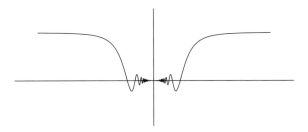

---

[5] Continuous functions are Riemann integrable, so Fejér's theorem applies to all continuous periodic functions.

2. Find the Fourier series for the periodic function $f(\theta) = |\sin\theta|$ (the absolute value of $\sin\theta$). In other words, find the coefficients a_m and b_m using equations (2.2.5) and (2.2.6). You will need to divide the interval from 0 to 2π into two subintervals in order to evaluate the integral.

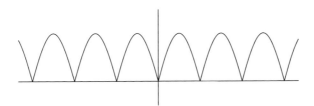

3. Let $\phi(\theta)$ be the periodic sawtooth function with period 2π defined by $\phi(\theta) = (\pi - \theta)/2$ for $0 < \theta < 2\pi$ and $\phi(0) = \phi(2\pi) = 0$. Find the Fourier series for $\phi(\theta)$.[6]

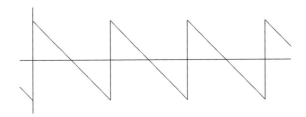

4. Find the Fourier series of the continuous periodic triangular wave function defined by

$$f(\theta) = \begin{cases} \frac{\pi}{2} - \theta & 0 \le \theta \le \pi, \\ \theta - \frac{3\pi}{2} & \pi \le \theta \le 2\pi, \end{cases}$$

and $f(\theta + 2\pi) = f(\theta)$.

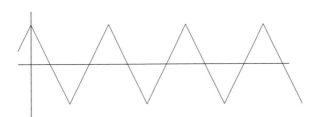

5. (a) Show that if $f(\theta)$ is a bounded (and Riemann integrable) periodic function with period 2π then the Fourier coefficients a_m and b_m defined by (2.2.5)–(2.2.7) are bounded.

---

[6] The sawtooth waveform is approximately what is produced by a violin or other bowed string instrument. This is because the bow pulls the string, and then suddenly releases it when the coefficient of static friction is exceeded. The coefficient of dynamic friction is smaller, so once the string is released by the bow, it will tend to continue moving rapidly until the other extreme of its trajectory is reached. See Section 3.4.

(b) If $f(\theta)$ is a differentiable periodic function with period 2π, find the relationship between the Fourier coefficients $a_m(f)$, $b_m(f)$ for $f(\theta)$ and the Fourier coefficients $a_m(f')$, $b_m(f')$ for the derivative $f'(\theta)$. (Hint: use integration by parts.)

(c) Show that if $f(\theta)$ is a k times differentiable periodic function with period 2π, and the kth derivative $f^{(k)}(\theta)$ is bounded, then the Fourier coefficients a_m and b_m of $f(\theta)$ are bounded by a constant multiple of $1/m^k$.

   We see from this question that smoothness of $f(\theta)$ is reflected in rapidity of decay of the Fourier coefficients.

6. Find the Fourier series for the function $f(\theta)$ defined by $f(\theta) = \theta^2$ for $-\pi \le \theta \le \pi$ and then extended to all values of θ by periodicity, $f(\theta + 2\pi) = f(\theta)$. Evaluate your answer at $\theta = 0$ and at $\theta = \pi$, and use your answer to find $\sum_{n=1}^{\infty} \frac{(-1)^n}{n^2}$ and $\sum_{n=1}^{\infty} \frac{1}{n^2}$.

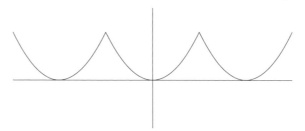

2.5 The Gibbs phenomenon

A function defined on a closed interval is said to be *piecewise continuous* if it is continuous except at a finite set of points, and at those points the left limit and right limit exist although they may not be equal. When we talk of the *size of* a discontinuity of a piecewise continuous function $f(\theta)$ at $\theta = a$, we mean the difference $f(a^+) - f(a^-)$, where

$$f(a^+) = \lim_{\theta \to a^+} f(\theta), \qquad f(a^-) = \lim_{\theta \to a^-} f(\theta)$$

denote the left limit and the right limit at that point. A periodic function is said to be piecewise continuous if it is so on a closed interval forming a period of the function.

   Many of the functions encountered in the theory of synthesized sound are piecewise continuous but not continuous. These include waveforms such as the square wave and the sawtooth function.

   Denote by $\phi(\theta)$ the piecewise continuous periodic sawtooth function defined by $\phi(\theta) = (\pi - \theta)/2$ for $0 < \theta < 2\pi$, $\phi(0) = 0$, and $\phi(\theta + 2\pi) = \phi(\theta)$. See Figure 2.5. Then, given any piecewise continuous periodic function $f(\theta)$, we may add some finite set of functions of the form $C\phi(\theta + \alpha)$ (with C and α constants) to make the left limits and right limits at the discontinuities agree. We can then just change the values of the function at the discontinuities, which will not affect the

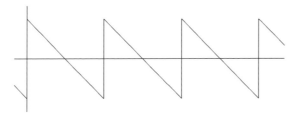

Figure 2.5 Sawtooth function.

Fourier series, to make the function continuous. It follows that in order to understand the Fourier series for piecewise continuous functions in general, it suffices
to understand the Fourier series of continuous functions together with the Fourier
series of the single function $\phi(\theta)$. The Fourier series of this function (see Exercise 3
of Section 2.4) is

$$\phi(\theta) = \sum_{n=1}^{\infty} \frac{\sin n\theta}{n}.$$

At the discontinuity ($\theta = 0$), this series converges to zero because all the terms are
zero. This is the average of the left limit and the right limit at this point. It follows
that for any piecewise continuous periodic function, the Cesàro sums σ_m described
in Section 2.4 converge everywhere, and at the points of discontinuity σ_m converges
to the average of the left and right limit at the point:

$$\lim_{m \to \infty} \sigma_m(a) = \tfrac{1}{2}(f(a^+) + f(a^-)).$$

A further examination of the function $\phi(\theta)$ shows that the convergence around
the point of discontinuity is not as straightforward as one might suppose. Namely,
setting

$$\phi_m(\theta) = \sum_{n=1}^{m} \frac{\sin n\theta}{n}, \qquad (2.5.1)$$

although it is true that we have *pointwise convergence*, in the sense that for each
point a we have $\lim_{m \to \infty} \phi_m(a) = \phi(a)$, this convergence is not uniform.

The definition in analysis of pointwise convergence is that, given a value a of
θ and given $\varepsilon > 0$, there exists N such that $m \geq N$ implies $|\phi_m(a) - \phi(a)| < \varepsilon$.
Uniform convergence means that, given $\varepsilon > 0$, there exists N (independent of a)
such that for all values a of θ, $m \geq N$ implies $|\phi_m(a) - \phi(a)| < \varepsilon$. What happens
with the Fourier series for the above function ϕ is that there is an overshoot, the
size of which does not tend to zero as m gets larger. The peak of the overshoot gets
closer and closer to the discontinuity, though, so that for any particular value a of
θ, convergence holds. But choosing ε smaller than the size of the overshoot shows

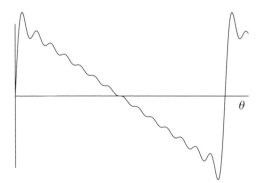

Figure 2.6 $\sin\theta + \frac{1}{2}\sin 2\theta + \cdots + \frac{1}{14}\sin 14\theta$.

that uniform convergence fails. This overshoot is called the Gibbs phenomenon.[7]
See Figure 2.6.

To demonstrate the reality of the overshoot, we shall compute its size in the
limit. The first step is to differentiate $\phi_m(\theta)$ to find its local maxima and minima. We concentrate on the interval $0 \le \theta \le \pi$, since $\phi_m(2\pi - \theta) = -\phi_m(\theta)$. We
have

$$\phi'_m(\theta) = \sum_{n=1}^{m} \cos n\theta = \frac{\sin\frac{1}{2}m\theta \, \cos\frac{1}{2}(m+1)\theta}{\sin\frac{1}{2}\theta}$$

(see Exercise 6 of Section 1.8). So the zeros of $\phi'_m(\theta)$ occur at $\theta = \frac{(2k+1)\pi}{m+1}$ and
$\theta = \frac{2k\pi}{m}, 0 \le k \le \lfloor\frac{m-1}{2}\rfloor$.[8]

Now $\sin\frac{1}{2}\theta$ is positive throughout the interval $0 \le \theta \le 2\pi$. At $\theta = \frac{(2k+1)\pi}{m+1}$,
$\sin\frac{1}{2}m\theta$ has sign $(-1)^k$ while $\cos\frac{1}{2}(m+1)\theta$ changes sign from $(-1)^k$ to $(-1)^{k+1}$,
so that $\phi'_m(\theta)$ changes from positive to negative. It follows that $\theta = \frac{(2k+1)\pi}{m+1}$ is a
local maximum of ϕ_m. Similarly, at $\theta = \frac{2k\pi}{m}$, $\cos\frac{1}{2}(m+1)\theta$ has sign $(-1)^k$ while
$\sin\frac{1}{2}m\theta$ changes sign from $(-1)^{k-1}$ to $(-1)^k$, so that $\phi'_m(\theta)$ changes from negative to positive. It follows that $\theta = \frac{2k\pi}{m}$ is a local minimum of $\phi_m(\theta)$. These local
maxima and minima alternate.

The first local maximum value of $\phi_m(\theta)$ for $0 \le \theta \le 2\pi$ happens at $\frac{\pi}{m+1}$. The
value of $\phi_m(\theta)$ at this maximum is

$$\phi_m\left(\frac{\pi}{m+1}\right) = \sum_{n=1}^{m} \frac{1}{n}\sin\left(\frac{n\pi}{m+1}\right) = \frac{\pi}{m+1}\sum_{n=1}^{m} \frac{\sin\left(\frac{n\pi}{m+1}\right)}{\left(\frac{n\pi}{m+1}\right)}.$$

[7] Josiah Willard Gibbs described this phenomenon in a series of letters to *Nature* in 1898 in correspondence with
A. E. H. Love. He seems to have been unaware of the previous treatment of the subject by Henry Wilbraham in
his article On a certain periodic function, *Cambridge & Dublin Math. J.* **3** (1848), 198–201.
[8] The notation $\lfloor x \rfloor$ denotes the largest integer less than or equal to x.

This is the Riemann sum for

$$\int_0^\pi \frac{\sin\theta}{\theta}\,d\theta$$

with $m+1$ equal intervals of size $\frac{\pi}{m+1}$ (note that $\lim\limits_{\theta\to 0}\frac{\sin\theta}{\theta}=1$ so that we should define the integrand to be 1 when $\theta=0$ to make a continuous function on the closed interval $0\le\theta\le\pi$). Therefore the limit as m tends to infinity of the height of the first maximum point of the sum of the first m terms in the Fourier series for $\phi(\theta)$ is

$$\int_0^\pi \frac{\sin\theta}{\theta}\,d\theta\approx 1.851\,937\,0.$$

This overshoots the maximum value $\frac{\pi}{2}\approx 1.570\,7963$ of the function $\phi(\theta)$ by a factor of $1.178\,9797$. Of course, the size of the discontinuity is not $\frac{\pi}{2}$ but π, so that as a proportion of the size of the discontinuity, the overshoot is about 8.9490%.[9] It follows that, for any piecewise continuous function, the overshoot of the Fourier series just after a discontinuity is this proportion of the size of the discontinuity.

After the function overshoots, it then returns to undershoot, then overshoot again, and so on, each time with a smaller value than before. An argument similar to the above shows that the value at the kth critical point of $\phi_m(\theta)$ tends to $\int_0^{k\pi}\frac{\sin\theta}{\theta}\,d\theta$ as m tends to infinity. Thus, for example, the first undershoot ($k=2$) has a value with a limit of about $1.418\,1516$, which undershoots $\frac{\pi}{2}$ by a factor of $0.902\,8233$. The undershoot is therefore about 4.8588% of the size of the discontinuity.

The Gibbs phenomenon can be interpreted in terms of the response of an amplifier as follows. No matter how good your amplifier is, if you feed it a square wave, the output will overshoot at the discontinuity by roughly 9%. This is because any amplifier has a frequency beyond which it does not respond. Improving the amplifier can only increase this frequency, but cannot get rid of the limitation altogether.

Manufacturers of cathode ray tubes also have to contend with this problem. The beam is being made to run across the tube from left to right linearly and then switch back suddenly to the left. Much effort goes into preventing the inevitable overshoot from causing problems.

As mentioned above, the Gibbs phenomenon is a good example to illustrate the distinction between pointwise convergence and uniform convergence. For pointwise convergence of a sequence of functions $f_n(\theta)$ to a function $f(\theta)$, it is required that for each value of θ, the values $f_n(\theta)$ must converge to $f(\theta)$. For uniform convergence, it is required that the distance between $f_n(\theta)$ and $f(\theta)$ is bounded by a quantity which depends on n and not on θ, and which tends to zero as n tends to infinity. In

---

[9] This value was first computed by Maxime Bôcher, Introduction to the theory of Fourier's series. *Ann. Math.* (2) **7** (1905–6), 81–152. A number of otherwise reputable sources overstate the size of the overshoot by a factor of two for some reason probably associated with uncritical copying.

the above example, the distance between the nth partial sum of the Fourier series and the original function can at best be bounded by a quantity which depends on n and not on θ, but which tends to roughly 0.281 14. So this Fourier series converges pointwise, but not uniformly.

Exercise

Show that

$$\int_0^x \frac{\sin\theta}{\theta}\, d\theta = \sum_{n=0}^{\infty} \frac{(-1)^n x^{2n+1}}{(2n+1)(2n+1)!}.$$

Use this formula to verify the approximate value of $\int_0^\pi \frac{\sin\theta}{\theta}\, d\theta$ given in the text.

2.6 Complex coefficients

The theory of Fourier series is considerably simplified by the introduction of complex exponentials. The relationships

$$e^{i\theta} = \cos\theta + i\sin\theta, \qquad \cos\theta = \frac{e^{i\theta} + e^{-i\theta}}{2}, \tag{2.6.1}$$

$$e^{-i\theta} = \cos\theta - i\sin\theta, \qquad \sin\theta = \frac{e^{i\theta} - e^{-i\theta}}{2i}$$

mean that equation (2.2.1) can be rewritten as[10]

$$f(\theta) = \sum_{n=-\infty}^{\infty} \alpha_n e^{in\theta}, \tag{2.6.2}$$

where $\alpha_0 = a_0/2$, and for $m > 0$, $\alpha_m = a_m/2 + b_m/(2i)$ and $\alpha_{-m} = a_m/2 - b_m/(2i)$. Conversely, given a series of the form (2.6.2) we can reconstruct the series (2.2.1) using $a_0 = 2\alpha_0$, $a_m = \alpha_m + \alpha_{-m}$ and $b_m = i(\alpha_m - \alpha_{-m})$ for $m > 0$. Equations (2.2.2)–(2.2.4) are replaced by the single equation[11]

$$\int_0^{2\pi} e^{im\theta} e^{in\theta}\, d\theta = \begin{cases} 2\pi & \text{if } m = -n, \\ 0 & \text{if } m \neq -n, \end{cases}$$

and equations (2.2.5)–(2.2.7) are replaced by

$$\alpha_m = \frac{1}{2\pi} \int_0^{2\pi} e^{-im\theta} f(\theta)\, d\theta. \tag{2.6.3}$$

---

[10] Note that we are dealing with complex valued functions of a real periodic variable, and not with functions of a complex variable here.

[11] Over the complex numbers, to interpret this equation as an orthogonality relation (see the footnote on page 40), the inner product needs to be taken to be $\langle f, g \rangle = \frac{1}{2\pi} \int_0^{2\pi} f(\theta)\overline{g(\theta)}\, d\theta$.

Exercise

For the square wave example discussed in Section 2.2, show that

$$\alpha_m = \begin{cases} 2/im\pi & m \text{ odd}, \\ 0 & m \text{ even}. \end{cases}$$

so that the Fourier series is

$$\sum_{n=-\infty}^{\infty} \frac{2}{i(2n+1)\pi} e^{i(2n+1)\theta}.$$

2.7 Proof of Fejér's theorem

We are now in a position to prove Fejér's theorem 2.4.2. This section may safely be skipped on first reading.

In terms of the complex form of the Fourier series, the partial sums (2.4.1) become

$$S_m = \sum_{n=-m}^{m} \alpha_n e^{in\theta}, \tag{2.7.1}$$

and so the Cesàro sums σ_m are given by

$$
\begin{aligned}
\sigma_m(\theta) &= \frac{s_0 + \cdots + s_m}{m+1} \\
&= \frac{1}{m+1} \sum_{j=0}^{m} \sum_{n=-j}^{j} \alpha_n e^{in\theta} \\
&= \frac{1}{m+1} \left(\alpha_{-m} e^{-im\theta} + 2\alpha_{-(m-1)} e^{-i(m-1)\theta} + 3\alpha_{-(m-2)} e^{-i(m-2)\theta} + \cdots \right. \\
&\qquad \left. + \cdots + m\alpha_{-1} e^{-i\theta} + (m+1)\alpha_0 e^{0} + m\alpha_1 e^{i\theta} + \cdots + \alpha_m e^{im\theta} \right) \\
&= \sum_{n=-m}^{m} \frac{m+1-|n|}{m+1} \alpha_n e^{in\theta} \\
&= \sum_{n=-m}^{m} \frac{m+1-|n|}{m+1} \left(\frac{1}{2\pi} \int_{0}^{2\pi} e^{-inx} f(x)\,dx \right) e^{in\theta} \\
&= \frac{1}{2\pi} \int_{0}^{2\pi} f(x) \left(\sum_{n=-m}^{m} \frac{m+1-|n|}{m+1} e^{in(\theta-x)} \right) dx \\
&= \frac{1}{2\pi} \int_{0}^{2\pi} f(x) K_m(\theta - x)\,dx,
\end{aligned}
$$

where

$$K_m(y) = \sum_{n=-m}^{m} \frac{m+1-|n|}{m+1} e^{iny}.$$

The functions K_m are called the Fejér kernels.

The substitution $y = \theta - x$ shows that

$$\frac{1}{2\pi} \int_0^{2\pi} f(x) K_m(\theta - x)\, dx = \frac{1}{2\pi} \int_0^{2\pi} f(\theta - y) K_m(y)\, dy.$$

By examining what happens when a geometric series is squared, for $y \neq 0$ we have

$$K_m(y) = \frac{1}{m+1}\left(e^{-imy} + 2e^{-i(m-1)y} + \cdots + (m+1)e^0 + \cdots + e^{imy}\right)$$

$$= \frac{1}{m+1}\left(e^{-i\frac{m}{2}y} + e^{-i(\frac{m}{2}-1)y} + \cdots + e^{i\frac{m}{2}y}\right)^2 \qquad (2.7.2)$$

$$= \frac{1}{m+1}\left(\frac{e^{i\frac{m+1}{2}y} - e^{-i\frac{m+1}{2}y}}{e^{i\frac{1}{2}y} - e^{-i\frac{1}{2}y}}\right)^2$$

$$= \frac{1}{m+1}\left(\frac{\sin\frac{m+1}{2}y}{\sin\frac{1}{2}y}\right)^2,$$

and $K_m(0) = m + 1$ can also be read off from (2.7.2). Figure 2.7 shows the graphs of $K_m(y)$ for some small values of m.

The function $K_m(y)$ satisfies $K_m(y) \geq 0$ for all y; for any $\delta > 0$, $K_m(y) \to 0$ uniformly as $m \to \infty$ on $[\delta, 2\pi - \delta]$; and $\int_0^{2\pi} K_m(y)\, dy = 2\pi$. So

$$\sigma_m(\theta) = \frac{1}{2\pi} \int_0^{2\pi} f(\theta - y) K_m(y)\, dy \approx \frac{1}{2\pi} \int_{-\delta}^{\delta} f(\theta - y) K_m(y)\, dy$$

$$\approx f(\theta)\left(\frac{1}{2\pi} \int_{-\delta}^{\delta} K_m(y)\, dy\right) \approx f(\theta).$$

If $f(\theta)$ is continuous at θ, then by choosing δ small enough, the second approximation may be made as close as desired (independently of m). Then by choosing m large enough, the first and third approximations may be made as close as desired. This completes the proof of Fejér's theorem.

Exercise

(i) Substitute equation (2.6.3) in equation (2.7.1) to show that

$$s_m(\theta) = \frac{1}{2\pi} \int_0^{2\pi} f(x) D_m(\theta - x)\, dx,$$

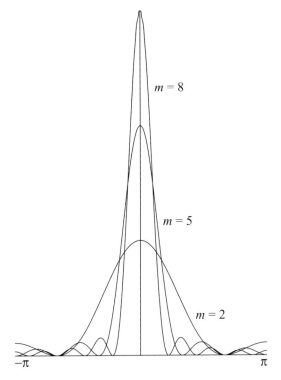

$m = 8$

$m = 5$

$m = 2$

$-\pi$ π

Figure 2.7 $K_m(y)$.

where

$$D_m(y) = \sum_{n=-m}^{m} e^{iny}.$$

The functions D_m are called the *Dirichlet kernels*.

(ii) Use a substitution to show that

$$s_m(\theta) = \frac{1}{2\pi} \int_0^{2\pi} f(\theta - y) D_m(y) \, dy.$$

(iii) By regarding the formula for $D_m(y)$ as a geometric series, show that

$$D_m(y) = \frac{\sin\left(m + \frac{1}{2}\right)y}{\sin\frac{1}{2}y}.$$

(iv) Show that $|D_m(y)| \leq |\operatorname{cosec}\frac{1}{2}y|$

(v) Sketch the graphs of the Dirichlet kernels for small values of m. What happens as m gets large?

2.8 Bessel functions

Bessel functions[12] are the result of applying the theory of Fourier series to the functions $\sin(z \sin \theta)$ and $\cos(z \sin \theta)$ as functions of θ, where z is to be thought of at first as a real (or complex) constant, and later it will be allowed to vary. We shall have two uses for the Bessel functions. One is understanding the vibrations of a drum in Section 3.6, and the other is understanding the amplitudes of side bands in FM synthesis in Section 8.8.

Now $\sin(z \sin \theta)$ is an odd periodic function of θ, so its Fourier coefficients a_n (2.2.1) are zero for all n (see Section 2.3). Since

$$\sin(z \sin(\pi + \theta)) = -\sin(z \sin \theta),$$

the Fourier coefficients b_{2n} are also zero (see Section 2.3 again). The coefficients b_{2n+1} depend on the parameter z, and so we write $2J_{2n+1}(z)$ for this coefficient. The factor of two simplifies some later calculations. So the Fourier expansion (2.2.1) is

$$\sin(z \sin \theta) = 2 \sum_{n=0}^{\infty} J_{2n+1}(z) \sin(2n + 1)\theta. \qquad (2.8.1)$$

Similarly, $\cos(z \sin \theta)$ is an even periodic function of θ, so the coefficients b_n are zero. Since

$$\cos(z \sin(\pi + \theta)) = \cos(z \sin \theta)$$

we also have $a_{2n+1} = 0$, and we write $2J_{2n}(z)$ for the coefficient a_{2n} to obtain

$$\cos(z \sin \theta) = J_0(z) + 2 \sum_{n=1}^{\infty} J_{2n}(z) \cos 2n\theta. \qquad (2.8.2)$$

The functions $J_n(z)$ giving the Fourier coefficients in these expansions are called the *Bessel functions* of the first kind.

Equations (2.2.5) and (2.2.6) allow us to find the Fourier coefficients $J_n(z)$ in the above expansions as integrals. We obtain

$$2J_{2n+1}(z) = \frac{1}{\pi} \int_0^{2\pi} \sin(2n + 1)\theta \sin(z \sin \theta) \, d\theta.$$

---

[12] Friedrich Wilhelm Bessel was a German astronomer and a friend of Gauss. He was born in Minden on July 22, 1784. His working life started as a ship's clerk. But in 1806, he became an assistant at an astronomical observatory in Lilienthal. In 1810 he became director of the then new Prussian Observatory in Königsberg, where he remained until he died on March 17, 1846. The original context (around 1824) of his investigations of the functions that bear his name was the study of planetary motion, as we shall describe in Section 2.11.

The integrand is an even function of θ, so the integral from 0 to 2π is twice the integral from 0 to π,

$$J_{2n+1}(z) = \frac{1}{\pi} \int_0^\pi \sin(2n+1)\theta \sin(z \sin \theta) \, d\theta.$$

Now the function $\cos(2n+1)\theta \cos(z \sin \theta)$ is negated when θ is replaced by $\pi - \theta$, so

$$\frac{1}{\pi} \int_0^\pi \cos(2n+1)\theta \cos(z \sin \theta) \, d\theta = 0.$$

Adding this into the above expression for $J_{2n+1}(z)$, we obtain

$$J_{2n+1}(z) = \frac{1}{\pi} \int_0^\pi [\cos(2n+1)\theta \cos(z \sin \theta) + \sin(2n+1)\theta \sin(z \sin \theta)] \, d\theta$$

$$= \frac{1}{\pi} \int_0^\pi \cos((2n+1)\theta - z \sin \theta) \, d\theta.$$

In a similar way, we have

$$2 J_{2n}(z) = \frac{1}{\pi} \int_0^{2\pi} \cos 2n\theta \cos(z \sin \theta) \, d\theta,$$

which a similar manipulation puts in the form

$$J_{2n}(z) = \frac{1}{\pi} \int_0^\pi \cos(2n\theta - z \sin \theta) \, d\theta.$$

This means that we have the single equation for all values of n, even or odd,

$$\boxed{J_n(z) = \frac{1}{\pi} \int_0^\pi \cos(n\theta - z \sin \theta) \, d\theta,} \qquad (2.8.3)$$

which can be taken as a definition for the Bessel functions for integers $n \geq 0$. In fact, this definition also makes sense when n is a negative integer,[13] and gives

$$J_{-n}(z) = (-1)^n J_n(z). \qquad (2.8.4)$$

This means that (2.8.1) and (2.8.2) can be rewritten as

$$\sin(z \sin \theta) = \sum_{n=-\infty}^\infty J_{2n+1}(z) \sin(2n+1)\theta, \qquad (2.8.5)$$

$$\cos(z \sin \theta) = \sum_{n=-\infty}^\infty J_{2n}(z) \cos 2n\theta. \qquad (2.8.6)$$

---

[13] For non-integer values of n, the above formula is not the correct definition of $J_n(z)$. Rather, one uses the differential equation (2.10.1). See, for example, Whittaker and Watson, *A Course in Modern Analysis*, Cambridge University Press, 1927, p. 358.

We also have

$$\sum_{n=-\infty}^{\infty} J_{2n}(z) \sin 2n\theta = 0,$$

$$\sum_{n=-\infty}^{\infty} J_{2n+1}(z) \cos(2n+1)\theta = 0,$$

because the terms with positive subscript cancel with the corresponding terms with negative subscript. So we can rewrite equations (2.8.5) and (2.8.6) as

$$\sin(z \sin \theta) = \sum_{n=-\infty}^{\infty} J_n(z) \sin n\theta, \qquad (2.8.7)$$

$$\cos(z \sin \theta) = \sum_{n=-\infty}^{\infty} J_n(z) \cos n\theta. \qquad (2.8.8)$$

So, using equation (1.7.2), we have

$$\sin(\phi + z \sin \theta) = \sin \phi \cos(z \sin \theta) + \cos \phi \sin(z \sin \theta)$$

$$= \sin \phi \sum_{n=-\infty}^{\infty} J_n(z) \cos n\theta + \cos \phi \sum_{n=-\infty}^{\infty} J_n(z) \sin n\theta$$

$$= \sum_{n=-\infty}^{\infty} J_n(z)(\sin \phi \cos n\theta + \cos \phi \sin n\theta).$$

Finally, recombining the terms using equation (1.7.2), we obtain

$$\boxed{\sin(\phi + z \sin \theta) = \sum_{n=-\infty}^{\infty} J_n(z) \sin(\phi + n\theta).} \qquad (2.8.9)$$

This equation will be of fundamental importance for FM synthesis in Section 8.8. A similar argument gives

$$\cos(\phi + z \sin \theta) = \sum_{n=-\infty}^{\infty} J_n(z) \cos(\phi + n\theta), \qquad (2.8.10)$$

which can also be obtained from equation (2.8.9) by replacing ϕ by $\phi + \frac{\pi}{2}$, or by differentiating with respect to ϕ, keeping z and θ constant.

Figure 2.8 shows graphs of the first few Bessel functions:

Exercises

1. Replace θ by $\frac{\pi}{2} - \theta$ in equations (2.8.1) and (2.8.2) to obtain the Fourier series for $\sin(z \cos \theta)$ and $\cos(z \cos \theta)$.
2. Deduce equation (2.8.10) from equation (2.8.9).

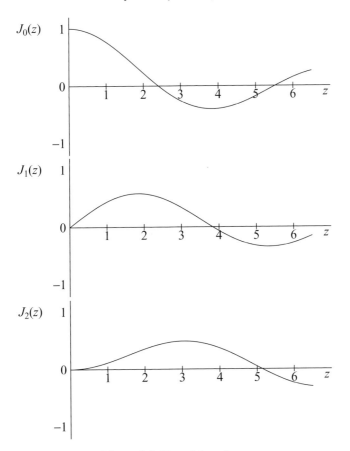

Figure 2.8 Bessel functions.

2.9 Properties of Bessel functions

From equation (2.8.9), we can obtain relationships between the Bessel functions and their derivatives, as follows. Differentiating (2.8.9) with respect to z, keeping θ and ϕ constant, we obtain

$$\sin\theta \, \cos(\phi + z\sin\theta) = \sum_{n=-\infty}^{\infty} J_n'(z)\sin(\phi + n\theta). \qquad (2.9.1)$$

On the other hand, multiplying equation (2.8.10) by $\sin\theta$ and using (1.8.4), we have

$$\sin\theta \, \cos(\phi + z\sin\theta) = \sum_{n=-\infty}^{\infty} J_n(z)\cdot\tfrac{1}{2}(\sin(\phi + (n+1)\theta) - \sin(\phi + (n-1)\theta))$$

$$= \sum_{n=-\infty}^{\infty} \tfrac{1}{2}(J_{n-1}(z) - J_{n+1}(z))\sin(\phi + n\theta). \qquad (2.9.2)$$

In the last step, we have split the sum into two parts, reindexed by replacing n by $n-1$ and $n+1$ respectively in the two parts, and then recombined the parts.

We would like to compare formulas (2.9.1) and (2.9.2) and deduce that

$$J_n'(z) = \tfrac{1}{2}(J_{n-1}(z) - J_{n+1}(z)). \tag{2.9.3}$$

In order to do this, we need to know that the functions $\sin(\phi + n\theta)$ are independent. This can be seen using Fourier series as follows.

Lemma 2.9.1 *If*

$$\sum_{n=-\infty}^{\infty} a_n \sin(\phi + n\theta) = \sum_{n=-\infty}^{\infty} a_n' \sin(\phi + n\theta),$$

as an identity between functions of ϕ and θ, where a_n and a_n' do not depend on θ and ϕ, then each coefficient $a_n = a_n'$.

Proof Subtracting one side from the other, we see that we must prove that if $\sum_{n=-\infty}^{\infty} c_n \sin(\phi + n\theta) = 0$ (where $c_n = a_n - a_n'$) then each $c_n = 0$. To prove this, we expand using (1.7.2) to give

$$\sum_{n=-\infty}^{\infty} c_n \sin\phi \cos n\theta + \sum_{n=-\infty}^{\infty} c_n \cos\phi \sin n\theta = 0.$$

Putting $\phi = 0$ and $\phi = \frac{\pi}{2}$ in this equation, we obtain

$$\sum_{n=-\infty}^{\infty} c_n \cos n\theta = 0, \tag{2.9.4}$$

$$\sum_{n=-\infty}^{\infty} c_n \sin n\theta = 0. \tag{2.9.5}$$

Multiply equation (2.9.4) by $\cos m\theta$, integrate from 0 to 2π and divide by π. Using equation (2.2.3), we get $c_m + c_{-m} = 0$. Similarly, from equations (2.9.5) and (2.2.4), we get $c_m - c_{-m} = 0$. Adding and dividing by two, we get $c_m = 0$. $\quad\square$

This completes the proof of equation (2.9.3). As an example, setting $n = 0$ in (2.9.3) and using (2.8.4), we obtain

$$J_1(z) = -J_0'(z). \tag{2.9.6}$$

In a similar way, we can differentiate (2.8.9) with respect to θ, keeping z and ϕ constant, to obtain

$$z \cos\theta \, \cos(\phi + z\sin\theta) = \sum_{n=-\infty}^{\infty} n J_n(z) \cos(\phi + n\theta). \tag{2.9.7}$$

On the other hand, multiplying equation (2.8.10) by $z\cos\theta$ and using (1.8.7), we obtain

$$z\cos\theta\,\cos(\phi + z\sin\theta)$$

$$= \sum_{n=-\infty}^{\infty} J_n(z).\tfrac{z}{2}(\cos(\phi + (n+1)\theta) + \cos(\phi + (n-1)\theta))$$

$$= \sum_{n=-\infty}^{\infty} \tfrac{z}{2}(J_{n-1}(z) + J_{n+1}(z))\cos(\phi + n\theta). \tag{2.9.8}$$

Comparing equations (2.9.7) and (2.9.8) and using Lemma 2.9.1, we obtain the recurrence relation

$$J_n(z) = \frac{z}{2n}(J_{n-1}(z) + J_{n+1}(z)). \tag{2.9.9}$$

Exercise

Show that $\displaystyle\int_0^\infty J_1(z)\,dz = 1$.

(You may use the fact that $\displaystyle\lim_{z\to\infty} J_0(z) = 0$.)

2.10 Bessel's equation and power series

Using equations (2.9.3) and (2.9.9), we can now derive the differential equation (2.10.1) for the Bessel functions $J_n(z)$. Using (2.9.3) twice, we obtain

$$J_n''(z) = \tfrac{1}{2}(J_{n-1}'(z) - J_{n+1}'(z))$$

$$= \tfrac{1}{4}J_{n-2}(z) - \tfrac{1}{2}J_n(z) + \tfrac{1}{4}J_{n+2}(z).$$

On the other hand, substituting (2.9.9) into (2.9.3), we obtain

$$J_n'(z) = \tfrac{1}{2}\left(\tfrac{z}{2(n-1)}(J_{n-2}(z) + J_n(z)) - \tfrac{z}{2(n+1)}(J_n(z) + J_{n+2}(z))\right)$$

$$= \tfrac{z}{4(n-1)}J_{n-2}(z) + \tfrac{z}{2(n^2-1)}J_n(z) - \tfrac{z}{4(n-1)}J_{n+2}(z).$$

In a similar way, using (2.9.9) twice gives

$$J_n(z) = \tfrac{z}{2n}\left(\tfrac{z}{2(n-1)}(J_{n-2}(z) + J_n(z)) + \tfrac{z}{2(n+1)}(J_n(z) + J_{n+2}(z))\right)$$

$$= \tfrac{z^2}{4n(n-1)}J_{n-2}(z) + \tfrac{z^2}{n^2-1}J_n(z) + \tfrac{z^2}{4n(n+1)}J_{n+2}(z).$$

Combining these three formulae, we obtain

$$J_n''(z) + \tfrac{1}{z}J_n'(z) - \tfrac{n^2}{z^2}J_n(z) = -J_n(z),$$

or

$$J_n''(z) + \frac{1}{z}J_n'(z) + \left(1 - \frac{n^2}{z^2}\right)J_n(z) = 0. \qquad (2.10.1)$$

We now discuss the general solution to *Bessel's equation*, namely the differential equation

$$f''(z) + \frac{1}{z}f'(z) + \left(1 - \frac{n^2}{z^2}\right)f(z) = 0. \qquad (2.10.2)$$

This is an example of a second order linear differential equation, and once one solution is known, there is a general procedure for obtaining all solutions. In this case, this consists of substituting $f(z) = J_n(z)g(z)$, and finding the differential equation satisfied by the new function $g(z)$. We find that

$$f'(z) = J_n'(z)g(z) + J_n(z)g'(z),$$
$$f''(z) = J_n''(z)g(z) + 2J_n'(z)g'(z) + J_n(z)g''(z).$$

So, substituting into Bessel's equation (2.10.2), we obtain

$$\left(J_n''(z) + \frac{1}{z}J_n'(z) + \left(1 - \frac{n^2}{z^2}\right)J_n(z)\right)g(z) +$$

$$\left(2J_n'(z) + \frac{1}{z}J_n(z)\right)g'(z) + J_n(z)g''(z) = 0.$$

The coefficient of $g(z)$ vanishes by equation (2.10.1), and so we are left with

$$\left(2J_n'(z) + \frac{1}{z}J_n(z)\right)g'(z) + J_n(z)g''(z) = 0. \qquad (2.10.3)$$

This is a separable first order equation for $g'(z)$, so we separate the variables,

$$\frac{g''(z)}{g'(z)} = -2\frac{J_n'(z)}{J_n(z)} - \frac{1}{z},$$

and integrate to obtain

$$\ln|g'(z)| = -2\ln|J_n(z)| - \ln|z| + C,$$

where C is the constant of integration. Exponentiating, we obtain

$$g'(z) = \frac{B}{zJ_n(z)^2},$$

where $B = \pm e^C$. Alternatively, we could have obtained this directly from equation (2.10.3) by multiplying by $zJ_n(z)$ to see that the derivative of $zJ_n(z)^2g'(z)$ is zero.

Integrating again, we obtain

$$g(z) = A + B \int \frac{dz}{z J_n(z)^2},$$

where the integral sign denotes a chosen antiderivative. Finally, it follows that the general solution to Bessel's equation is given by

$$f(z) = A J_n(z) + B J_n(z) \int \frac{dz}{z J_n(z)^2}. \qquad (2.10.4)$$

The function

$$Y_n(z) = \frac{2}{\pi} J_n(z) \int \frac{dz}{z J_n(z)^2},$$

for a suitable choice of constant of integration, is called Neumann's Bessel function of the second kind, or Weber's function. The factor of $2/\pi$ is introduced (by most, but not all authors) so that formulae involving $J_n(z)$ and $Y_n(z)$ look similar; we shall not go into the details. From the above integral, it is not hard to see that $Y_n(z)$ tends to $-\infty$ as z tends to zero from above; we shall be more explicit about this towards the end of this section.

Next, we develop the power series for $J_n(z)$. We begin with $J_0(z)$. Putting $z = \theta = 0$ in equation (2.8.2), we see that $J_0(0) = 1$. By (2.8.4), $J_0(z)$ is an even function of z, so we look for a power series of the form

$$J_0(z) = 1 + a_2 z^2 + a_4 z^4 + \cdots = \sum_{k=0}^{\infty} a_{2k} z^{2k},$$

where $a_0 = 1$. Then

$$J_0'(z) = 2a_2 z + 4a_4 z^3 + \cdots = \sum_{k=1}^{\infty} 2k a_{2k} z^{2k-1},$$

$$J_0''(z) = 2 \cdot 1\, a_2 + 4 \cdot 3\, a_4 z^2 + \cdots = \sum_{k=1}^{\infty} 2k(2k-1) a_{2k} z^{2k-2}.$$

Putting $n = 0$ in equation (2.10.1) and comparing coefficients of a_{2k-2}, we obtain

$$2k(2k-1)a_{2k} + 2k a_{2k} + a_{2k-2} = 0,$$

or

$$(2k)^2 a_{2k} = -a_{2k-2}.$$

So, starting with $a_0 = 1$, we obtain $a_2 = -1/2^2$, $a_4 = 1/(2^2 \cdot 4^2)$, ..., and by induction on k, we have

$$a_{2k} = \frac{(-1)^k}{2^2 \cdot 4^2 \dots (2k)^2} = \frac{(-1)^k}{2^k (k!)^2}.$$

So we have

$$J_0(z) = 1 - \frac{z^2}{2^2} + \frac{z^4}{2^2 \cdot 4^2} - \frac{z^6}{2^2 \cdot 4^2 \cdot 6^2} + \cdots = \sum_{k=0}^{\infty} \frac{(-1)^k \left(\frac{z}{2}\right)^{2k}}{(k!)^2}. \qquad (2.10.5)$$

Since the coefficients in this power series are tending to zero very rapidly, it has an infinite radius of convergence.[14] So it is uniformly convergent, and can be differentiated term by term. It follows that the sum of the power series satisfies Bessel's equation, because that's how we chose the coefficients. We have already seen that there is only one solution of Bessel's equation with value 1 at $z = 0$, which completes the proof that the sum of the power series is indeed $J_0(z)$.

Differentiating equation (2.10.5) term by term and using (2.9.6), we see that

$$J_1(z) = \frac{z}{2} - \frac{z^3}{2^2 \cdot 4} + \frac{z^5}{2^2 \cdot 4^2 \cdot 6} - \cdots = \sum_{k=0}^{\infty} \frac{(-1)^k \left(\frac{z}{2}\right)^{1+2k}}{k!(1+k)!}.$$

Now using equation (2.9.9) and induction on n, we find that

$$\boxed{J_n(z) = \sum_{k=0}^{\infty} \frac{(-1)^k \left(\frac{z}{2}\right)^{n+2k}}{k!(n+k)!},} \qquad (2.10.6)$$

with infinite radius of convergence.

From the power series, we can get information about $Y_n(z)$ as $z \to 0^+$. For small positive values of z, $J_n(z)$ is equal to $z^n/2^n n!$ plus much smaller terms. So $\frac{1}{z J_n(z)^2}$ is equal to $2^{2n}(n!)^2 z^{-2n-1}$ plus much smaller terms, and $\int \frac{1}{z J_n(z)^2} dz$ is equal to $-2^{2n-1} n!(n-1)! z^{-2n}$ plus much smaller terms. Finally, $Y_n(z)$ is equal to $-2^n(n-1)! z^{-n}/\pi$ plus much smaller terms. In particular, this shows that $Y_n(z) \to -\infty$ as $z \to 0^+$.

Exercises

1. Show that $y = J_n(\alpha x)$ is a solution of the differential equation

$$\frac{d^2 y}{dx^2} + \frac{1}{x}\frac{dy}{dx} + \left(\alpha^2 - \frac{n^2}{x^2}\right) y = 0.$$

   Show that the general solution to this equation is given by $y = A J_n(\alpha x) + B Y_n(\alpha x)$.
2. Show that $y = \sqrt{x} J_n(x)$ is a solution of the differential equation

$$\frac{d^2 y}{dx^2} + \left(1 + \frac{\frac{1}{4} - n^2}{x^2}\right) y = 0.$$

   Find the general solution of this equation.

---

[14] For any value of z, the ratio of successive terms tends to zero, so by the ratio test the series converges.

3. Show that $y = J_n(e^x)$ is a solution of the differential equation

$$\frac{d^2 y}{dx^2} + (e^{2x} - n^2)y = 0.$$

Find the general solution of this equation.

4. The following exercise is another route to Bessel's differential equation (2.10.1).
   (a) Differentiate equation (2.8.9) twice with respect to z, keeping ϕ and θ constant.
   (b) Differentiate equation (2.8.9) twice with respect to θ, keeping z and ϕ constant.
   (c) Divide the result of (b) by z^2 and add to the result of (a), and use the relation $\sin^2 \theta + \cos^2 \theta = 1$. Deduce that

$$\sum_{n=-\infty}^{\infty} \left(J_n''(z) + \frac{1}{z} J_n'(z) + \left(1 - \frac{n^2}{z^2} \right) J_n(z) \right) \sin(\phi + z\theta) = 0.$$

   (d) Finally, use Lemma 2.9.1 to show that Bessel's equation (2.8.9) holds.
   (The following exercises suppose some knowledge of complex analysis in order to give an alternative development of the power series and recurrence relations for the Bessel functions)

5. Show that

$$J_n(z) = \frac{1}{2\pi} \int_0^\pi e^{i(n\theta - z \sin \theta)} \, d\theta + \frac{1}{2\pi} \int_0^\pi e^{-i(n\theta - z \sin \theta)} \, d\theta$$

$$= \frac{1}{2\pi} \int_{-\pi}^\pi e^{-i(n\theta - z \sin \theta)} \, d\theta.$$

Substitute $t = e^{i\theta}$ (so that $\frac{1}{2i}(t - \frac{1}{t}) = \sin \theta$) to obtain

$$J_n(z) = \frac{1}{2\pi i} \oint t^{-n-1} e^{\frac{1}{2} z(t - \frac{1}{t})} \, dt, \tag{2.10.7}$$

where the contour of integration goes counterclockwise once around the unit circle. Use Cauchy's integral formula to deduce that $J_n(z)$ is the coefficient of t^n in the Laurent expansion of $e^{\frac{1}{2} z(t - \frac{1}{t})}$:

$$e^{\frac{1}{2} z(t - \frac{1}{t})} = \sum_{n=-\infty}^{\infty} J_n(z) t^n.$$

6. Substitute $t = 2s/z$ in (2.10.7) to obtain

$$J_n(z) = \frac{1}{2\pi i} \left(\frac{z}{2} \right)^n \oint s^{-n-1} e^{s - \frac{z^2}{4s}} \, ds.$$

Discuss the contour of integration. Expand the integrand in powers of z to give

$$J_n(z) = \frac{1}{2\pi i} \sum_{k=0}^{\infty} \frac{(-1)^k}{k!} \left(\frac{z}{2} \right)^{n+2k} \oint s^{-n-k-1} e^s \, ds$$

and justify the term by term integration. Show that the residue of the integrand at $s = 0$ is $1/(n + k)!$ when $n + k \geq 0$ and is zero when $n + k < 0$ (note that $0! = 1$). Deduce the power series (2.10.6).

7. (a) Use the power series (2.10.6) to show that

$$J_n(z) = \tfrac{z}{2n}(J_{n-1}(z) + J_{n+1}(z)).$$

(b) Differentiate the power series (2.10.6) term by term to show that

$$J'_n(z) = \tfrac{1}{2}(J_{n-1}(z) - J_{n+1}(z)).$$

Further reading on Bessel functions

Milton Abramowitz and Irene A. Stegun, *Handbook of Mathematical Functions*, National Bureau of Standards, 1964, reprinted by Dover in 1965 and still in print. This contains extensive tables of many mathematical functions including $J_n(z)$ and $Y_n(z)$.

Frank Bowman, *Introduction to Bessel Functions*, reprinted by Dover in 1958 and still in print.

G. N. Watson (1922), *A Treatise on the Theory of Bessel Functions* is an 800 page tome on the theory of Bessel functions. This work contains essentially everything that was known in 1922 about these functions, and is still pretty much the standard reference.

E. T. Whittaker and G. N. Watson, *Modern Analysis*, Cambridge University Press, 1927, Chapter XVII.

See also Appendix B for some tables and a summary of some properties of Bessel functions, as well as a C++ program for calculation.

2.11 Fourier series for FM feedback and planetary motion

We shall see in Section 8.9 that in the theory of FM synthesis, feedback is represented by an equation of the form

$$\phi = \sin(\omega t + z\phi), \qquad (2.11.1)$$

where ω and z are constants with $|z| \le 1$, and the equation implicitly defines ϕ as a function of t. With an equation like this, we should regard it as extraordinary that we can explicitly find ϕ as a function of t.

In the theory of planetary motion, Kepler's laws imply that the angle θ subtended at the centre (not the focus) of the elliptic orbit by the planet, measured from the major axis of the ellipse, satisfies

$$\omega t = \theta - z \sin\theta, \qquad (2.11.2)$$

where z is the eccentricity[15] of the ellipse, a number in the range $0 \le z \le 1$, and $\omega = 2\pi\nu$ is a constant which plays the role of average angular velocity. Again, this equation implicitly defines θ as a function of t.

---

[15] The eccentricity of an ellipse is defined to be the distance from the centre to the focus, as a proportion of the major radius.

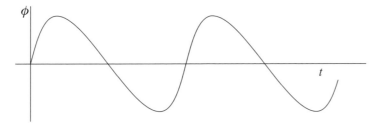

Figure 2.9 $Z = 1/2$.

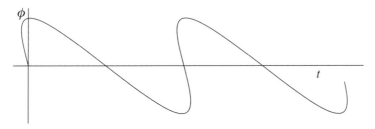

Figure 2.10 $Z = 3/2$.

Both of these equations define periodic functions of t, namely ϕ in the first case and $\sin\theta = (\theta - \omega t)/z$ in the second. In fact, they are really just different ways of writing the same equation. To get from equation (2.11.2) to (2.11.1), we use the substitution $\theta = \omega t + z\phi$. To go the other way, we use the inverse substitution $\phi = (\theta - \omega t)/z$.

The same functions turn up in other places too. In an exercise at the end of this section, we describe the relevance to nonlinear acoustics.

To graph ϕ as a function of t, it is best to use θ as a parameter and set $t = (\theta - z\sin\theta)/\omega$, $\phi = \sin\theta$. Figure 2.9 shows the result when $z = \frac{1}{2}$.

When $|z| > 1$, the parametrized form of the equation still makes sense, but it is easy to see that the resulting graph does not define ϕ uniquely as a function of t. Figure 2.10 shows the result when $z = \frac{3}{2}$.

In this section, we examine equation (2.11.2), and find the Fourier coefficients of $\phi = \sin\theta$ as a function of t, regarding z as a constant. The answer is given in terms of Bessel functions. In fact, the solution of this equation in the context of planetary motion was the original motivation for Bessel to introduce his functions $J_n(z)$.[16]

First, for convenience we write $T = \omega t$. Next, we observe that provided $|z| \leq 1$, $\theta - z\sin\theta$ is a strictly increasing fuction of θ whose domain and range are the

[16] Bessel, Untersuchung der Theils der planetarischen Störungen, welcher aus der Bewegung der Sonne entsteht, *Berliner Abh.* (1826), 1–52.

whole real line. It follows that solving equation (2.11.2) gives a unique value of θ for each T, so that θ may be regarded as a continuous function of T. Furthermore, adding 2π to both θ and T, or negating both θ and T, does not affect equation (2.11.2), so $z\phi = z\sin\theta = \theta - T$ is an odd periodic function of T with period 2π. So it has a Fourier expansion

$$z\phi = \sum_{n=1}^{\infty} b_n \sin nT. \tag{2.11.3}$$

The coefficients b_n can be calculated directly using equation (2.2.6):

$$b_n = \frac{1}{\pi} \int_0^{2\pi} z\phi \sin nT \, dT = \frac{2}{\pi} \int_0^{\pi} z\phi \sin nT \, dT.$$

Integrating by parts gives

$$b_n = \frac{2}{\pi} \left[-z\phi \frac{\cos nT}{n} \right]_0^{\pi} + \frac{2}{\pi} \int_0^{\pi} z \frac{d\phi}{dT} \frac{\cos nT}{n} \, dT.$$

We have $\phi = 0$ when $T = 0$ or $T = \pi$, so the first term vanishes. Rewriting the second term, we obtain

$$b_n = \frac{2}{n\pi} \int_0^{\pi} \cos nT \frac{d(z\phi)}{dT} \, dT.$$

Now $\int_0^{\pi} \cos nT \, dT = 0$, so we can rewrite this as

$$b_n = \frac{2}{n\pi} \int_0^{\pi} \cos nT \frac{d(z\phi + T)}{dT} \, dT = \frac{2}{n\pi} \int_0^{\pi} \cos nT \frac{d\theta}{dT} \, dT$$

$$= \frac{2}{n\pi} \int_0^{\pi} \cos nT \, d\theta.$$

In the last step, we have used the fact that as T increases from 0 to π, so does θ. Substituting $T = \theta - z\sin\theta$ now gives

$$b_n = \frac{2}{n\pi} \int_0^{\pi} \cos(n\theta - nz\sin\theta) \, d\theta.$$

Comparing with equation (2.8.3) finally gives

$$b_n = \frac{2}{n} J_n(nz).$$

Substituting back into equation (2.11.3) gives

$$\phi = \sin\theta = \sum_{n=1}^{\infty} \frac{2 J_n(nz)}{nz} \sin n\omega t. \tag{2.11.4}$$

So this equation gives the Fourier series relevant to feedback in FM synthesis (2.11.1), planetary motion (2.11.2) and nonlinear acoustics (2.11.5).

Exercise

Show that if a function ϕ satisfying equation (2.11.1) is regarded as a function of z and t, and ω is regarded as a constant, then ϕ is a solution of the partial differential equation

$$\frac{\partial \phi}{\partial z} = \frac{\phi}{\omega} \frac{\partial \phi}{\partial t}. \tag{2.11.5}$$

Show that if α is a nonzero constant, then $\psi(z, t) = \alpha\phi(\alpha z, t)$ is another solution to this equation.

(Warning: this equation is *nonlinear*: adding solutions does not give another solution, and multiplying a solution by a scalar does not give another solution.)

This equation turns out to be relevant to nonlinear acoustics. In this context, the solutions given by applying the above dilation to equation (2.11.4) are called *Fubini solutions*,[17] in spite of the fact that they were described by Bessel more than a century earlier. Figure 2.11 now represents the solution for $|\alpha z| > 1$, and describes an acoustic shock wave (in this context, αz is called the *distortion range variable*).

2.12 Pulse streams

In this section, we examine streams of square pulses. The purpose of this is twofold. First, we wish to prepare for a discussion of analogue synthesizers in Chapter 8. One method for obtaining a time varying frequency spectrum in analogue synthesis is to use a technique called *pulse width modulation* (PWM).[18] For this purpose, a *low frequency oscillator* (LFO, Section 8.2) is used to control the pulse width of a square wave, while keeping the fundamental frequency constant.

The second purpose of looking at pulse streams is that by keeping the pulse width constant and decreasing the frequency, we motivate the definition of Fourier transform, to be introduced in Section 2.13.

Let us investigate the frequency spectrum of the square wave given by

$$f(t) = \begin{cases} 1 & 0 \le t \le \rho/2, \\ 0 & \rho/2 < t < T - \rho/2, \\ 1 & T - \rho/2 \le t < T, \end{cases}$$

where ρ is some number between 0 and T, and $f(t + T) = f(t)$. See Figure 2.11.

---

[17] Eugene Fubini, Anomalies in the propagation of acoustic waves at great amplitude (in Italian), *Alta Frequenza* **4** (1935), 530–581. Eugene Fubini (1913–1997) was son of the mathematician Guido Fubini (1879–1943), after whom Fubini's theorem is named.

[18] This is also used in some of the more modern analogue modelling synthesizers such as the Roland JP-8000/JP-8080.

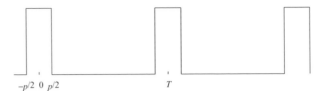

$-p/2$ 0 $p/2$ T

Figure 2.11 Square wave.

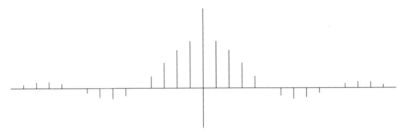

Figure 2.12 Frequency spectrum for $T = 5\rho$.

Figure 2.13 .

The Fourier coefficients are given by

$$\alpha_m = \frac{1}{T} \int_{-\rho/2}^{\rho/2} e^{-2m\pi it/T} \, dt = \frac{1}{m\pi} \sin(m\pi\rho/T).$$

For example, if $T = 5\rho$, the frequency spectrum is as shown in Figure 2.12.

If we keep ρ constant and increase T, the shape of the spectrum stays the same, but vertically scaled down in proportion, so as to keep the energy density along the horizontal axis constant. It makes sense to rescale to take account of this, and to plot $T\alpha_m$ instead of α_m. If we do this, and increase T while keeping ρ constant, all that happens is that the graph fills in. So, for example, removing every second peak from the original square wave, as in Figure 2.13, the spectrum fills in as shown in Figure 2.14.

Letting T tend to infinity while keeping ρ constant, we obtain the *Fourier transform* of a single square pulse, which (after suitable scaling) is the function $\sin(v)/v$. Here, v is a continuously variable quantity representing frequency.

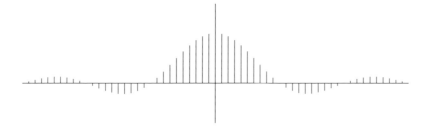

Figure 2.14.

2.13 The Fourier transform

The theory of Fourier series, as described in Sections 2.2–2.4, decomposes periodic waveforms into infinite sums of sines and cosines, or equivalently (Section 2.6) complex exponential functions of the form e^{int}. It is often desirable to analyze nonperiodic functions in a similar way. This leads to the theory of Fourier transforms. The theory is more beset with conditions than the theory of Fourier series. In particular, without the introduction of *generalized functions* or *distributions*, the theory only describes functions which tend to zero for large positive or negative values of the time variable t. To deal with this from a musical perspective, we introduce the theory of windowing. The point is that any actual sound is not really periodic, since periodic functions have no starting point and no end point. Moreover, we don't really want a frequency analysis of, for example, the whole of a symphony, because the answer would be dominated by extremely phase sensitive low frequency information. We'd really like to know at each instant what the frequency spectrum of the sound is, and to plot this frequency spectrum against time. Now, it turns out that it doesn't really make sense to ask for the instantaneous frequency spectrum of a sound, because there's not enough information. We really need to know the waveform for a time window around each point, and analyze that. Small window sizes give information which is more localized in time, but the frequency components are smeared out along the spectrum. Large window sizes give information in which the frequency components are more accurately described, but more smeared out along the time axis. This limitation is inherent to the process, and has nothing to do with how accurately the waveform is measured. In this respect, it resembles the Heisenberg uncertainty principle.[19]

---

[19] In fact, this is more than just an analogy. In quantum mechanics, the probability distributions for position and momentum of a particle are related by the Fourier transform, with an extra factor of Planck's constant $\hbar$. The Heisenberg uncertainty principle applies to the expected deviation from the average value of any two quantities related by the Fourier transform, and says that the product of these expected deviations is at least $\frac{1}{2}$. So in the quantum mechanical context the product is at least $\hbar/2$, because of the extra factor.

If $f(t)$ is a real or complex valued function of a real variable t, then its *Fourier transform* $\hat{f}(v)$ is the function of a real variable v defined by[20]

$$\hat{f}(v) = \int_{-\infty}^{\infty} f(t)e^{-2\pi i v t}\, dt. \tag{2.13.1}$$

The interpretation of this formula is that t represents time and v represents frequency. So $\hat{f}(v)$ should be thought of as measuring how much of $f(t)$ there is at frequency v. Somehow we've broken up the signal $f(t)$ into periodic components, but at all possible frequencies; reassembling the signal from its Fourier transform is given by means of the *inverse Fourier transform*, described below in equation (2.13.3).

Existence of a Fourier transform for a function assumes convergence of the above integral, and this already puts restrictions on the function $f(t)$. A reasonable condition which ensures convergence is the following. A function $f(t)$ is said to be L^1, or *absolutely integrable* on $(-\infty, \infty)$ if the integral $\int_{-\infty}^{\infty} |f(t)|\, dt$ converges. In particular, this forces $f(t)$ to tend to zero as $|t| \to \infty$ (except possibly on a set of measure zero, which may be ignored), which makes integrating by parts easier. In Section 2.17, we shall see how to extend the definition to a much wider class of functions using the theory of distributions. For example, we would at least like to be able to take the Fourier tranform of a sine wave.

Calculating the Fourier transform of a function is usually a difficult process. As an example, we now calculate the Fourier transform of $e^{-\pi t^2}$. This function is unusual, in that it turns out to be its own Fourier transform.

Theorem 2.13.1 *The Fourier transform of* $e^{-\pi t^2}$ *is* $e^{-\pi v^2}$.

---

[20] There are a number of variations on this definition to be found in the literature, depending mostly on the placement of the factor of 2π. The way we have set it up means that the variable v directly represents frequency. Most authors delete the 2π from the exponential in this definition, which amounts to using the angular velocity ω instead. This means that they either have a factor of $1/2\pi$ appearing in formula (2.13.3), causing an annoying asymmetry, or an even more annoying factor of $1/\sqrt{2\pi}$ in both (2.13.1) and (2.13.3).

Strictly speaking, the meaning of equation (2.13.1) should be

$$\lim_{a \to -\infty} \lim_{b \to \infty} \int_a^b f(t)e^{-2\pi i v t}\, dt.$$

However, under some conditions this double limit may not exist, while

$$\lim_{R \to \infty} \int_{-R}^{R} f(t)e^{-2\pi i v t}\, dt$$

may exist. This weaker symmetric limit is called the *Cauchy principal value* of the integral. Principal values are often used in the theory of Fourier transforms.

Proof Let $f(t) = e^{-\pi t^2}$. Then

$$\hat{f}(v) = \int_{-\infty}^{\infty} e^{-\pi t^2} e^{-2\pi i v t} \, dt$$

$$= \int_{-\infty}^{\infty} e^{-\pi(t^2 + 2ivt)} \, dt$$

$$= \int_{-\infty}^{\infty} e^{-\pi((t+iv)^2 + v^2)} \, dt.$$

Substituting $x = t + iv$, $dx = dt$, we obtain

$$\hat{f}(v) = \int_{-\infty}^{\infty} e^{-\pi(x^2 + v^2)} \, dx. \tag{2.13.2}$$

This form of the integral makes it obvious that $\hat{f}(v)$ is positive and real, but it is not obvious how to evaluate the integral. It turns out that it can be evaluated using a trick. The trick is to square both sides, and then regard the right hand side as a double integral.

$$\hat{f}(v)^2 = \int_{-\infty}^{\infty} e^{-\pi(x^2 + v^2)} \, dx \int_{-\infty}^{\infty} e^{-\pi(y^2 + v^2)} \, dy$$

$$= \int_{-\infty}^{\infty} \int_{-\infty}^{\infty} e^{-\pi(x^2 + y^2 + 2v^2)} \, dx \, dy.$$

We now convert this double integral over the (x, y) plane into polar coordinates (r, θ). Remembering that the element of area in polar coordinates is $r \, dr \, d\theta$, we get

$$\hat{f}(v)^2 = \int_{0}^{2\pi} \int_{0}^{\infty} e^{-\pi(r^2 + 2v^2)} r \, dr \, d\theta.$$

We can easily perform the integration with respect to θ, since the integrand is constant with respect to θ. And then the other integral can be carried out explicitly.

$$\hat{f}(v)^2 = \int_{0}^{\infty} 2\pi r e^{-\pi(r^2 + 2v^2)} \, dr$$

$$= \left[-e^{-\pi(r^2 + 2v^2)} \right]_{0}^{\infty}$$

$$= e^{-2\pi v^2}.$$

Finally, since equation (2.13.2) shows that $\hat{f}(v)$ is positive, taking square roots gives $\hat{f}(v) = e^{-\pi v^2}$ as desired. $\qquad\square$

The following gives a formula for the Fourier transform of the derivative of a function.

Theorem 2.13.2 *The Fourier transform of $f'(t)$ is $2\pi i v \hat{f}(v)$.*

Proof Integrating by parts, we have

$$\int_{-\infty}^{\infty} f'(t)e^{-2\pi i v t} \, dt = \left[f(t)e^{-2\pi i v t} \right]_{0}^{\infty} - \int_{-\infty}^{\infty} f(t)(-2\pi i v)e^{-2\pi i v t} \, dt$$

$$= 0 + 2\pi i v \hat{f}(v).$$

$\square$

The inversion formula is the following, which should be compared with Theorem 2.4.1.

Theorem 2.13.3 *Let $f(t)$ be a piecewise C^1 function (i.e., on any finite interval, $f(t)$ is C^1 except at a finite set of points) which is also L^1. Then, at points where $f(t)$ is continuous, its value is given by the* **inverse Fourier transform**

$$f(t) = \int_{-\infty}^{\infty} \hat{f}(v)e^{2\pi i v t} \, dv. \tag{2.13.3}$$

(Note the change of sign in the exponent from equation (2.13.1).)

At discontinuities, the expression on the right of this formula gives the average of the left limit and the right limit, $\frac{1}{2}(f(t^{+}) + f(t^{-}))$, just as in Section 2.5.

Just as in the case of Fourier series, it is not true that a piecewise continuous L^1 function satisfies the conclusions of the above theorem. But a device analogous to Cesàro summation works equally well here. The analogue of averaging the first n sums is to introduce a factor of $1 - |v|/R$ into the integral defining the inverse Fourier transform, before taking principal values.

Theorem 2.13.4 *Let $f(t)$ be a piecewise continuous L^1 function. Then, at points where $f(t)$ is continuous, its value is given by*

$$f(t) = \lim_{R \to \infty} \int_{-R}^{R} \left(1 - \frac{|v|}{R} \right) \hat{f}(v)e^{2\pi i v t} \, dv.$$

At discontinuities, this formula gives $\frac{1}{2}(f(t^{+}) + f(t^{-}))$.

Exercises

1. (a) This part of the exercise is for people who run the Mac OS X operating system. Go to

   www.dr-lex.34sp.com/software/spectrograph.html

   and download the SpectroGraph plugin for iTunes, a frequency analyzing program.
   (b) This part of the exercise is for people who run the Windows operating system. Download a copy of Sound Frequency Analyzer from

   www.relisoft.com/freeware/index.htm

   This is a freeware realtime audio frequency analyzing programme for a PC running

Windows 95 or higher. Plug a microphone into the audio card on your PC, if there isn't one built in.

In both cases, use the programme to watch a windowed frequency spectrum analysis of sounds such as any musical instruments you may have around, bells, whistles, and so on. Experiment with various vowel sounds such as 'ee', 'oo', 'ah', and try varying the pitch of your voice. Both programmes use the fast Fourier transform, see Section 7.10.

The Windows Media Player contains an elementary oscilloscope. Use 'Windows Update' to make sure you have at least version 7 of the Media Player, play your favourite CD, and under View → Visualizations, choose Bars and Waves → Scope. Notice how it is almost impossible to get much meaningful information about how the waveform will sound, just by seeing the oscilloscope trace.

2. Find $\int_{-\infty}^{\infty} e^{-x^2} dx$.

(Hint: square the integral and convert to polar coordinates, as in the proof of Theorem 2.13.1.)

3. Show that if a is a constant then the Fourier transform of $f(at)$ is $\frac{1}{a} \hat{f}(\frac{v}{a})$.

4. Show that if a is a constant then the Fourier transform of $f(t - a)$ is $e^{-2\pi i v a} \hat{f}(v)$.

5. Find the Fourier transform of the square wave pulse of Section 2.12,

$$f(t) = \begin{cases} 1 & \text{if } -\rho/2 \leq t \leq \rho/2, \\ 0 & \text{otherwise.} \end{cases}$$

6. Using Theorem 2.13.1 and integration by parts, show that the Fourier transform of $2\pi t^2 e^{-\pi t^2}$ is $(1 - 2\pi v^2)e^{-\pi v^2}$.

(Hint: substitute $x = t + iv$ in the integral.)

2.14 Proof of the inversion formula

The purpose of this section is to prove the Fourier inversion formula, Theorem 2.13.3. This says that under suitable conditions, if a function $f(t)$ has Fourier transform

$$\hat{f}(v) = \int_{-\infty}^{\infty} f(t)e^{-2\pi i v t} dt, \tag{2.14.1}$$

then the original function $f(t)$ can be reconstructed as the Cauchy principal value of the integral

$$f(t) = \int_{-\infty}^{\infty} \hat{f}(v)e^{2\pi i v t} dv. \tag{2.14.2}$$

First of all, we have the same difficulty here as we did with Fourier series. Namely, if we change the value of $f(t)$ at just one point, then $\hat{f}(v)$ will not change. So the best we can hope for is to reconstruct the average of the left and right limits, if this exists, $\frac{1}{2}(f(t^+) + f(t^-))$.

To avoid using t both as a variable of integration and the independent variable, let us use τ instead of t in (2.14.2). Then the Cauchy principal value of the right hand side of (2.14.2) becomes

$$\lim_{A \to \infty} \int_{-A}^{A} \left(\int_{-\infty}^{\infty} f(t) e^{-2\pi i \nu t}\, dt \right) e^{2\pi i \nu \tau}\, d\nu.$$

So this is the expression we must compare with $f(\tau)$, or rather with $\frac{1}{2}(f(\tau^+) + f(\tau^-))$. Since the outer integral just involves a finite interval, and the inner integral is absolutely convergent, we may reverse the order of integration to see that (2.14.2) is equal to

$$\lim_{A \to \infty} \int_{-\infty}^{\infty} f(t) \int_{-A}^{A} e^{2\pi i \nu (\tau - t)}\, d\nu\, dt$$

$$= \lim_{A \to \infty} \int_{-\infty}^{\infty} f(t) \left[\frac{1}{2\pi i (\tau - t)} e^{2\pi i \nu (\tau - t)} \right]_{\nu = -A}^{A}\, dt$$

$$= \lim_{A \to \infty} \int_{-\infty}^{\infty} f(t) \frac{\sin 2\pi A(\tau - t)}{\pi(\tau - t)}\, dt,$$

where we've used the equation $\sin \theta = \frac{1}{2i}(e^{i\theta} - e^{-i\theta})$ to rewrite the complex exponentials in terms of sines.

Substituting $x = t - \tau$, $t = \tau + x$ in the $\int_0^\infty$ part, and substituting $x = \tau - t$, $t = \tau - x$ in the $\int_{-\infty}^0$ part of the above integral, we find that (2.14.2) is equal to

$$\lim_{A \to \infty} \int_0^\infty (f(\tau + x) + f(\tau - x)) \frac{\sin 2\pi A x}{\pi x}\, dx. \qquad (2.14.3)$$

So we really need to understand the behavior of $\frac{\sin 2\pi A x}{\pi x}$ and its integral, as A gets large. We do this in the following theorem.

Theorem 2.14.1 (i) *For $A > 0$, we have* $\displaystyle\int_0^\infty \frac{\sin 2\pi A x}{\pi x}\, dx = \frac{1}{2}$,

(ii) *For any $\varepsilon > 0$, we have*

$$\lim_{A \to \infty} \int_0^\varepsilon \frac{\sin 2\pi A x}{\pi x}\, dx = \frac{1}{2} \quad and \quad \lim_{A \to \infty} \int_\varepsilon^\infty \frac{\sin 2\pi A x}{\pi x}\, dx = 0.$$

Proof To see that the integral converges, write

$$I_n = \int_{n/2A}^{(n+1)/2A} \frac{\sin 2\pi A x}{\pi x}\, dx.$$

Then the I_n alternate in sign and monotonically decrease to zero, so their sum converges. To find the value of the integral, we first find

$$\int_0^{\frac{\pi}{2}} \frac{\sin(2n+1)u}{\sin u} \, du = \int_0^{\frac{\pi}{2}} \frac{e^{(2n+1)iu} - e^{-(2n+1)iu}}{e^{iu} - e^{-iu}} \, du$$

$$= \int_0^{\frac{\pi}{2}} \left(e^{2niu} + e^{2(n-1)iu} + \cdots + e^{-2niu} \right) du$$

$$= \frac{\pi}{2}. \tag{2.14.4}$$

For the last step, the terms in the integral cancel out in pairs, so that the only term giving a nonzero contribution is the middle one, which is $e^0 = 1$.

Now $\frac{1}{\sin u} - \frac{1}{u} \to 0$ as $u \to 0$ (combine and use l'Hôpital's rule, for example), so this expression defines a nonnegative, uniformly continuous function on $[0, \frac{\pi}{2}]$. An elementary estimate of the difference between consecutive positive and negative areas then shows that

$$\lim_{n\to\infty} \int_0^{\frac{\pi}{2}} \left(\frac{1}{\sin u} - \frac{1}{u} \right) \sin(2n+1)u \, du = 0.$$

Combining with (2.14.4) gives

$$\lim_{n\to\infty} \int_0^{\frac{\pi}{2}} \frac{\sin(2n+1)u}{u} \, du = \frac{\pi}{2}.$$

Now substitute $(2n+1)u = 2\pi A x$ and divide by π to get

$$\frac{1}{\pi} \int_0^{\frac{\pi}{2}} \frac{\sin(2n+1)u}{u} \, du = \int_0^{\frac{2n+1}{4A}} \frac{\sin 2\pi A x}{\pi x} \, dx \to \frac{1}{2}$$

as $n \to \infty$. For any given $A > 0$, letting $n \to \infty$ gives (i). Given $\varepsilon > 0$, set $A = \frac{2n+1}{4\varepsilon}$ and let $n \to \infty$ to get (ii). □

To prove Theorem 2.13.3, we first note that if $f(t)$ is L^1 then the Fourier integral makes sense, and our task is to understand (2.14.2), or equivalently (2.14.3). The idea is to use the above theorem to say that, for any $\varepsilon > 0$,

$$\lim_{A\to\infty} \int_\varepsilon^\infty (f(\tau + x) + f(\tau - x)) \frac{\sin 2\pi A x}{\pi x} \, dx = 0,$$

so that (2.14.3) is equal to

$$\lim_{A\to\infty} \int_0^\varepsilon (f(\tau + x) + f(\tau - x)) \frac{\sin 2\pi A x}{\pi x} \, dx.$$

So at any point where $\lim_{x\to 0}(f(\tau + x) + f(\tau - x))$ exists, the theorem shows that the above integral is equal to $\frac{1}{2} \lim_{x\to 0}(f(\tau + x) + f(\tau - x))$. In particular, this holds for piecewise continuous functions.

2.15 Spectrum

How does the Fourier transform tell us about the frequency distribution in the original function? Well, just as in Section 2.6, the relations (2.6.1) tell us how to rewrite complex exponentials in terms of sines and cosines, and vice versa. So the values of $\hat{f}$ at v and at $-v$ tell us not only about the magnitude of the frequency component with frequency v, but also the phase. If the original function $f(t)$ is real valued, then $\hat{f}(-v)$ is the complex conjugate $\hat{f}(v)$. The *energy density* at a particular value of v is defined to be the square of the amplitude $|\hat{f}(v)|$,

$$\text{energy density} = |\hat{f}(v)|^2.$$

Integrating this quantity over an interval will measure the total energy corresponding to frequencies in this interval. But note that both v and $-v$ contribute to energy, so if only positive values of v are used, we must remember to double the answer.

The usual way to represent the *frequency spectrum* of a real valued signal is to represent the amplitude and the phase of $\hat{f}(v)$ separately for positive values of v. In polar coordinates, we can write $\hat{f}(v)$ as $re^{i\theta}$, where $r = |\hat{f}(v)|$ is the amplitude of the corresponding frequency component and θ is the phase. So r is always nonnegative, and we take θ to lie between $-\pi$ and π. Then $\hat{f}(-v) = \overline{\hat{f}(v)} = re^{-i\theta}$, so we have already represented the information about negative values of v if we have given both amplitude and phase for positive values of v. Phase is often regarded as less important than amplitude, and so the frequency spectrum is often displayed just as a graph of $|\hat{f}(v)|$ for $v > 0$. For example, if we look at the frequency spectrum of the square wave pulse described in Section 2.12 and we ignore phase information (which is just a sign in this case), we get the following picture.

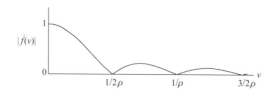

In this graph, we represented frequency linearly along the horizontal axis. But since our perception of frequency is logarithmic, the horizontal axis is often represented logarithmically. With this convention each octave, representing a doubling of the frequency, is represented by the same distance along the axis.

Parseval's identity states that the total energy of a signal is equal to the total energy in its spectrum:

$$\int_{-\infty}^{\infty} |f(t)|^2 \, dt = \int_{-\infty}^{\infty} |\hat{f}(v)|^2 \, dv.$$

More generally, if $f(t)$ and $g(t)$ are two functions, it states that

$$\int_{-\infty}^{\infty} f(t)\overline{g(t)} \, dt = \int_{-\infty}^{\infty} \hat{f}(v)\overline{\hat{g}(v)} \, dv. \tag{2.15.1}$$

The term *white noise* refers to a waveform whose spectrum is flat; for pink noise, the spectrum level decreases by 3dB per octave, while for brown noise (named after Brownian motion), the spectrum level decreases by 6dB per octave.

The windowed Fourier transform was introduced by Gabor,[21] and is described as follows. Given a windowing function $\psi(t)$ and a waveform $f(t)$, the windowed Fourier transform is the function of two variables

$$\mathcal{F}_\psi(f)(p, q) = \int_{-\infty}^{\infty} f(t) e^{-2\pi i q t} \psi(t - p) \, dt,$$

for p and q real numbers. This may be thought of as using all possible time translations of the windowing function, and pulling out the frequency components of the result. The typical windowing function might look as follows.

It's a good idea for the window to have smooth edges, and not just be a simple rectangular pulse, since corners in the windowing function tend to introduce extraneous high frequency artifacts in the windowed signal.

2.16 The Poisson summation formula

When we come to study digital music in Chapter 7, we shall need to use the Poisson summation formula.

Theorem 2.16.1 (Poisson's summation formula)

$$\sum_{n=-\infty}^{\infty} f(n) = \sum_{n=-\infty}^{\infty} \hat{f}(n). \tag{2.16.1}$$

Proof Define

$$g(\theta) = \sum_{n=-\infty}^{\infty} f\left(\frac{\theta}{2\pi} + n\right).$$

[21] D. Gabor, Theory of communication, *J. Inst. Electr. Eng.* **93** (1946), 429–457.

Then the left hand side of the desired formula is $g(0)$. Furthermore, $g(\theta)$ is periodic with period 2π, $g(\theta + 2\pi) = g(\theta)$. So we may apply the theory of Fourier series to $g(\theta)$. By equation (2.6.2), we have

$$g(\theta) = \sum_{n=-\infty}^{\infty} \alpha_n e^{in\theta},$$

and by equation (2.6.3), we have

$$
\begin{aligned}
\alpha_m &= \frac{1}{2\pi} \int_0^{2\pi} g(\theta) e^{-im\theta}\, d\theta \\
&= \frac{1}{2\pi} \int_0^{2\pi} \sum_{n=-\infty}^{\infty} f\left(\frac{\theta}{2\pi} + n\right) e^{-im\theta}\, d\theta \\
&= \frac{1}{2\pi} \sum_{n=-\infty}^{\infty} \int_0^{2\pi} f\left(\frac{\theta}{2\pi} + n\right) e^{-im\theta}\, d\theta \\
&= \frac{1}{2\pi} \int_{-\infty}^{\infty} f\left(\frac{\theta}{2\pi}\right) e^{-im\theta}\, d\theta \\
&= \int_{-\infty}^{\infty} f(t) e^{-2\pi imt}\, dt \\
&= \hat{f}(m).
\end{aligned}
$$

The third step above consists of piecing together the real line from segments of length 2π. The fourth step is given by the substitution $\theta = 2\pi t$. Finally, we have

$$\sum_{n=-\infty}^{\infty} f(n) = g(0) = \sum_{n=-\infty}^{\infty} \alpha_n = \sum_{n=-\infty}^{\infty} f(n). \qquad \square$$

Warning There are limitations on the applicability of the Poisson summation formula, coming from the limitations on applying Fourier inversion (2.6.2). For a discussion of this point, see Y. Katznelson, *An Introduction to Harmonic Analysis*. Dover (1976), p. 129.

2.17 The Dirac delta function

Dirac's delta function $\delta(t)$ is defined by the following properties:

(i) $\delta(t) = 0$ for $t \neq 0$, and

(ii) $\int_{-\infty}^{\infty} \delta(t)\, dt = 1$.

Think of $\delta(t)$ as being zero except for a spike at $t = 0$, so large that the area under it is equal to one. The awake reader will immediately notice that these properties are contradictory. This is because changing the value of a function at a single point does not change the value of an integral, and the function is zero except at one point, so the integral should be zero. Later in this section, we'll explain the resolution of this problem, but for the moment, let's continue as though there were no problem, and as though equations (2.13.1) and (2.13.3) work for functions involving $\delta(t)$.

It is often useful to shift the spike in the definition of the delta function to another value of t, say $t = t_0$, by using $\delta(t - t_0)$ instead of $\delta(t)$. The fundamental property of the delta function is that it can be used to pick out the value of another function at a desired point by integrating. Namely, if we want to find the value of $f(t)$ at $t = t_0$, we notice that $f(t)\delta(t - t_0) = f(t_0)\delta(t - t_0)$, because $\delta(t - t_0)$ is only nonzero at $t = t_0$. So

$$\int_{-\infty}^{\infty} f(t)\delta(t - t_0)\,dt = \int_{-\infty}^{\infty} f(t_0)\delta(t - t_0)\,dt = f(t_0)\int_{-\infty}^{\infty} \delta(t - t_0)\,dt = f(t_0).$$

Next, notice what happens if we take the Fourier transform of a delta function. If $f(t) = \delta(t - t_0)$, then by equation (2.13.1)

$$\hat{f}(v) = \int_{-\infty}^{\infty} \delta(t - t_0) e^{-2\pi i v t}\,dt = e^{-2\pi i v t_0}.$$

In other words, the Fourier transform of a delta function $\delta(t - t_0)$ is a complex exponential $e^{-2\pi i v t_0}$. In particular, in the case $t_0 = 0$, we find that the Fourier transform of $\delta(t)$ is the constant function 1. The Fourier transform of $\frac{1}{2}(\delta(t - t_0) + \delta(t + t_0))$ is

$$\frac{1}{2}\left(e^{-2\pi i v t_0} + e^{2\pi i v t_0}\right) = \cos(2\pi v t_0).$$

Conversely, if we apply the inverse Fourier transform (2.13.3) to the function $\hat{f}(v) = \delta(v - v_0)$, we obtain $f(t) = e^{2\pi i v_0 t}$. So we can think of the Dirac delta function concentrated at a frequency v_0 as the Fourier transform of a complex exponential. Similarly, $\frac{1}{2}(\delta(v - v_0) + \delta(v + v_0))$ is the Fourier transform of a cosine wave $\cos(2\pi v_0 t)$ with frequency v_0. We shall justify these manipulations towards the end of this section.

The relationship between Fourier series and the Fourier transform can be made more explicit in terms of the delta function. Suppose that $f(t)$ is a periodic function of t of the form $\sum_{n=-\infty}^{\infty} \alpha_n e^{in\theta}$ (see equation (2.6.2)) where $\theta = 2\pi v_0 t$. Then we have

$$\hat{f}(v) = \sum_{n=-\infty}^{\infty} \alpha_n \delta(v - n v_0).$$

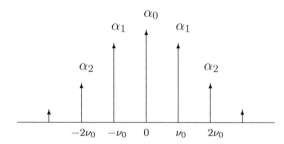

Figure 2.15 Fourier transform of a periodic function.

So the Fourier transform of a real valued periodic function has a spike at plus and minus each frequency component, consisting of a delta function multiplied by the amplitude of that frequency component.

So what kind of a function is $\delta(t)$? The answer is that it really isn't a function at all, it's a *distribution*, sometimes also called a *generalized function*. A distribution is only defined in terms of what happens when we multiply by a function and integrate. Whenever a delta function appears, there is an implicit integration lurking in the background.

More formally, one starts with a suitable space of *test functions*,[22] and a distribution is defined as a continuous linear map from the space of test functions to the complex numbers (or the real numbers, according to context).

A function $f(t)$ can be regarded as a distribution, namely, we identify it with the linear map taking $g(t)$ to $\int_{-\infty}^{\infty} f(t)g(t)\,dt$, as long as this makes sense. The delta function is the distribution which is defined as the linear map taking a test function $g(t)$ to $g(0)$. It is easy to see that this distribution does not come from an ordinary function in the above way. The argument is given at the beginning of this section. But we write distributions as though they were functions, and we write integration for the value of a distribution on a function. So, for example, the distribution $\delta(t)$ is defined by $\int_{-\infty}^{\infty} \delta(t)g(t)\,dt = g(0)$, and this just means that the value of the distribution $\delta(t)$ on the test function $g(t)$ is $g(0)$, nothing more nor less.

There is one warning that must be stressed at this stage. Namely, it does not make sense to multiply distributions. So, for example, the square of the delta function

---

[22] In the context of the theory of Fourier transforms, it is usual to start with the *Schwartz space* $\mathcal{S}$ consisting of infinitely differentiable functions $f(t)$ with the property that there is a bound not depending on m and n for the value of any derivative $f^{(m)}(t)$ times any power t^n of t ($m, n \geq 0$). So these functions are very smooth and all their derivatives tend to zero very rapidly as $|t| \to \infty$. An example of a function in $\mathcal{S}$ is the function e^{-t^2}. The sum, product and Fourier transform of functions in $\mathcal{S}$ are again in $\mathcal{S}$. For the purpose of saying what it means for a linear map on $\mathcal{S}$ to be continuous, the distance between two functions $f(t)$ and $g(t)$ in $\mathcal{S}$ is defined to be the largest distance between the values of $t^n f^{(m)}(t)$ and $t^n g^{(m)}(t)$ as m and n run through the nonnegative integers. The space of distributions defined on $\mathcal{S}$ is written $\mathcal{S}'$. Distributions in $\mathcal{S}'$ are called *tempered* distributions.

does not make sense as a distribution. After all, what would $\int_{-\infty}^{\infty} \delta(t)^2 g(t) \, dt$ be? It would have to be $\delta(0)g(0)$, which isn't a number!

However, distributions can be multiplied by functions. The value of a distribution times $f(t)$ on $g(t)$ is equal to the value of the original distribution on $f(t)g(t)$. As long as $f(t)$ has the property that whenever $g(t)$ is a test function then so is $f(t)g(t)$, this makes sense. Test functions and polynomials satisfy this condition, for example.

Distributions can also be differentiated. The way this is done is to use integration by parts to give the *definition* of differentiation. So if $f(t)$ is a distribution and $g(t)$ is a test function then $f'(t)$ is defined via

$$\int_{-\infty}^{\infty} f'(t)g(t) \, dt = - \int_{-\infty}^{\infty} f(t)g'(t) \, dt.$$

So, for example, the value of the distribution $\delta'(t)$ on the test function $g(t)$ is $-g'(0)$.

To illustrate how to manipulate distributions, let us find $t\delta'(t)$. Integration by parts shows that if $g(t)$ is a test function, then

$$\int_{-\infty}^{\infty} t\delta'(t)g(t) \, dt = - \int_{-\infty}^{\infty} \delta(t)\frac{d}{dt}(tg(t)) \, dt = - \int_{-\infty}^{\infty} \delta(t)(tg'(t) + g(t)) \, dt.$$

Now $t\delta(t) = 0$, so this gives $-g(0)$. If two distributions take the same value on all test functions, they are by definition the same distribution. So we have

$$\boxed{t\delta'(t) = -\delta(t).}$$

The reader should be warned, however, that extreme caution is necessary when playing with equations of this kind. For example, dividing the above equation by t to get $\delta'(t) = -\delta(t)/t$ makes no sense at all. After all, what if we were to apply the same logic to the equation $t\delta(t) = 0$?

It is also useful at this stage to go back to the proof of Fejér's theorem give in Section 2.7. Basically, the reason why this proof works is that the functions $K_m(y)$ are finite approximations to the distribution $2\pi\delta(y)$. Approximations to delta functions, used in this way, are called *kernel functions*, and they play a very important role in the theory of partial differential equations, analogous to the role they play in the proof of Fejér's theorem.

The Fourier transform of a distribution is defined using Parseval's identity (2.15.1). Namely, if $f(t)$ is a distribution, then for any function $g(t)$ the quantity $\int_{-\infty}^{\infty} f(t)\overline{g(t)} \, dt$ denotes the value of the distribution on $\overline{g(t)}$. We define $\hat{f}(v)$ to be the distribution whose value on $\overline{\hat{g}(v)}$ is the same quantity. In other words, the *definition* of $\hat{f}(v)$ is

$$\int_{-\infty}^{\infty} \hat{f}(v)\overline{\hat{g}(v)} \, dv = \int_{-\infty}^{\infty} f(t)\overline{g(t)} \, dt.$$

Notice that even if we are only interested in *functions*, this considerably extends the definition of Fourier transforms, and that the Fourier transform of a function can easily end up being a distribution which is not a function. For example, we saw earlier that the Fourier transform of the function $e^{2\pi i\nu_0 t}$ is the distribution $\delta(\nu - \nu_0)$.

Exercises

1. Find the Fourier transform of the sine wave $f(t) = \sin(2\pi \nu_0 t)$ in terms of the Dirac delta function.

2. Show that if C is a constant then
$$\delta(Ct) = \frac{1}{|C|}\delta(t).$$

3. The Heaviside function $H(t)$ is defined by
$$H(t) = \begin{cases} 1 & \text{if } t \geq 0, \\ 0 & \text{if } t < 0. \end{cases}$$

   Prove that the derivative of $H(t)$ is equal to the Dirac delta function $\delta(t)$.
   (Hint: use integration by parts.)

4. Show that $t\delta(t) = 0$.

5. Using Theorem 2.13.2, show that the Fourier transform of t^n is $\left(\frac{-1}{2\pi i}\right)^n \delta^{(n)}(\nu)$, where $\delta^{(n)}$ is the nth derivative of the Dirac delta function.

Further reading

F. G. Friedlander and M. Joshi, *Introduction to the Theory of Distributions*, second edition, Cambridge University Press, 1998.
A. H. Zemanian, *Distribution Theory and Transform Analysis*, Dover, 1987.

2.18 Convolution

The Fourier transform does not preserve multiplication. Instead, it turns it into *convolution*. If $f(t)$ and $g(t)$ are two test functions, their convolution $f * g$ is defined by
$$(f * g)(t) = \int_{-\infty}^{\infty} f(s)g(t - s)\, ds.$$

The corresponding verb is to *convolve* the function f with the function g. The formula also makes sense if one of f and g is a distribution and the other is a test function. The result is a function, but not necessarily a test function. The convolution of two distributions sometimes but not always makes sense; for example, the convolution of two constant functions is not defined but the convolution of two Dirac delta functions is defined.

It is easy to check that the following properties of convolution hold whenever both sides make sense:

(i) (commutativity) $f * g = g * f$;
(ii) (associativity) $(f * g) * h = f * (g * h)$;
(iii) (distributivity) $f * (g + h) = f * g + f * h$;
(iv) (identity element) $\delta * f = f * \delta = f$.

Here, δ denotes the Dirac delta function.

Theorem 2.18.1 (i) $\widehat{f * g}(v) = \hat{f}(v)\hat{g}(v)$,
(ii) $\widehat{fg}(v) = (\hat{f} * \hat{g})(v)$.

Proof To prove part (i), from the definition of convolution we have

$$\widehat{f * g}(v) = \int_{-\infty}^{\infty} \int_{-\infty}^{\infty} f(s)g(t - s)e^{-2\pi i v t} \, ds \, dt$$

$$= \int_{-\infty}^{\infty} \int_{-\infty}^{\infty} f(s)g(u)e^{-2\pi i v(s+u)} \, ds \, du$$

$$= \left(\int_{-\infty}^{\infty} f(s)e^{-2\pi i v s} \, ds \right) \left(\int_{-\infty}^{\infty} g(u)e^{-2\pi i v u} \, du \right)$$

$$= \hat{f}(v)\hat{g}(v).$$

Here, we have made the substitution $u = t - s$. Part (ii) follows from part (i) by the Fourier inversion formula (2.13.3); in other words, by reversing the roles of t and v. □

Part (i) of this theorem can be interpreted in terms of frequency filters. Applying a frequency filter to an audio signal is supposed to have the effect of multiplying the frequency distribution of the signal, $\hat{f}(v)$, by a filter function $\hat{g}(v)$. So, in the time domain, this corresponds to convolving the signal $f(t)$ with $g(t)$, the inverse Fourier transform of the filter function.

The output of a filter is usually taken to depend only on the input at the current and previous times. Looking at the formula for convolution, this corresponds to the statement that $g(t)$, the inverse Fourier transform of the filter function, should be zero for negative values of its argument.

The function $g(t)$ for the filter is called the *impulse response*, because it represents the output when a delta function is present at the input. The statement that $g(t) = 0$ for $t < 0$ is a manifestation of *causality*.

For example, let $g(t)$ be a delta function at zero plus a hump a little later.

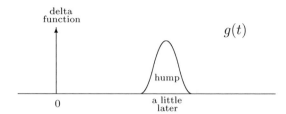

Then convolving a signal $f(t)$ with $g(t)$ will give $f(t)$ plus a smeared echo of $f(t)$ a short time later. The graph of $g(t)$ is interpreted as the impulse response, namely what comes out when a delta function is put in (in this case, crack – thump). These days, effects are often added to sound using a *digital filter*, which uses a discrete version of this process of convolution. See Section 7.8 for a brief description of the theory.

Exercises

1. Show that $\delta' * f = -f'$. Find a formula for $\delta^{(n)} * f$.
2. Prove the associativity formula $(f * g) * h = f * (g * h)$.

Further reading

Curtis Roads (1997), *Sound transformation by convolution*, appears as article 12 of Roads *et al.* (1997), pages 411–438.

2.19 Cepstrum

The idea of *cepstrum* is to look for periodicity in the Fourier transform of a signal, but measured on a logarithmic scale. So, for example, this would pick out a series of frequency components separated by octaves. So the definition of the cepstrum of a signal is

$$\widehat{\ln \hat{f}}(\rho) = \int_{-\infty}^{\infty} e^{-2\pi i \rho \nu} \ln \hat{f}(\nu)\, d\nu.$$

This gives a sort of twisted up, backwards spectrum. The idea was first introduced by Bogert, Healy and Tukey, who introduced the terminology. The variable ρ is called *quefrency*, to indicate that it is a twisted version of frequency. Peaks of quefrency are called *rahmonics*.

   If filtering a signal corresponds to multiplying its Fourier transform by a function, then *liftering* a signal is achieved by finding the cepstrum, multiplying by a function, and then undoing the cepstrum process. This process is often used in the analysis of vocal signals, in order to locate and extract formants.

Further reading

B. P. Bogert, M. J. R. Healy and J. W. Tukey, Quefrency analysis of time series for echoes: cepstrum, pseudo-autocovariance, cross-cepstrum and saphe cracking. In *Proceedings of the Symposium on Time Series Analysis*, Wiley 1963, pages 209–243.

Judith C. Brown, Computer identification of wind instruments using cepstral coefficients, *Proceedings of the 16th International Congress on Acoustics and 135th Meeting of the Acoustical Society of America, Seattle, Washington* (1998), 1889–1890.

Judith C. Brown, Computer identification of musical instruments using pattern recognition with cepstral coefficients as features, *J. Acoust. Soc. Amer.* **105** (3) (1999), 1933–1941.

M. R. Schroeder, *Computer Speech*, Springer Series in Information Sciences, Springer-Verlag, 1999, Section 10.14 and Appendix B.

Stan Tempelaars (1996), *Signal processing, speech and music*, Section 7.2.

2.20 The Hilbert transform and instantaneous frequency

Although the notion of instantaneous *frequency spectrum* of a signal makes no sense (because of the Heisenberg uncertainty principle), there is a notion of *instantaneous frequency* of a signal at a point in time. The idea is to use the Hilbert transform. If $f(t)$ is the signal, its Hilbert transform $g(t)$ is defined to be the Cauchy principal value[23] of the integral

$$g(t) = \frac{1}{\pi} \int_{-\infty}^{\infty} \frac{f(\tau)}{t - \tau} \, d\tau.$$

This makes an *analytic* signal $f(t) + ig(t)$.

For example, if $f(t) = c\cos(\omega t + \phi)$ then $g(t) = c\sin(\omega t + \phi)$ and $f(t) + ig(t) = ce^{i(\omega t + \phi)}$. In this case, $f(t) + ig(t)$ is rotating counterclockwise around the origin of the complex plane at a rate of ω radians per unit time. This suggests that the instantaneous angular frequency $\omega(t)$ is defined as the rate at which $f(t) + ig(t)$ is rotating around the origin. The angle $\theta(t)$ satisfies[24]

$$\tan\theta = g(t)/f(t),$$

so, differentiating, we obtain

$$\sec^2\theta \, \frac{d\theta}{dt} = \frac{f(t)g'(t) - g(t)f'(t)}{f(t)^2}.$$

---

[23] i.e., $g(t) = \lim_{A \to \infty} \lim_{\varepsilon \to 0} \frac{1}{\pi} \left(\int_{-A}^{-\varepsilon} \frac{f(\tau)}{t-\tau} \, d\tau + \int_{\varepsilon}^{A} \frac{f(\tau)}{t-\tau} \, d\tau \right)$.

[24] The formula $\theta = \tan^{-1}(g(t)/f(t))$ is incorrect. Why?

Using the relation

$$\sec^2 \theta = 1 + \tan^2 \theta = \frac{f(t)^2 + g(t)^2}{f(t)^2},$$

we obtain

$$\omega(t) = \frac{d\theta}{dt} = \frac{f(t)g'(t) - g(t)f'(t)}{f(t)^2 + g(t)^2}.$$

So the instantaneous frequency is given by

$$v(t) = \frac{\omega(t)}{2\pi} = \frac{1}{2\pi} \frac{f(t)g'(t) - g(t)f'(t)}{f(t)^2 + g(t)^2}.$$

The same reasoning also leads to the notion of *instantaneous amplitude* whose value is $\sqrt{f(t)^2 + g(t)^2}$. This is not the same as $|f(t)|$, which fails to capture the notion of instantaneous amplitude of a signal even for a sine wave.

From the formula for Hilbert transform, it can be seen that the definitions of instantaneous frequency and amplitude depend mostly on information about the signal close to the point being considered, but they do also have small contributions from the behaviour far away.

Further reading

B. Boashash, Estimating and interpreting the instantaneous frequency of a signal – Part I: Fundamentals, *Proc. IEEE* **80** (1992), 520–538.

L. Rossi and G. Girolami, Instantaneous frequency and short term Fourier transforms: applications to piano sounds, *J. Acoust. Soc. Amer.* **110** (5) (2001), 2412–2420.

Zachary M. Smith, Bertrand Delgutte and Andrew J. Oxenham, Chimaeric sounds reveal dichotomies in auditory perception, *Nature* **416**, 7 March 2002, 87–90. This article discusses the Fourier transform and Hilbert transform as models for auditory perception of music and speech, and concludes that both play a role.

3

A mathematician's guide to the orchestra

3.1 Introduction

Ethnomusicologists classify musical instruments into five main categories, which correspond reasonably well to the mathematical description of the sound they produce.[1]

1. **Idiophones,** where sound is produced by the body of a vibrating instrument. This category includes percussion instruments other than drums. It is divided into four subcategories: struck idiophones such as xylophones and cymbals, plucked idiophones (lamellophones) such as the mbira and the balafon, friction idiophones such as the (bowed) saw, and blown idiophones such as the aeolsklavier (a nineteenth century German instrument in which wooden rods are blown by bellows).
2. **Membranophones,** where the sound is produced by the vibration of a stretched membrane; for example, drums are membranophones. This category also has four subdivisions: struck drums, plucked drums, friction drums, and singing membranes such as the kazoo.
3. **Chordophones,** where the sound is produced by one or more vibrating strings. This category includes not only stringed instruments such as the violin and harp, but also keyboard instruments such as the piano and harpsichord.
4. **Aerophones,** where the sound is produced by a vibrating column of air. This category includes woodwind instruments such as the flute, clarinet and oboe, brass instruments such as the trombone, trumpet and French horn, and also various more exotic instruments such as the bullroarer and the conch shell.
5. **Electrophones,** where the sound is produced primarily by electrical or electronic means. This includes the modern electronic synthesizer (analogue or digital) as well as sound

---

[1] This classification was due to Hornbostel and Sachs (*Zeitschrift für Musik*, 1914), who omitted the fifth category of electrophones. This last category was added in 1961 by Anthony Baines and Klaus P. Wachsmann in their translation of the article of Hornbostel and Sachs.

The Hornbostel–Sachs system had antecedents. A Hindu system dating back more than two thousand years divides instruments into four similar groups. Victor Mahillon, curator of the collection of musical instruments of the Brussels conservatoire, used a similar classification in his 1888 catalogue of the collection.

Figure 3.3 Jean-le-Rond d'Alembert (1717–1783).

The mass of the segment of string will be approximately $\rho \Delta x$. So Newton's law $(F = ma)$ for the acceleration $a = \dfrac{\partial^2 y}{\partial t^2}$ gives

$$T \Delta x \frac{\partial^2 y}{\partial x^2} \approx (\rho \Delta x) \frac{\partial^2 y}{\partial t^2}.$$

Cancelling a factor of Δx on both sides gives

$$T \frac{\partial^2 y}{\partial x^2} \approx \rho \frac{\partial^2 y}{\partial t^2}.$$

In other words, as long as $\theta(x)$ never gets large, the motion of the string is essentially determined by the wave equation

$$\boxed{\frac{\partial^2 y}{\partial t^2} = c^2 \frac{\partial^2 y}{\partial x^2},} \qquad (3.2.2)$$

where $c = \sqrt{T/\rho}$.

D'Alembert[2] discovered a strikingly simple method for finding the general solution to equation (3.2.2). Roughly speaking, his idea is to factorize the differential

---

[2] Jean-le-Rond d'Alembert was born in Paris on November 16, 1717, and died there on October 29, 1783. He was the illegitimate son of a chevalier by the name of Destouches, and was abandoned by his mother on the steps of a small church called St. Jean-le-Rond, from which his first name is taken. He grew up in the family of a glazier and his wife, and lived with his adoptive mother until she died in 1757. But his father paid for his education, which allowed him to be exposed to mathematics. Two essays written in 1738 and 1740 drew attention to his mathematical abilities, and he was elected to the French Academy in 1740. Most of his mathematical works were written there in the years 1743–1754, and his solution of the wave equation appeared in his paper: *Recherches sur la courbe que forme une corde tendue mise en vibration, Hist. Acad. Sci. Berlin* **3** (1747), 214–219.

Figure 3.4 19th century lyre found in Nuba Hills, Sudan. British Museum, London.
ⓒ The Trustees of the British Museum.

operator

$$\frac{\partial^2}{\partial t^2} - c^2 \frac{\partial^2}{\partial x^2}$$

as

$$\left(\frac{\partial}{\partial t} + c\frac{\partial}{\partial x}\right)\left(\frac{\partial}{\partial t} - c\frac{\partial}{\partial x}\right).$$

More precisely, we make a change of variables

$$u = x + ct, \qquad v = x - ct.$$

Then, by the multivariable form of the chain rule, we have

$$\frac{\partial y}{\partial t} = \frac{\partial y}{\partial u}\frac{\partial u}{\partial t} + \frac{\partial y}{\partial v}\frac{\partial v}{\partial t} = c\frac{\partial y}{\partial u} - c\frac{\partial y}{\partial v}.$$

Differentiating again, we have

$$\frac{\partial^2 y}{\partial t^2} = \frac{\partial}{\partial u}\left(\frac{\partial y}{\partial t}\right)\frac{\partial u}{\partial t} + \frac{\partial}{\partial v}\left(\frac{\partial y}{\partial t}\right)\frac{\partial v}{\partial t}$$

$$= c\left(c\frac{\partial^2 y}{\partial u^2} - c\frac{\partial^2 y}{\partial u\partial v}\right) - c\left(c\frac{\partial^2 y}{\partial v\partial u} - c\frac{\partial^2 y}{\partial v^2}\right)$$

$$= c^2\left(\frac{\partial^2 y}{\partial u^2} - 2\frac{\partial^2 y}{\partial u\partial v} + \frac{\partial^2 y}{\partial v^2}\right).$$

Similarly,

$$\frac{\partial y}{\partial x} = \frac{\partial y}{\partial u}\frac{\partial u}{\partial x} + \frac{\partial y}{\partial v}\frac{\partial v}{\partial x} = \frac{\partial y}{\partial u} + \frac{\partial u}{\partial x},$$

$$\frac{\partial^2 y}{\partial x^2} = \frac{\partial^2 y}{\partial u^2} + 2\frac{\partial^2 y}{\partial u\partial v} + \frac{\partial^2 y}{\partial v^2}.$$

Then equation (3.2.2) becomes

$$c^2\left(\frac{\partial^2 y}{\partial u^2} - 2\frac{\partial^2 y}{\partial u\partial v} + \frac{\partial^2 y}{\partial v^2}\right) = c^2\left(\frac{\partial^2 y}{\partial u^2} + 2\frac{\partial^2 y}{\partial u\partial v} + \frac{\partial^2 y}{\partial v^2}\right)$$

or

$$\boxed{\frac{\partial^2 y}{\partial u\partial v} = 0.}$$

This equation may be integrated directly to see that the general solution is given by $y = f(u) + g(v)$ for suitably chosen functions f and g. Substituting back, we obtain

$$\boxed{y = f(x + ct) + g(x - ct).}$$

This represents a superposition of two waves, one travelling to the left and one travelling to the right, each with velocity c.

Now the boundary conditions tell us that the left and right ends of the string are fixed, so that when $x = 0$ or $x = \ell$ (the length of the string), we have $y = 0$ (independent of t). The condition with $x = 0$ gives

$$0 = f(ct) + g(-ct)$$

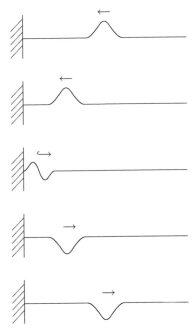

Figure 3.5 Reflected wave.

for all t, so that

$$g(\lambda) = -f(-\lambda) \qquad (3.2.3)$$

for any value of λ. Thus

$$y = f(x + ct) - f(ct - x).$$

Physically, this means that the wave travelling to the left hits the end of the string and returns inverted as a wave travelling to the right. This is called the 'principle of reflection'. See Figure 3.5.

Substituting the other boundary condition $x = \ell$, $y = 0$ gives $f(\ell + ct) = f(ct - \ell)$ for all t, so that

$$f(\lambda) = f(\lambda + 2\ell) \qquad (3.2.4)$$

for all values of λ. We summarize all the above information in the following theorem.

Theorem 3.2.1 (d'Alembert) *The general solution of the wave equation*

$$\frac{\partial^2 y}{\partial t^2} = c^2 \frac{\partial^2 y}{\partial x^2}$$

is given by

$$y = f(x + ct) + g(x - ct).$$

The solutions satisfying the boundary conditions $y = 0$ for $x = 0$ and for $x = \ell$, for all values of t, are of the form

$$y = f(x + ct) - f(-x + ct), \tag{3.2.5}$$

where f satisfies $f(\lambda) = f(\lambda + 2\ell)$ for all values of λ.

One interesting feature of d'Alembert's solution to the wave equation is worth emphasizing. Although the wave equation only makes sense for functions with second order partial derivatives, the solutions make sense for any *continuous* periodic function f. (Discontinuous functions cannot represent displacement of an unbroken string!) This allows us, for example, to make sense of the plucked string, where the initial displacement is continuous, but not even once differentiable. This is a common phenomenon when solving partial differential equations. A technique which is very often used is to rewrite the equation as an integral equation, meaning an equation involving integrals rather than derivatives. Integrable functions are much more general than differentiable functions, so one should expect a more general class of solutions.

Equation (3.2.4) means that the function f appearing in d'Alembert's solution is periodic with period 2ℓ, so that f has a Fourier series expansion. So, for example, if only the fundamental frequency is present, then the function $f(x)$ takes the form $f(x) = C \cos((\pi x/\ell) + \phi)$. If only the nth harmonic is present, then we have $f(x) = C \cos((n\pi x/\ell) + \phi)$,

$$y = C \cos\left(\frac{n\pi(x + ct)}{\ell} + \phi\right) - C \cos\left(\frac{n\pi(-x + ct)}{\ell} + \phi\right). \tag{3.2.6}$$

The theory of Fourier series allows us to write the general solution as a combination of the above harmonics, as long as we take care of the details of what sort of functions are allowed and what sort of convergence is intended.

Using equation (1.8.9), we can rewrite the nth harmonic solution (3.2.6) as

$$y = 2C \sin\left(\frac{n\pi x}{\ell}\right) \sin\left(\frac{n\pi ct}{\ell} + \phi\right). \tag{3.2.7}$$

This is Bernoulli's solution to the wave equation.[3] Thus the frequency of the nth harmonic is given by $2\pi v = n\pi c/\ell$, or, replacing c by its value $\sqrt{T/\rho}$,

$$\boxed{v = (n/2\ell)\sqrt{T/\rho}.}$$

[3] Daniel Bernoulli, *Réflections et éclairissements sur les nouvelles vibrations des cordes, Exposées dans les Mémoires de l'Academie de 1747 et 1748*, Royal Academy, Berlin, (1755), 147ff.

Figure 3.6 Marin Mersenne (1588–1648).

This formula for frequency was essentially discovered by Marin Mersenne[4] (see Figure 3.6) as his 'laws of stretched strings'. These say that the frequency of a stretched string is inversely proportional to its length, directly proportional to the square root of its tension, and inversely proportional to the square root of the linear density.

Exercises

1. Piano wire is manufactured from steel of density approximately 5900 kg/m^3. The manufacturers recommend a stress of approximately 1.1×10^9 newtons/m^2. What is the speed of propagation of waves along the wire? Does it depend on cross-sectional area? How long does the string need to be to sound middle C (262 Hz)?

2. By what factor should the tension on a string be increased, to raise its pitch by a perfect fifth? Assume that the length and linear density remain constant.
   (A perfect fifth represents a frequency ratio of 3:2.)

3. Read the beginning of Appendix F on music theory, and then explain why the back of a grand piano is shaped in a good approximation to an exponential curve.

---

[4] Marin Mersenne, *Harmonie Universelle*, Sebastien Cramoisy, 1636–37. Translated by R. E. Chapman as *Harmonie Universelle: The Books on Instruments*, Martinus Nijhoff, The Hague, 1957. Also republished in French by the CNRS in 1975 from a copy annotated by Mersenne.

3.3 Initial conditions

In this section, we see that in the analysis of the wave equation (3.2.2) described in the last section, specifying the initial position and velocity of each point on the string uniquely determines the subsequent motion.

Let $s_0(x)$ and $v_0(x)$ be the initial vertical displacement and velocity of the string as functions of the horizontal coordinate x, for $0 \leq x \leq \ell$. These must satisfy $s_0(0) = s_0(\ell) = 0$ and $v_0(0) = v_0(\ell) = 0$ to fit with the boundary conditions at the two ends of the string.

The first step is to extend the definitions of s_0 and v_0 to all values of x using the reflection principle. If we specify that $s_0(-x) = -s_0(x)$ and $v_0(-x) = -v_0(x)$, so that s_0 and v_0 are odd functions of x, this extends the domain of definition to the values $-\ell \leq x \leq \ell$. The values match up at $-\ell$ and ℓ, so we can extend to all values of x by specifying periodicity with period 2ℓ; namely that $s_0(x + 2\ell) = s_0(x)$ and $v_0(x + 2\ell) = v_0(x)$.

Now we simply substitute into the solution given by d'Alembert's theorem. Namely, we know that

$$y = f(x + ct) - f(-x + ct), \tag{3.3.1}$$

where f is periodic with period 2ℓ. Differentiating with respect to t gives the formula for velocity,

$$\frac{\partial y}{\partial t} = cf'(x + ct) - cf'(x - ct).$$

Substituting $t = 0$ in both the equation and its derivative gives the following equations:

$$f(x) - f(-x) = s_0(x), \tag{3.3.2}$$
$$cf'(x) - cf'(-x) = v_0(x). \tag{3.3.3}$$

Integrating equation (3.3.3) and noting that $v_0(0) = 0$, we obtain

$$cf(x) + cf(-x) = \int_0^x v_0(u)\,du.$$

We divide this equation by c to obtain a formula for $f(x) + f(-x)$. So we can then add equation (3.3.2) and divide by two to obtain $f(x)$. This gives

$$f(x) = \tfrac{1}{2}s_0(x) + \frac{1}{2c}\int_0^x v_0(u)\,du.$$

Putting this back into equation (3.3.1) gives

$$y = \tfrac{1}{2}(s_0(x + ct) - s_0(-x + ct)) + \frac{1}{2c}\left(\int_0^{x+ct} v_0(u)\,du - \int_0^{-x+ct} v_0(u)\,du\right).$$

Figure 3.7 Initial displacement.

Figure 3.8 Extended to a periodic function.

Using the fact that v_0 is an odd function, we have

$$\int_{x-ct}^{-x+ct} v_0(u)\,du = 0.$$

So we can rewrite the solution as

$$y = \tfrac{1}{2}(s_0(x+ct) + s_0(x-ct)) + \frac{1}{2c}\int_{x-ct}^{x+ct} v_0(u)\,du.$$

It is now easy to check that this is the unique solution satisfying both the initial conditions and the boundary conditions.

So, for example, if the initial velocity is zero, as is the case for a plucked string, then the solution is given by

$$y = \tfrac{1}{2}(s_0(x+ct) + s_0(x-ct)).$$

In other words, the initial displacement moves both ways along the string, with velocity c, and the displacement at time t is the average of the two travelling waves.

Let's see how this works in practice. Choose a satisfying $0 < a < 1$, and set

$$s_0(x) = \begin{cases} x/a & 0 \le x \le a, \\ (\ell - x)/(\ell - a) & a \le x \le 1. \end{cases}$$

See Figure 3.7.

We use the reflection principle to extend this to a periodic function of period 2ℓ as described above and shown in Figure 3.8.

Now we let this wave travel both left and right, and average the two resulting functions. Figure 3.9 shows the resulting motion of the plucked string.

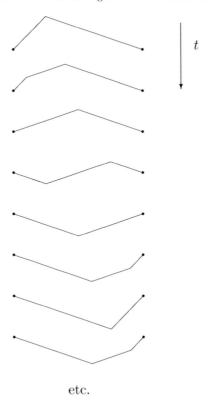

etc.

Figure 3.9 Motion of plucked string.

Exercise

(Effect of errors in initial conditions) Consider two sets of initial conditions for the wave equation (3.2.2), $s_0(x)$ and $v_0(x)$, $s_0'(x)$ and $v_0'(x)$, and let y and y' be the corresponding solutions. If we have bounds (not depending on x) on the distance between these initial conditions,

$$|s_0(x) - s_0'(x)| < \varepsilon_s, \qquad |v_0(x) - v_0'(x)| < \varepsilon_v,$$

show that the distance between y and y' satisfies

$$|y - y'| < \varepsilon_s + \frac{\ell \varepsilon_v}{2c}$$

(independently of x and t). This means, in particular, that the solution to the wave equation (3.2.2) depends *continuously* on the initial conditions.

Further reading

J. Beament (1997), *The Violin Explained: Components, Mechanism, and Sound.*
R. Courant and D. Hilbert, *Methods of Mathematical Physics, I*, Interscience, 1953,
        Section V.3.

Figure 3.10 Ousainou Chaw on the *riti*, from Jacqueline Cogdell DjeDje, *Turn up the Volume! A Celebration of African Music*, UCLA 1999, p. 105.

L. Cremer (1984), *The Physics of the Violin*.
Neville H. Fletcher and Thomas D. Rossing (1991), *The Physics of Musical Instruments*. Part III, String instruments.
T. D. Rossing (1990), *The Science of Sound*, Section 10.

3.4 The bowed string

Helmholtz[5] carried out experiments on bowed violins, using a *vibration microscope* to produce *Lissajous* figures. He discovered that the motion of the string at every point describes a triangular pattern, but with slopes which depend on the point of observation. Near the bow, the displacement is as follows:

whereas nearer the bridge it looks as follows.

[5] See Section V.4 of Helmholtz (1877).

This means that the graph of velocity against time has the following form:

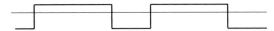

where the area under the axis equals the area above, and the width of the trough decreases towards the bridge.

The interpretation of this motion is that the bowing action alternates between two distinct phases. In one phase, the bow sticks to the string and pulls it with it. In the other phase, the bow slides against the string. This form of motion reflects the fact that the coefficient of static friction is higher than the coefficient of dynamic friction.

The resulting motion of the entire string has the following form. The envelope of the motion is described by two parabolas, a lower one and an inverted upper one. Inside this envelope, at any point of time the string has two straight segments from the two ends to a point on the envelope. This point circulates around the envelope as shown in Figure 3.11.

To understand this behaviour mathematically, we must solve the following problem. What are the solutions to the wave equation (3.2.2) satisfying not only the boundary conditions $y = 0$ for $x = 0$ and for $x = \ell$, for all values of t, but also the condition that the value of y as a function of t is prescribed for a particular value x_0 of x and for all t? Of course, the prescribed motion at $x = x_0$ must have the right periodicity, because all solutions of the wave equation do:

$$y(x_0, t + 2\ell/c) = y(x_0, t).$$

It is tempting to try to solve this problem using d'Alembert's solution of the wave equation (Theorem 3.2.1). The problems we run into when we try to do this are interesting. For example, let's suppose that $x_0 = \ell/2$. Then we have

$$f(\ell/2 + ct) - f(-\ell/2 + ct) = y(\ell/2, t).$$

Replacing t by $t + \ell/c$ in this equation, we get

$$f(3\ell/2 + ct) - f(\ell/2 + ct) = y(\ell/2, t + \ell/c).$$

Adding, we get

$$f(3\ell/2 + ct) - f(-\ell/2 + ct) = y(\ell/2, t) + y(\ell/2, t + \ell/c).$$

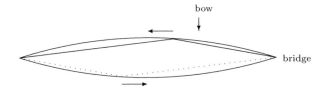

Figure 3.11 Motion of a bowed string.

But f is supposed to be periodic with period 2ℓ, so

$$f(3\ell/2 + ct) = f(-\ell/2 + ct).$$

This means that we have

$$y(\ell/2, t + \ell/c) = -y(\ell/2, t).$$

So not every periodic function with period $2\ell/c$ will work as the function $y(\ell/2, t)$. The function is forced to be half-period antisymmetric, so that only odd harmonics are present (see Section 2.3). This is only to be expected. After all, the even harmonics have a node at $x = \ell/2$, so how could we expect to involve even harmonics in the value of $y(x, t)$ at $x = \ell/2$?

Similar problems occur at $x = \ell/3$. The harmonics divisible by three are not allowed to occur in $y(\ell/3, t)$, because they have a node at $x = \ell/3$. This is a problem at every *rational* proportion of the string length.

It is becoming clear that Bernoulli's form (3.2.7) of the solution of the wave equation is going to be easier to use for this problem than d'Alembert's.

Since we are interested in functions $y(x_0, t)$ of the form shown in the diagrams at the beginning of this section, we may choose to measure time in such a way that $y(x_0, t)$ is an odd function of t, so that only sine waves and not cosine waves come into the Fourier series. So we set

$$y(x_0, t) = \sum_{n=1}^{\infty} b_n \sin\left(\frac{n\pi ct}{\ell}\right).$$

Since the wave equation is linear, we can work with one frequency component at a time. So we set $y(x_0, t) = b_n \sin(n\pi ct/\ell)$. We look for solutions of the form

$$f(x) = c_n \cos\left(\frac{n\pi x}{\ell} + \phi_n\right),$$

and we want to determine c_n and ϕ_n in terms of b_n. We plug into d'Alembert's equation (3.2.5)

$$y(x_0, t) = f(x_0 + ct) - f(-x_0 + ct)$$

to get

$$b_n \sin\left(\frac{n\pi ct}{\ell}\right) = c_n \cos\left(\frac{n\pi(x_0 + ct)}{\ell} + \phi_n\right) + c_n \cos\left(\frac{n\pi(-x_0 + ct)}{\ell} + \phi_n\right).$$

Using equation (1.8.11), this becomes

$$b_n \sin\left(\frac{n\pi ct}{\ell}\right) = 2c_n \sin\left(\frac{n\pi x_0}{\ell}\right) \sin\left(\frac{n\pi ct}{\ell} + \phi_n\right).$$

Since this is supposed to be an identity between functions of t, we get $\phi_n = 0$ and

$$b_n = 2c_n \sin\left(\frac{n\pi x_0}{\ell}\right).$$

We now have a problem, very similar to the problem we ran into when we tried to use d'Alembert's solution. Namely, if $\sin(n\pi x_0/\ell)$ happens to be zero and $b_n \neq 0$, then there is no solution. So if x_0 is a rational multiple of ℓ then some frequency components are forced to be missing from $y(x_0, t)$. Apart from that, we have almost solved the problem. The value of c_n is

$$c_n = \frac{b_n}{2 \sin(n\pi x_0/\ell)},$$

and so

$$f(x) = \sum_{n=1}^{\infty} \frac{b_n \sin(n\pi x/\ell)}{2 \sin(n\pi x_0/\ell)}. \tag{3.4.1}$$

The solution of the wave equation is then given by plugging this into the formula (3.2.5). Using equation (1.8.9), we get

$$y = f(x + ct) - f(-x + ct) = \sum_{n=1}^{\infty} b_n \frac{\sin(n\pi x/\ell)\cos(n\pi ct/\ell)}{\sin(n\pi x_0/\ell)}.$$

The only thing that isn't clear so far is when the sum (3.4.1) converges. This is a point that we shall finesse by using Helmholtz's observation that in the case of a bowed string, for any chosen value of x_0 we have a triangular waveform

$$y(x_0, t) = \begin{cases} A\frac{t}{\alpha} & -\alpha \leq t \leq \alpha, \\ A\dfrac{\ell - ct}{\ell - c\alpha} & \alpha \leq t \leq \frac{2\ell}{c} - \alpha, \end{cases}$$

where α is some number depending on x_0, determining how long the leading edge of the triangular waveform lasts at the position x_0 along the string. The quantity A also depends on x_0, and represents the maximum amplitude of the vibration at that point. Using equation (2.2.9), we then calculate

$$b_n = \frac{c}{\ell} \int_{-\alpha}^{\alpha} A\frac{t}{\alpha} \sin\left(\frac{n\pi ct}{\ell}\right) dt + \frac{c}{\ell} \int_{\alpha}^{\frac{2\ell}{c} - \alpha} A\frac{\ell - ct}{\ell - c\alpha} \sin\left(\frac{n\pi ct}{\ell}\right) dt$$

$$= \frac{2A\ell^2}{n^2\pi^2 c\alpha(\ell - c\alpha)} \sin\left(\frac{n\pi c\alpha}{\ell}\right),$$

so that

$$c_n = \frac{A\ell^2}{n^2\pi^2 c\alpha(\ell - c\alpha)} \frac{\sin(n\pi c\alpha/\ell)}{\sin(n\pi x_0/\ell)}.$$

Since the ratios of the c_n can't depend on the value of x_0 which we chose for our initial measurements, the only way this can work is if the two sine terms in this equation are equal, namely if

$$\frac{\pi c\alpha}{\ell} = \frac{\pi x_0}{\ell},$$

or

$$\alpha = x_0/c.$$

So if we measure the vibration at x_0 then the proportion $\alpha/(\ell/c)$ of the cycle spent in the trailing part of the triangular wave is equal to x_0/ℓ. In particular, if we measure at the bowing point, we obtain the following principle.

The proportion of the cycle for which the bow slips on the string is the same as the proportion of the string between the bow and the bridge.

Now A is just some constant depending on x_0. Since c_n doesn't depend on x_0, the constant $A/c\alpha(\ell - c\alpha) = A/x_0(\ell - x_0)$ must be independent of x_0. If we write K for this quantity, we obtain a formula for amplitude in terms of position along the string,

$$A = K x_0(\ell - x_0).$$

This formula explains the parabolic amplitude envelope for the vibration of the bowed string.

Further reading

L. Cremer (1984), *The Physics of the Violin*.

Joseph B. Keller, Bowing of violin strings, *Comm. Pure and Appl. Math.* **6** (1953), 483–495.

B. Lawergren, On the motion of bowed violin strings, *Acustica* **44** (1980), 194–206.

C. V. Raman, On the mechanical theory of the vibrations of bowed strings and of musical instruments of the violin family, with experimental verification of the results: Part I, *Indian Assoc. Cultivation Sci. Bull.* **15** (1918), 1–158.

J. C. Schelleng, The bowed string and the player, *J. Acoust. Soc. Amer.* **53** (1) (1973), 26–41.

J. C. Schelleng, The physics of the bowed string, *Scientific American* **235** (1) (1974), 87–95. Reproduced in Hutchins, *The Physics of Music*, W. H. Freeman and Co., 1978.

Lily M. Wang and Courtney B. Burroughs, Acoustic radiation from bowed violins, *J. Acoust. Soc. Amer.* **110** (1) (2001), 543–555.

3.5 Wind instruments

To understand the vibration of air in a tube or pipe, we introduce two variables, displacement and acoustic pressure. Both of these will end up satisfying the wave equation, but with different phases.

Figure 3.12 Bone flute from Henan province, China, 6000 B.C.E. Picture from J. F. So, *Music in the Age of Confucius*, Smithsonian Institution, 2000, p. 90. The oldest known flute is 35 000 years old, made from the tusk of the now extinct woolly mammoth. It was discovered in a German cave in December 2004. .

We consider the air in the tube to have a rest position, and the wave motion is expressed in terms of displacement from that position. So let x denote position along the tube, and let $\xi(x, t)$ denote the displacement of the air at position x at time t. The pressure also has a rest value, namely the ambient air pressure ρ. We measure the *acoustic pressure* $p(x, t)$ by subtracting ρ from the absolute pressure $P(x, t)$, so that

$$p(x, t) = P(x, t) - \rho.$$

Hooke's law in this situation states that

$$p = -B \frac{\partial \xi}{\partial x},$$

where B is the *bulk modulus* of air. Newton's second law of motion implies that

$$\frac{\partial p}{\partial x} = -\rho \frac{\partial^2 \xi}{\partial t^2}.$$

Combining these equations, we obtain the equations

$$\frac{\partial^2 \xi}{\partial x^2} = \frac{1}{c^2} \frac{\partial^2 \xi}{\partial t^2} \tag{3.5.1}$$

and

$$\frac{\partial^2 p}{\partial x^2} = \frac{1}{c^2} \frac{\partial^2 p}{\partial t^2}, \tag{3.5.2}$$

where $c = \sqrt{B/\rho}$. These equations are the wave equation for displacement and acoustic pressure respectively.

The boundary conditions depend upon whether the end of the tube is open or closed. For a closed end of a tube, the displacement ξ is forced to be zero for all values of t. For an open end of a tube, the acoustic pressure p is zero for all values of t.

So, for a tube open at both ends, such as the flute, the behaviour of the acoustic pressure p is determined by exactly the same boundary conditions as in the case of a vibrating string. It follows that d'Alembert's solution given in Section 3.2 works in

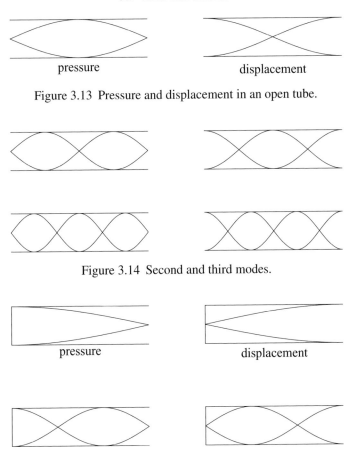

Figure 3.13 Pressure and displacement in an open tube.

Figure 3.14 Second and third modes.

Figure 3.15 Pressure and displacement in a tube closed at one end.

this case, and we again get integer multiples of a fundamental frequency. The basic mode of vibration is a sine wave, represented by Figure 3.13. The displacement is also a sine wave, but with a different phase.

Bear in mind that the vertical axis in this diagram actually represents *horizontal* displacement or pressure, and not vertical, because of the longitudinal nature of air waves. Furthermore, the two parts of the graphs only represent the two extremes of the motion. In these diagrams, the nodes of the pressure diagram correspond to the antinodes of the displacement diagram and vice versa. The second and third vibrational modes will be represented by the diagrams in Figure 3.14.

Tubes or pipes which are closed at one end behave differently, because the displacement is forced to be zero at the closed end. So the first two modes are as shown in Figure 3.15. In these diagrams, the left end of the tube is closed.

It follows that for closed tubes, odd multiples of the fundamental frequency dominate. For example, as mentioned above, the flute is an open tube, so all multiples

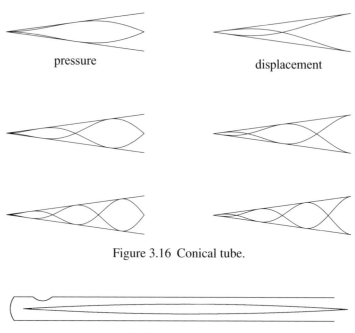

pressure   displacement

Figure 3.16 Conical tube.

Figure 3.17 Flute with all holes closed.

of the fundamental are present. The clarinet is a closed tube, so odd multiples predominate.

Conical tubes are equivalent to open tubes of the same length, as illustrated in Figure 3.16. These diagrams are obtained from the ones for the open tube, by squashing down one end.

The oboe has a conical bore, so again all multiples are present. This explains why the flute and oboe overblow at the octave, while the clarinet overblows at an octave plus a perfect fifth, which represents tripling the frequency. The odd multiples of the fundamental frequency dominate for a clarinet, although in practice there are small amplitudes present for the even ones from four times the fundamental upwards as well.

At this point, it should be mentioned that for an open end, $p = 0$ is really only an approximation, because the volume of air just outside of the tube is not infinite. A good way to adjust to make a more accurate representation of an actual tube is to work in terms of an *effective* length, and consider the tube to end a little beyond where it really does. Figure 3.17 shows the effective length for the fundamental vibrational mode of a flute, with all holes closed.

The *end correction* is the amount by which the effective length exceeds the actual length, and under normal conditions it is usually somewhere around three fifths of the width of the tube.

Figure 3.18 Flute with one hole open.

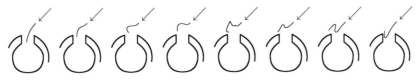

Figure 3.19 Airstream in a flute mouthpiece.

The effect of an open hole is to decrease the effective length of the tube. Figure 3.18 is a diagram of the first vibrational mode with one open hole.

The effective length of the tube can be seen by continuing the left part of the wave as though the hole doesn't exist, and seeing where the wave ends. This is represented by the dotted lines in the diagram. The larger the hole, the greater the effect on the effective length.

So what happens when the flutist blows into the mouthpiece of the flute? How does this cause a note to sound? The pictures in Figure 3.19, adapted from stroboscopic experiments of Coltman using smoke particles, show how the airstream varies with time. The arrow represents the incoming air stream.

This air pattern results in a series of vortices being sent down the tube. When the vortices get to the end of the tube, they are reflected back up. They reach the beginning of the tube and are reflected again. Some of these will be out of phase with the new vortices being generated, and some will be in phase. The ones that are in phase reinforce, and feed back to build up a coherent tone. This in turn makes it more favorable for vortices to be formed in synchronization with the tone.

Further reading

Giles Brindley, The standing wave-patterns of the flute, *Galpin Society Journal* **24** (1971), 5–15.

John W. Coltman (1988), Acoustics of the flute, *Physics Today* **21** (11) (1968), 25–32. Reprinted in Rossing.

Neville H. Fletcher and Thomas D. Rossing (1991), *The Physics of Musical Instruments.* Part IV, Wind instruments.

Ian Johnston (1989), *Measured Tones*, 207–233.

C. J. Nederveen (1998), *Acoustical Aspects of Woodwind Instruments.*

T. D. Rossing (1990), *The Science of Sound*, Section 12.

3.6 The drum

Consider a circular drum whose skin has area density (mass per unit area) ρ. If the boundary is under uniform tension T, this ensures that the entire surface is under the same uniform tension. The tension is measured in force per unit distance (newtons per metre).

To understand the wave equation in two dimensions, for a membrane such as the surface of a drum, the argument is analogous to the one dimensional case. We parametrize the surface with two variables x and y, and we use z to denote the displacement perpendicular to the surface. Consider a rectangular element of surface of width Δx and length Δy. Then the tension on the left and right sides is $T \Delta y$, and the argument which gave equation (3.2.1) in the one dimensional case shows in this case that the difference in vertical components is approximately

$$(T \Delta y) \left(\Delta x \frac{\partial^2 z}{\partial x^2} \right).$$

Similarly, the difference in vertical components between the front and back of the rectangular element is approximately

$$(T \Delta x) \left(\Delta y \frac{\partial^2 z}{\partial y^2} \right).$$

So the total upward force on the element of surface is approximately

$$T \Delta x \Delta y \left(\frac{\partial^2 z}{\partial x^2} + \frac{\partial^2 z}{\partial y^2} \right).$$

The mass of the element of surface is approximately $\rho \Delta x \Delta y$, so Newton's second law of motion gives

$$T \Delta x \Delta y \left(\frac{\partial^2 z}{\partial x^2} + \frac{\partial^2 z}{\partial y^2} \right) \approx (\rho \Delta x \Delta y) \frac{\partial^2 z}{\partial t^2}.$$

Dividing by $\Delta x \Delta y$, we obtain the wave equation in two dimensions, namely the partial differential equation

$$\rho \frac{\partial^2 z}{\partial t^2} = T \left(\frac{\partial^2 z}{\partial x^2} + \frac{\partial^2 z}{\partial y^2} \right).$$

As in the one dimensional case, we set $c = \sqrt{T/\rho}$, which will play the role of the speed of the waves on the membrane. So the wave equation becomes

$$\frac{\partial^2 z}{\partial t^2} = c^2 \left(\frac{\partial^2 z}{\partial x^2} + \frac{\partial^2 z}{\partial y^2} \right).$$

Converting to polar coordinates (r, θ) and using the fact that

$$\frac{\partial^2 z}{\partial r^2} + \frac{1}{r}\frac{\partial z}{\partial r} + \frac{1}{r^2}\frac{\partial^2 z}{\partial \theta^2} = \frac{\partial^2 z}{\partial x^2} + \frac{\partial^2 z}{\partial y^2}.$$

we obtain

$$\frac{\partial^2 z}{\partial t^2} = c^2 \left(\frac{\partial^2 z}{\partial r^2} + \frac{1}{r}\frac{\partial z}{\partial r} + \frac{1}{r^2}\frac{\partial^2 z}{\partial \theta^2} \right). \tag{3.6.1}$$

We look for *separable* solutions of this equation, namely solutions of the form

$$z = f(r)g(\theta)h(t).$$

The reason for looking for separable solutions will be explained further in the next section. Substituting this into the wave equation, we obtain

$$f(r)g(\theta)h''(t) = c^2 \left(f''(r)g(\theta)h(t) + \frac{1}{r}f'(r)g(\theta)h(t) + \frac{1}{r^2}f(r)g''(\theta)h(t) \right).$$

Dividing by $f(r)g(\theta)h''(t)$ gives

$$\frac{h''(t)}{h(t)} = c^2 \left(\frac{f''(r)}{f(r)} + \frac{1}{r}\frac{f'(r)}{f(r)} + \frac{1}{r^2}\frac{g''(\theta)}{g(\theta)} \right).$$

In this equation, the left hand side only depends on t, and is independent of r and θ, while the right hand side only depends on r and θ, and is independent of t. Since t, r and θ are three independent variables, this implies that the common value of the two sides is independent of t, r and θ, so that it has to be a constant. We shall see in the next section that this constant has to be a negative real number, so we shall write it as $-\omega^2$. So we obtain two equations,

$$h''(t) = -\omega^2 h(t), \tag{3.6.2}$$

$$\frac{f''(r)}{f(r)} + \frac{1}{r}\frac{f'(r)}{f(r)} + \frac{1}{r^2}\frac{g''(\theta)}{g(\theta)} = -\frac{\omega^2}{c^2}. \tag{3.6.3}$$

The general solution to equation (3.6.2) is a multiple of the solution

$$h(t) = \sin(\omega t + \phi),$$

where ϕ is a constant determined by the initial temporal phase. Multiplying equation (3.6.3) by r^2 and rearranging, we obtain

$$r^2\frac{f''(r)}{f(r)} + r\frac{f'(r)}{f(r)} + \frac{\omega^2}{c^2}r^2 = -\frac{g''(\theta)}{g(\theta)}.$$

The left hand side depends only on r, while the right hand side depends only on θ, so their common value is again a constant. This makes $g(\theta)$ either a sine function or an exponential function, depending on the sign of the constant. But the function $g(\theta)$ has to be periodic of period 2π since it is a function of angle. So the common value of the constant must be the square of an integer n, so that

$$g''(\theta) = -n^2 g(\theta)$$

and $g(\theta)$ is a multiple of $\sin(n\theta + \psi)$. Here, ψ is another constant representing spatial phase. So we obtain

$$r^2 \frac{f''(r)}{f(r)} + r\frac{f'(r)}{f(r)} + \frac{\omega^2}{c^2}r^2 = n^2.$$

Multiplying by $f(r)$, dividing by r^2 and rearranging, this becomes

$$f''(r) + \frac{1}{r}f'(r) + \left(\frac{\omega^2}{c^2} - \frac{n^2}{r^2}\right)f(r) = 0.$$

Now Exercise 2 in Section 2.10 shows that the general solution to this equation is a linear combination of $J_n(\omega r/c)$ and $Y_n(\omega r/c)$. But the function $Y_n(\omega r/c)$ tends to $-\infty$ as r tends to zero, so this would introduce a singularity at the centre of the membrane. So the only physically relevant solutions to the above equation are multiples of $J_n(\omega r/c)$. So we have shown that the functions

$$\boxed{z = A J_n(\omega r/c)\sin(\omega t + \phi)\sin(n\theta + \psi)}$$

are solutions to the wave equation.

If the radius of the drum is a, then the boundary condition which we must satisfy is that $z = 0$ when $r = a$, for all values of t and θ. So it follows that $J_n(\omega a/c) = 0$. This is a constraint on the value of ω. The function J_n takes the value zero for a discrete infinite set of values of its argument. So ω is also constrained to an infinite discrete set of values.

It turns out that linear combinations of functions of the above form uniformly approximate the general, twice continuously differentiable solution of (3.6.1) as closely as desired, so that these form the drum equivalent of the sine and cosine functions of Fourier series.

Here is a table of the first few zeros of the Bessel functions. For more, see Appendix A.

k	J_0	J_1	J_2	J_3	J_4
1	2.404 83	3.831 71	5.135 62	6.380 16	7.588 34
2	5.520 08	7.015 59	8.417 24	9.761 02	11.064 71
3	8.653 73	10.173 47	11.619 84	13.015 20	14.372 54

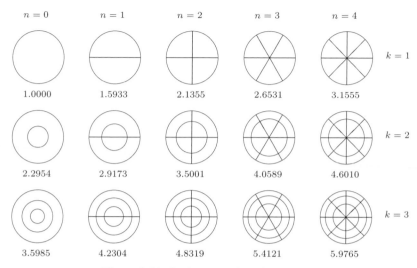

$n = 0$	$n = 1$	$n = 2$	$n = 3$	$n = 4$	
1.0000	1.5933	2.1355	2.6531	3.1555	$k = 1$
2.2954	2.9173	3.5001	4.0589	4.6010	$k = 2$
3.5985	4.2304	4.8319	5.4121	5.9765	$k = 3$

Figure 3.20 Stationary points for a drum.

We have seen that to choose a vibrational mode, we must choose a nonnegative integer n and we must choose a zero of $J_n(z)$. Denoting the kth zero of J_n by $j_{n,k}$, the corresponding vibrational mode has frequency $(cj_{n,k}/2\pi a)$, which is $j_{n,k}/j_{0,1}$ times the fundamental frequency. The stationary points have the pictures shown in Figure 3.20. Underneath each picture, we have recorded the value of $j_{n,k}/j_{0,1}$ for the relative frequency.

In the late eighteenth century, Chladni[6] discovered a way to see normal modes of vibration. He was interested in the vibration of plates, but the same technique can be used for drums and other instruments. He placed sand on the plate and then set it vibrating in one of its normal modes, using a violin bow. The sand collects on the stationary lines and gives a picture similar to the ones described above for the drum. A picture of Chladni patterns on a kettledrum can be found in Figure 3.21.

In practice, for a drum in which the air is confined (such as a kettledrum) the fundamental mode of the drum is heavily damped, because it involves compression and expansion of the air enclosed in the drum. So what is heard as the fundamental is really the mode with $n = 1, k = 1$, namely the second entry in the top row in the above diagram. The higher modes mostly involve moving the air from side to side. The inertia of the air has the effect of raising the frequency of the modes with $n = 0$, especially the fundamental, while the modes with $n > 0$ are lowered in frequency in such a way as to widen the frequency gaps. For an open drum, on the other hand, all the vibrational frequencies are lowered by the inertia of the air, but the ones of lower frequency are lowered the most.

---

[6] E. F. F. Chladni, *Entdeckungen über die Theorie des Klanges*, Weidmanns Erben und Reich 1787.

Figure 3.21 Chladni patterns on a kettledrum from Risset, *Les instruments de l'orchestre*, Bibliothèque pour la science, Diffusion Belin, 1987. Copyright B. Stark, Northern Illinois University.

The design of the orchestral kettledrum carefully utilises the inertia of the air to arrange for the modes with $n = 1, k = 1$ and $n = 2, k = 1$ to have frequency ratio approximating 3:2, so that what is perceived is a missing fundamental at half the actual fundamental frequency. Furthermore, the modes with $n = 3$, 4 and 5 (still with $k = 1$) are arranged to approximate frequency ratios of 4:2, 5:2 and 6:2 with the $n = 1, k = 1$ mode, thus accentuating the perception of the missing fundamental. The frequency of the $n = 1, k = 1$ mode is called the *nominal frequency* of the drum.

It is not true that the air in the kettle of a kettledrum acts as a resonator. A kettledrum can be retuned by a little more than a perfect fourth, whereas if the air were acting as a resonator, it could only do so for a small part of the frequency range. In fact, the resonances of the body of air are usually much higher in pitch, and do not have much effect on the overall sound. A more important effect is that the underside of the drum skin is prevented from radiating sound, and this makes the radiation of sound from the upper side more efficient.

Exercise

The women of Portugal (never the men) play a double sided square drum called an *adufe*. Find the separable solutions (i.e., the ones of the form $z = f(x)g(y)h(t)$) to the wave equation for a square drum. Write the answer in the form of an essay, with title: 'What does a square drum sound like?'. Try to integrate the words with the mathematics. Explain what you're doing at each step, and don't forget to answer the title question (i.e., describe the frequency spectrum).

Further reading

Murray Campbell and Clive Greated (1986), *The Musician's Guide to Acoustics*, Chapter 10.

R. Courant and D. Hilbert, *Methods of Mathematical Physics, I*, Interscience, 1953, Section V.5.

William C. Elmore and Mark A. Heald (1969), *Physics of Waves*, Chapter 2.

Neville H. Fletcher and Thomas D. Rossing (1991), *The Physics of Musical Instruments*, Section 18.

C. V. Raman (1988), The Indian musical drums, *Proc. Indian Acad. Sci.* **A1** (1934), 179–188. Reprinted in Rossing.

B. S. Ramakrishna and Man Mohan Sondhi, Vibrations of Indian musical drums regarded as composite membranes, *J. Acoust. Soc. Amer.* **26** (4) (1954), 523–529.

Thomas D. Rossing (2000), *Science of Percussion Instruments*.

3.7 Eigenvalues of the Laplace operator

In this section, we put the discussion of the vibrational modes of the drum into a broader context. Namely, we explain the relationship between the shape of a drum and its frequency spectrum, in terms of the eigenvalues of the Laplace operator. This discussion explains the connection between the uses of the word 'spectrum' in linear algebra, where it refers to the eigenvalues of an operator, and in music, where it refers to the distribution of frequency components. Parts of this discussion assume that the reader is familiar with elementary vector calculus and the divergence theorem.

We write ∇^2 for the operator $\frac{\partial^2}{\partial x^2} + \frac{\partial^2}{\partial y^2}$. This is known as the *Laplace operator* (in three dimensions the Laplace operator ∇^2 denotes $\frac{\partial^2}{\partial x^2} + \frac{\partial^2}{\partial y^2} + \frac{\partial^2}{\partial z^2}$; the analogous operator makes sense for any number of variables). In this notation, the wave equation becomes

$$\frac{\partial^2 z}{\partial t^2} = c^2 \nabla^2 z.$$

We consider the solutions to this equation on a closed and bounded region Ω. So for the drum of the last section, Ω was a disc in two dimensions.

A *separable solution* to the wave equation is one of the form

$$z = f(x, y)h(t).$$

Substituting into the wave equation, we obtain

$$f(x, y)h''(t) = c^2 \nabla^2 f(x, y) h(t)$$

or

$$\frac{h''(t)}{h(t)} = c^2 \frac{\nabla^2 f(x, y)}{f(x, y)}.$$

The left hand side is independent of x and y, while the right hand side is independent of t, so their common value is a constant. We write this constant as $-\omega^2$, because it will transpire that it has to be negative. Then we have

$$g''(t) = -\omega^2 g(t), \tag{3.7.1}$$

$$\nabla^2 f(x, y) = -\frac{\omega^2}{c^2} f(x, y). \tag{3.7.2}$$

The first of these equations is just the equation for simple harmonic motion with angular frequency ω, so the general solution is

$$g(t) = A \sin(\omega t + \phi).$$

A nonzero, twice differentiable function $f(x, y)$ satisfying the second equation is called an *eigenfunction* of the Laplace operator ∇^2 (or more accurately, of $-\nabla^2$), with *eigenvalue*

$$\lambda = \omega^2/c^2. \tag{3.7.3}$$

There are two important kinds of eigenfunctions and eigenvalues. The *Dirichlet spectrum* is the set of eigenvalues for eigenfunctions which vanish on the boundary of the region Ω. The *Neumann spectrum* is the set of eigenvalues for eigenfunctions with vanishing derivative normal (i.e., perpendicular) to the boundary. The latter functions are important when studying the wave equation for sound waves, where the dependent variable is acoustic pressure (i.e., pressure minus the average ambient pressure).

Here is a summary of some standard facts about the wave equation. We can choose Dirichlet eigenfunctions $f_1, f_2, \ldots$ of $-\nabla^2$ on Ω with eigenvalues $0 < \lambda_1 \le \lambda_2 \le \ldots$ with the following properties.

(i) Every eigenfunction is a finite linear combination of eigenfunctions f_i for which the λ_i are equal.
(ii) Each eigenvalue is repeated only a finite number of times.
(iii) $\lim_{n\to\infty} \lambda_n = \infty$.
(iv) (Completeness) Every continuous function can be written as a sum of an absolutely and uniformly convergent series of the form $f(x, y) = \sum_i a_i f_i(x, y)$.

The eigenvalue λ_i determines the frequency of the corresponding vibration via (3.7.3):

$$\omega_i = c\sqrt{\lambda_i}, \qquad \nu_i = c\sqrt{\lambda_i}/2\pi, \tag{3.7.4}$$

(recall that angular velocity ω is related to frequency ν by $\omega = 2\pi\nu$).

Initial conditions for the wave equation on Ω are specified by stipulating the values of z and $\frac{\partial z}{\partial t}$ for (x, y) in Ω, at $t = 0$. To solve the wave equation subject to these initial conditions, we use completeness to write $z = \sum_i a_i f_i(x, y)$ and $\frac{\partial z}{\partial t} = \sum_i b_i f_i(x, y)$ at $t = 0$. Then the unique solution is given by

$$z = \sum_\lambda f_i(x, y) \left(a_i \cos(c\sqrt{\lambda}\, t) + \frac{b_i}{c\sqrt{\lambda}} \sin(c\sqrt{\lambda}\, t) \right).$$

We have phrased the above discussion in terms of the two dimensional wave equation, but the same arguments work in any number of dimensions. For example, in one dimension it corresponds to the vibrational modes of a string, and we recover the theory of Fourier series.

An interesting problem, which was posed by Mark Kac in 1965 and solved by Gordon, Webb and Wolpert in 1991, is *whether one can hear the shape of a drum*. In other words, can one tell the shape of a simply connected closed region in two dimensions from its Dirichlet spectrum? Simply connected just means there are no holes in the region. Based on a method developed by Sunada a few years previously, Gordon, Webb and Wolpert found examples of pairs of regions with the same Dirichlet spectrum. The example which appears in their paper is Figure 3.22.

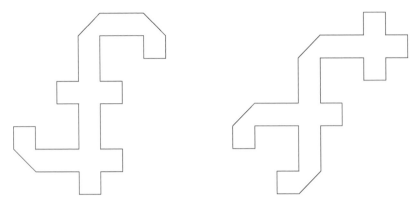

Figure 3.22 Two regions with the same Dirichlet spectrum.

Admittedly, it had probably not occurred to anyone to make drums using vibrating surfaces of these shapes, prior to this investigation. Many other pairs of regions with the same Dirichlet spectrum have been found. This and many more can be found in the paper of Buser, Conway, Doyle and Semmler listed below. But it is still not known whether there are any *convex* examples.

Further reading

P. Buser, J. H. Conway, P. Doyle and K.-D. Semmler, Some planar isospectral domains, *International Mathematics Research Notices* (1994), 391–400.

S. J. Chapman, Drums that sound the same, *Amer. Math. Monthly* **102** (2) (1995), 124–138.

Tobin Driscoll, Eigenmodes of isospectral drums. *SIAM Rev.* **39** (1997), 1–17.

Carolyn Gordon, David L. Webb and Scott Wolpert, One cannot hear the shape of a drum, *Bull. Amer. Math. Soc.* **27** (1992), 134–138.

Carolyn Gordon, David L. Webb and Scott Wolpert, Isospectral plane domains and surfaces via Riemannian orbifolds, *Invent. Math.* **110** (1992), 1–22.

V. E. Howle and Lloyd N. Trefethen, Eigenvalues and musical instruments, *J. Comp. Appl. Math.* **135** (2001), 23–40.

Mark Kac, Can one hear the shape of a drum?, *Amer. Math. Monthly* **73**, (1966), 1–23.

M. H. Protter, Can one hear the shape of a drum? Revisited. *SIAM Rev.* **29** (1987), 185–197.

K. Stewartson and R. T. Waechter, On hearing the shape of a drum: further results, *Proc. Camb. Phil. Soc.* **69** (1971), 353–363.

T. Sunada, Riemannian coverings and isospectral manifolds, *Ann. Math.* **121** (1985), 169–186.

3.8 The horn

The horn, and other instruments of the brass family, can be regarded as a hard walled tube of varying cross-section. Fortunately, the cross-section matters more than the exact shape and curvature of the tube.

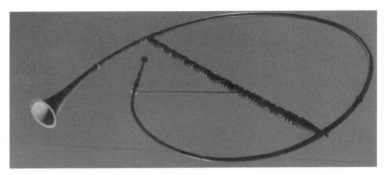

Figure 3.23 Cornu, Pompeii, first century CE (copy, XIXth century) Musical Instruments Museum, Brussels (inv. 464).

If $A(x)$ represents the cross-section as a function of position x along the tube, then assuming that the wavefronts are approximately planar and propagate along the direction of the horn, equation (3.5.2) can be modified to *Webster's horn equation*

$$\frac{1}{A(x)}\frac{\partial}{\partial x}\left(A(x)\frac{\partial p}{\partial x}\right) = \frac{1}{c^2}\frac{\partial^2 p}{\partial t^2}, \tag{3.8.1}$$

or equivalently

$$\frac{\partial^2 p}{\partial x^2} + \frac{1}{A}\frac{dA}{dx}\frac{\partial p}{\partial x} = \frac{1}{c^2}\frac{\partial^2 p}{\partial t^2}.$$

Solutions of this equation can be described using the theory of *Sturm–Liouville equations*. The theory of Sturm–Liouville equations is described in many standard texts on partial differential equations.

There is one particular form of $A(x)$ which is of physical importance because it gives a good approximation to the shape of actual brass instruments while at the same time giving an equation with relatively simple solutions. Namely, the *Bessel horn*, with cross section of radius and area

$$R(x) = bx^{-\alpha}, \qquad A(x) = \pi R(x)^2 = Bx^{-2\alpha}.$$

Here, the origin of the x coordinate and the constant b are chosen to give the correct radius at the two ends of the horn, and $B = \pi b^2$. Notice that the constant B disappears when $A(x)$ is put into equation (3.8.1). The parameter α is the 'flare parameter' that determines the shape of the flare of the horn. The case $\alpha = 0$ gives a conical tube, and we shall usually assume that $\alpha \geq 0$. The solutions are sums of ones of the form

$$p(x,t) = x^{\alpha+\frac{1}{2}} J_{\alpha+\frac{1}{2}}(\omega x/c)(a \cos \omega t + b \sin \omega t). \tag{3.8.2}$$

Here, as usual, the angular frequency ω must be chosen so that the boundary conditions are satisfied at the ends of the horn.

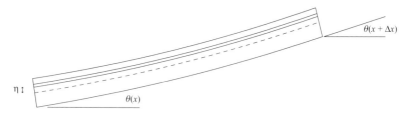

Figure 3.26 Bending a rod.

by $-\eta$ and integrating:[8]

$$M = \frac{E}{R} \int \eta^2 \, dA.$$

The quantity $I = \int \eta^2 \, dA$ is called the *sectional moment* of the cross-section of the rod. So we obtain $M = -EI/R$. Now the formula for radius of curvature is $R = \left(1 + \left(\frac{dy}{dx}\right)^2\right)^{\frac{3}{2}} / \frac{d^2 y}{dx^2}$. Assuming that $\frac{dy}{dx}$ is small, this can be approximated by the formula $1/R = \frac{d^2 y}{dx^2}$, so that

$$M(x, t) = EI \frac{\partial^2 y}{\partial x^2}.$$

Combining this with equation (3.9.1) gives

$$\frac{\partial^2 y}{\partial t^2} + \frac{EI}{\rho} \frac{\partial^4 y}{\partial x^4} = 0. \tag{3.9.2}$$

This is the differential equation which governs the transverse waves on the rod. It is known as the Euler–Bernoulli beam equation.

We look for separable solutions to equation (3.9.2). Setting

$$y = f(x)g(t),$$

we obtain

$$f(x)g''(t) + \frac{EI}{\rho} f^{(4)}(x)g(t) = 0$$

or

$$\frac{g''(t)}{g(t)} = -\frac{EI}{\rho} \frac{f^{(4)}(x)}{f(x)}.$$

Since the left hand side does not depend on x and the right hand side does not depend on t, both sides are constant. So

$$g''(t) = -\omega^2 g(t), \tag{3.9.3}$$

$$f^{(4)}(x) = \frac{\omega^2 \rho}{EI} f(x). \tag{3.9.4}$$

---

[8] The minus sign comes from the fact that counterclockwise moment is positive.

Equation (3.9.3) says that $g(t)$ is a multiple of $\sin(\omega t + \phi)$, while equation (3.9.4) has solutions

$$f(x) = A \sin \kappa x + B \cos \kappa x + C \sinh \kappa x + D \cosh \kappa x,$$

where

$$\kappa = \sqrt[4]{\frac{\omega^2 \rho}{EI}}. \qquad (3.9.5)$$

The general solution then decomposes as a sum of the normal modes

$$y = (A \sin \kappa x + B \cos \kappa x + C \sinh \kappa x + D \cosh \kappa x) \sin(\omega t + \phi). \qquad (3.9.6)$$

The boundary conditions depend on what happens at the end of the rod. It is these boundary conditions which constrain ω to a discrete set of values. If an end of the rod is free, then the quantities $V(x, t)$ and $M(x, t)$ have to vanish for all t, at the value of x corresponding to the end of the rod. So $\partial^2 y/\partial x^2 = 0$ and $\partial^3 y/\partial x^3 = 0$. If an end of the rod is clamped, then the displacement and slope vanish, so $y = 0$ and $\partial y/\partial x = 0$ for all t at the value of x corresponding to the end of the rod.

We calculate

$$\partial y/\partial x = \kappa(A \cos \kappa x - B \sin \kappa x + C \cosh \kappa x + D \sinh \kappa x) \sin(\omega t + \phi),$$
$$\partial^2 y/\partial x^2 = \kappa^2(-A \sin \kappa x - B \cos \kappa x + C \sinh \kappa x + D \cosh \kappa x) \sin(\omega t + \phi),$$
$$\partial^3 y/\partial x^3 = \kappa^3(-A \cos \kappa x + B \sin \kappa x + C \cosh \kappa x + D \sinh \kappa x) \sin(\omega t + \phi).$$

In the case of the xylophone or tubular bell, both ends are free. We take the two ends to be at $x = 0$ and $x = \ell$. The conditions $\partial^2 y/\partial x^2 = 0$ and $\partial^3 y/\partial x^3 = 0$ at $x = 0$ give $B = D$ and $A = C$. These conditions at $x = \ell$ give

$$A(\sinh \kappa \ell - \sin \kappa \ell) + B(\cosh \kappa \ell - \cos \kappa \ell) = 0,$$
$$A(\cosh \kappa \ell - \cos \kappa \ell) + B(\sinh \kappa \ell + \sin \kappa \ell) = 0.$$

These equations admit a nonzero solution in A and B exactly when the determinant

$$(\sinh \kappa \ell - \sin \kappa \ell)(\sinh \kappa \ell + \sin \kappa \ell) - (\cosh \kappa \ell - \cos \kappa \ell)^2$$

vanishes. Using the relations $\cosh^2 \kappa \ell - \sinh^2 \kappa \ell = 1$ and $\sin^2 \kappa \ell + \cos^2 \kappa \ell = 1$, this condition becomes

$$\cosh \kappa \ell \cos \kappa \ell = 1.$$

The values of $\kappa \ell$ for which this equation holds determine the allowed frequencies via the formula (3.9.5).

Set $\lambda = \kappa \ell$, so that λ has to be a solution of the equation

$$\cosh \lambda \cos \lambda = 1. \qquad (3.9.7)$$

Then equation (3.9.5) shows that the angular frequency and the frequency are given by

$$\omega = \sqrt{\frac{EI}{\rho}\frac{\lambda^2}{\ell^2}}; \qquad v = \frac{\omega}{2\pi} = \sqrt{\frac{EI}{\rho}\frac{\lambda^2}{2\pi\ell^2}}. \qquad (3.9.8)$$

Numerical computations for the positive solutions to equation (3.9.7) give the following values, with more accuracy than is strictly necessary:

$$\lambda_1 = \ 4.730\,040\,744\,862\,704\,026\,024\,0481,$$
$$\lambda_2 = \ 7.853\,204\,624\,095\,837\,556\,477\,0667,$$
$$\lambda_3 = 10.995\,607\,838\,001\,670\,906\,669\,0325,$$
$$\lambda_4 = 14.137\,165\,491\,257\,464\,177\,105\,9179,$$
$$\lambda_5 = 17.278\,759\,657\,399\,481\,438\,091\,0740,$$
$$\lambda_6 = 20.420\,352\,245\,626\,061\,090\,936\,4112.$$

As n increases, $\cosh \lambda_n$ increases exponentially, and so $\cos \lambda_n$ has to be very small and positive. So λ_n is close to $(n + \frac{1}{2})\pi$, the nth zero of the cosine function. For $n \geq 5$, the approximation

$$\lambda_n \approx \left(n + \tfrac{1}{2}\right)\pi - (-1)^n 2e^{-(n+\frac{1}{2})\pi} - 4e^{-2(n+\frac{1}{2})\pi} \qquad (3.9.9)$$

holds to at least 20 decimal places.[9]

Using equation (3.9.8), we find that the frequency ratios as multiples of the fundamental are given by the quantities λ_n^2/λ_1^2:

n	λ_n^2/λ_1^2
1	1.000\,000\,000\,000\,00
2	2.756\,538\,507\,099\,96
3	5.403\,917\,632\,383\,32
4	8.932\,950\,352\,381\,93
5	13.344\,286\,693\,666\,89
6	18.637\,887\,886\,581\,19

The resulting set of frequencies is certainly inharmonic, just as in the case of the drum. But as n increases, equation (3.9.9) shows that the higher partials have ratios approximating those of the squares of odd integers.

---

[9] This series continues as follows:

$$\lambda_n \approx \left(n + \tfrac{1}{2}\right)\pi - (-1)^n 2e^{-(n+\frac{1}{2})\pi} - 4e^{-2(n+\frac{1}{2})\pi} - (-1)^n \tfrac{34}{3}e^{-3(n+\frac{1}{2})\pi} - \tfrac{112}{3}e^{-4(n+\frac{1}{2})\pi} - \cdots$$

The (difficult) challenge to the reader is to compute the next few terms! As a check, $m!$ times the fraction in front of the mth exponential term should be an integer.

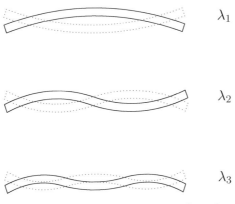

Figure 3.27 Vibrational modes of a rod.

The vibrational modes described by the above values of λ correspond to the pictures in Figure 3.27.

For actual instruments, rather than the idealized bar described above, the series of partials is somewhat different. Tubular bells are the closest to the ideal situation described above, with second and third partials at frequency ratios of 2.76:1 and 5.40:1 to the fundamental.

The bars of an orchestral xylophone are made of rosewood, or sometimes of more modern materials which are more durable and keep their pitch under more extreme conditions. There is a shallow arch cut out from the underside, with the intention of producing frequency ratios of 3:1 and 6:1 for the second and third partials with respect to the fundamental. These partials correspond to tones an octave and a perfect fifth, and two octaves and a perfect fifth above the fundamental, respectively.

The marimba is also made of rosewood, and the vibe is made from aluminium. For these instruments, a deeper arch is cut out from the underside, with the intention of producing frequency ratios of 4:1 and (usually) 10:1 with respect to the fundamental. These represent tones two octaves, and three octaves and a major third above the fundamental, respectively.

The tuning of the second partial can be made quite precise, because material removed from different parts of the bar affects different partials. Removing material from the end increases the fundamental and the partials. Taking material away from the sides of the arch lowers the second partial, while taking it from the centre of

the arch lowers the fundamental frequency. The third partial is harder to make accurate; this and the higher partials are part of the artistic expression of the maker.
Tuning can be carried out using stroboscopic equipment, which allows for tuning
of the fundamental and second partial to within plus or minus one cent (a cent is a
hundredth of a semitone).

Further reading

Antoine Chaigne and Vincent Doutaut, Numerical simulations of xylophones. I.
    Time-domain modeling of the vibrating bars, *J. Acoust. Soc. Amer.* **101** (1) (1997),
    539–557.
R. Courant and D. Hilbert, *Methods of Mathematical Physics, I*, Interscience, 1953,
    Section V.4.
William C. Elmore and Mark A. Heald (1969), *Physics of Waves*, Chapter 3.
Neville H. Fletcher and Thomas D. Rossing (1991), *The Physics of Musical Instruments*,
    Section 19.
D. Holz, Investigations on acoustically important qualities of xylophone-bar materials:
    can we substitute any tropical woods by European species?, *Proc. Int. Symp. Musical
    Acoustics*, Jouve (1995), 351–357.
A. M. Jones, *Africa and Indonesia: the Evidence of the Xylophone and Other Musical and
    Cultural Factors*, E. J. Brill, 1964. This book contains a large number of
    measurements of the tuning of African and Indonesian xylophones. The author
    argues the hypothesis that there was Indonesian influence on African music, and
    therefore visitations to Africa by Indonesians, long before the Portuguese
    colonization of Indonesia.
James L. Moore, *Acoustics of Bar Percussion Instruments*, Permus Publications, 1978.
Thomas D. Rossing (2000), *Science of Percussion Instruments*, Chapters 5–7.

3.10 The mbira

At a lecture demonstration I once attended in Seattle, Washington, Dumisani Maraire, a
visiting artist from Zimbabwe, walked onto the stage carrying a round-box resonator with a
fifteen-key instrument inside. He turned toward the audience and raised the round-box over
his head. 'What is this?' he called out.

There was no response.

'All right,' he said, 'it is an mbira; M-B-I-R-A. Now what did I say it was?'

A few people replied, 'Mbira; it is an mbira.' Most of the audience sat still in puzzlement.

'What is it?' Maraire repeated, as if slightly annoyed.

More people called out, 'Mbira.'

'Again,' Maraire insisted.

'Mbira!' returned the audience.

'Again!' he shouted. When the auditorium echoed with 'Mbira,' Maraire laughed out
loud. 'All right,' he said with good-natured sarcasm, 'that is the way the Christian mission-
aries taught me to say 'piano.''

(Paul F. Berliner, The Soul of Mbira, University of California Press)

Figure 3.28 Picture of mbira from Zimbabwe, from Jacqueline Cogdell DjeDje, 'Turn up the volume! A celebration of African music', UCLA 1999, p. 240.

The *mbira* is a popular melodic instrument of Africa, especially the Shona people of Zimbabwe. Other names for the instrument are *sanzhi*, *likembe* and *kalimba*; the general ethnomusicological category is the *lamellophone*. It consists of a set of keys on a soundboard, usually with some kind of resonator such as a gourd for amplifying and transmitting the sound. The keys are usually metal, clamped at one end and free at the other. They are depressed with the finger or thumb and suddenly released to produce the vibration.

The method of Section 3.9 can be used to analyze the resonant modes of the keys of the mbira. There is no change up to the point where the boundary conditions are applied to equation (3.9.6). We take the clamped end to be at $x = 0$ and the free end at $x = \ell$. At $x = 0$, the condition $y = 0$ gives $D = -B$ and $\partial y/\partial x = 0$ gives

$C = -A$. The conditions $\partial^2 y/\partial x^2 = 0$ and $\partial^3 y/\partial x^3 = 0$ at $x = \ell$ then give

$$-A(\sin \kappa \ell + \sinh \kappa \ell) - B(\cos \kappa \ell + \cosh \kappa \ell) = 0,$$
$$-A(\cos \kappa \ell + \cosh \kappa \ell) + B(\sin \kappa \ell - \sinh \kappa \ell) = 0.$$

These equations admit a nonzero solution in A and B exactly when the determinant

$$-(\sin \kappa \ell + \sinh \kappa \ell)(\sin \kappa \ell - \sinh \kappa \ell) - (\cos \kappa \ell + \cosh \kappa \ell)^2$$

vanishes. This time, the equation reduces to

$$\cosh \kappa \ell \cos \kappa \ell = -1.$$

Setting $\lambda = \kappa \ell$ as before, we find that λ has to be a solution of the equation

$$\cosh \lambda \cos \lambda = -1. \tag{3.10.1}$$

Then the angular frequency and the frequency are again given by equation (3.9.8). The following are the first few solutions of equation (3.10.1):

$$\lambda_1 = 1.875\,104\,068\,711\,961\,166\,445\,3082,$$
$$\lambda_2 = 4.694\,091\,132\,974\,174\,576\,436\,3918,$$
$$\lambda_3 = 7.854\,757\,438\,237\,612\,564\,861\,0086,$$
$$\lambda_4 = 10.995\,540\,734\,875\,466\,990\,667\,3491,$$
$$\lambda_5 = 14.137\,168\,391\,046\,470\,580\,917\,0468,$$
$$\lambda_6 = 17.278\,759\,532\,088\,236\,333\,543\,9284.$$

Notice that these are approximately the same as the values found in the last section, except that there is one extra value playing the role of the fundamental. The analogue of equation (3.9.9) is

$$\lambda_n \approx \left(n - \tfrac{1}{2}\right)\pi - (-1)^n 2e^{-(n-\frac{1}{2})\pi} - 4e^{-2(n-\frac{1}{2})\pi},$$

which holds to at least 20 decimal places for $n \geq 6$. The frequency ratios as multiples of the fundamental are given by the quantities λ_n^2/λ_1^2.

n	λ_n^2/λ_1^2
1	1.000\,000\,000\,000\,00
2	6.266\,893\,025\,770\,67
3	17.547\,481\,936\,808\,44
4	34.386\,061\,157\,203\,00
5	56.842\,622\,928\,102\,01
6	84.913\,035\,970\,713\,18

Of course, the above figures are based on an idealized mbira with constant cross-section for the keys. The keys of an actual mbira are very far from constant in cross-section, and so the actual relative frequencies of the partials may be far from what is described by the above table. But the most prominent feature, namely that the frequencies of the partials increase quite rapidly, holds in actual instruments.

Further reading

Paul F. Berliner, *The Soul of Mbira; Music and Traditions of the Shona People of Zimbabwe*, University of California Press, 1978. Reprinted by University of Chicago Press, 1993.

3.11 The gong

As a first approximation, the gong can be thought of as a circular flat stiff metal plate of uniform thickness. In practice, the gong is slightly curved, and the thickness is not uniform, but for the moment we shall ignore this. The stiff metal plate behaves like a mixture of the drum and the stiff rod. So the partial differential equation governing its motion is fourth order, as in the case of the stiff rod, but there are two directions in which to take partial derivatives, as in the case of the drum. If z represents displacement, and x and y represent Cartesian coordinates on the gong, then the equation is

$$\frac{\partial^2 z}{\partial t^2} + \frac{Eh^2}{12\rho(1-s^2)}\nabla^4 z = 0. \tag{3.11.1}$$

This equation first appears (without the explicit value of the constant in front of the second term) in a paper of Sophie Germain.[10] In this equation, h is the thickness of the plate, and an easy calculation shows that $\frac{h^2}{12} = \frac{1}{h}\int_{-h/2}^{+h/2} z^2\,dz$ is the corresponding sectional moment in the one thickness direction (in the case of the stiff rod, there were two dimensions for the cross-section, so the case of the stiff plate is easier in this regard). The quantity E is the Young's modulus as before, ρ is area density, and s is *Poisson's ratio*. This is a measure of the ratio of sideways spreading to the compression. The extra factor of $(1-s^2)$ in the denominator on the right hand of the above equation does not correspond to any term in equation (3.9.2). It arises

---

[10] Sophie Germain's paper, 'Recherches sur la théorie des surfaces élastiques', written in 1815 and published in 1821, won her a prize of a kilogram of gold from the French Academy of Sciences in 1816. The paper contained some significant errors, but became the basis for work on the subject by Lagrange, Poisson, Kirchoff, Navier and others.

Sophie Germain is probably better known for having made one of the first significant breakthroughs in the study of Fermat's last theorem. She proved that if x, y and z are integers satisfying $x^5 + y^5 = z^5$, then at least one of x, y and z has to be divisible by 5. More generally, she showed that the same was true when 5 is replaced by any prime p such that $2p + 1$ is also a prime.

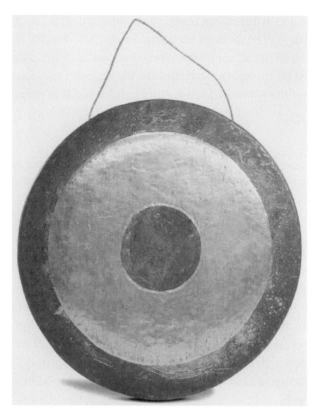

Figure 3.29 Gong from the Music Research Institute in Beijing. From *The Musical Arts of Ancient China*, exhibit 20.

from the fact that when the plate is bent downwards in one direction, it causes it to curl up in the perpendicular direction along the plate.

The term $\nabla^4 z$ denotes

$$\nabla^2 \nabla^2 z = \frac{\partial^4 z}{\partial x^4} + 2\frac{\partial^4 z}{\partial x^2 \partial y^2} + \frac{\partial^4 z}{\partial y^4}.$$

Observe the cross terms carefully. Without them, a rotational change of coordinates would not preserve this operation.

In the case of the stiff rod, we had to use the hyperbolic functions as well as the trigonometric functions. In this case, we are going to need to use the *hyperbolic Bessel functions*. These are defined by

$$I_n(z) = i^{-n} J_n(iz),$$

and bear the same relationship to the ordinary Bessel functions that the hyperbolic functions $\sinh x$ and $\cosh x$ do to the trigonometric functions $\sin x$ and $\cos x$.

Looking for separable solutions $z = Z(x, y)h(t) = f(r)g(\theta)h(t)$ to equation (3.11.1), we arrive at the equations

$$\nabla^4 Z = \kappa^4 Z \qquad (3.11.2)$$

and

$$\frac{\partial^2 h}{\partial t^2} = -\omega^2 h, \qquad (3.11.3)$$

where ω and κ are related by

$$\kappa^4 = \frac{12\rho(1 - s^2)\omega^2}{Eh^2}.$$

We factor equation (3.11.2) as

$$(\nabla^2 - \kappa^2)(\nabla^2 + \kappa^2)z = 0. \qquad (3.11.4)$$

So any solution to either the equation

$$\nabla^2 z = \kappa^2 z \qquad (3.11.5)$$

or the equation

$$\nabla^2 z = -\kappa^2 z \qquad (3.11.6)$$

is also a solution to (3.11.2).

Lemma 3.11.1 *Every solution z to equation (3.11.2) can be written uniquely as $z_1 + z_2$, where z_1 satisfies equation (3.11.5) and z_2 satisfies equation (3.11.6).*

Proof We use a variation of the even and odd function method. If $\nabla^4 z = \kappa^4 z$, we set

$$z_1 = \tfrac{1}{2}(z + \kappa^{-2}\nabla^2 z), \qquad z_2 = \tfrac{1}{2}(z - \kappa^{-2}\nabla^2 z).$$

Then

$$\nabla^2 z_1 = \tfrac{1}{2}(\nabla^2 z + \kappa^{-2}\nabla^4 z) = \tfrac{1}{2}(\nabla^2 z + \kappa^2 z) = \kappa^2 z_1,$$

$$\nabla^2 z_2 = \tfrac{1}{2}(\nabla^2 z - \kappa^{-2}\nabla^4 z) = \tfrac{1}{2}(\nabla^2 z - \kappa^2 z) = -\kappa^2 z_2$$

and $z_1 + z_2 = z$.

For the uniqueness, if z_1' and z_2' constitute another choice, then rearranging the equation $z_1 + z_2 = z_1' + z_2'$, we have $z_1 - z_1' = z_2' - z_2$. The common value z_3 of $z_1 - z_1'$ and $z_2' - z_2$ satisfies both equations (3.11.5) and (3.11.6). So $z_3 = \kappa^{-2}\nabla^2 z_3 = -z_3$, and hence $z_3 = 0$. It follows that $z_1 = z_1'$ and $z_2 = z_2'$. $\square$

Solving equation 3.11.6 is just the same as in the case of the drum, and the solutions are given as trigonometric functions of θ multiplied by Bessel functions of r. Equation 3.11.5 is similar, except that we must use the hyperbolic Bessel functions instead of the Bessel functions. We then have to combine the two classes of solutions in order to satisfy the boundary conditions, just as we did with the trigonometric and hyperbolic functions for the stiff rod. This leads us to solutions of the form

$$z = (A J_n(\kappa r) + B I_n(\kappa r)) \sin(\omega t + \phi) \sin(n\theta + \psi).$$

The boundary conditions for the gong require considerable care, and the first correct analysis was given by Kirchoff in 1850. His boundary conditions can be stated for any region with smooth boundary. Choosing coordinates in such a way that the element of boundary is a small segment of the y axis going through the origin, they are as follows:

$$\frac{\partial^2 z}{\partial x^2} + s \frac{\partial^2 z}{\partial y^2} = 0,$$

$$\frac{\partial^3 z}{\partial x^3} + (2 - s) \frac{\partial^3 w}{\partial x \partial y^2} = 0.$$

The derivation of these equations may be found in Chapter X of the first volume of Rayleigh's *The Theory of Sound* (1896), Section 216, where these boundary conditions appear as equations (6). He goes on to find the normal modes and eigenvalues by Fourier series methods. The results are similar to those for the drum in Section 3.6, but the modes with $k = 0$ and $n = 0$ or $n = 1$ are missing; it is easy to see why if we try to imagine the corresponding vibration of a gong. So the fundamental mode is $k = 0$ and $n = 2$. The relative frequencies are tabulated in Section 3.6 of Fletcher and Rossing, *The Physics of Musical Instruments* (1991), and reproduced in the following table with kind permission of Springer Science and Business Media.

Vibrational frequencies for a free circular plate

k	$n = 0$	$n = 1$	$n = 2$	$n = 3$	$n = 4$	$n = 5$
0	—	—	1.000	2.328	4.11	6.30
1	1.73	3.91	6.71	10.07	13.92	18.24
2	7.34	11.40	15.97	21.19	27.18	33.31

Actual gongs in real life are not perfect circular plates. Many designs feature circularly symmetric raised portions in the middle of the gong. This modifies the frequencies of the normal modes and the character of the sound. Often, eigenvalues become close enough together to degenerate, and then normal modes can mix. This

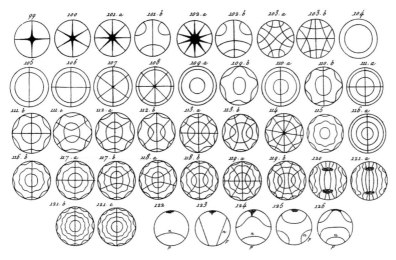

Figure 3.30 Normal modes of a gong. From E. F. F. Chladni, *Traité d'acoustique*, Courcier, Paris 1809.

seems to be in evidence in Chladni's original drawings (see Figure 3.30 and also Figure 3.20).

Cymbals are similar in design, and the theory works in a similar way. Because of the deviation from flatness, the normal modes again tend to combine in interesting ways. For example, the mode $(n, k) = (7, 0)$ and the mode $(2, 1)$ or $(3, 1)$ are often close enough in frequency to degenerate into a single compound mode (see Rossing and Peterson, 1982, below).

Further reading

R. C. Colwell and J. K. Stewart, The mathematical theory of vibrating membranes and plates, *J. Acoust. Soc. Amer.* **3** (4) (1932), 591–595.

R. C. Colwell, J. K. Stewart and H. D. Arnett, Symmetrical sand figures on circular plates, *J. Acoust. Soc. Amer.* **12** (2) (1940), 260–265.

R. Courant and D. Hilbert, *Methods of Mathematical Physics, I*, Interscience, 1953, Section V.6.

Neville H. Fletcher and Thomas D. Rossing (1991), *The Physics of Musical Instruments*, Sections 3.5–3.6 and Section 20.

Karl F. Graff (1975), *Wave Motion in Elastic Solids*.

Philip M. Morse and K. Uno Ingard (1968), *Theoretical Acoustics*, Section 5.3.

J. W. S. Rayleigh (1896), *The Theory of Sound*, Chapter X.

Thomas D. Rossing (2000), *Science of Percussion Instruments*, Chapters 8 and 9.

Thomas D. Rossing and Neville H. Fletcher, Nonlinear vibrations in plates and gongs, *J. Acoust. Soc. Amer.* **73** (1) (1983), 345–351.

Thomas D. Rossing and R. W. Peterson, Vibrations of plates, gongs and cymbals, *Percussive Notes* **19** (3) (1982), 31.

M. D. Waller, Vibrations of free circular plates. Part I: Normal modes, *Proc. Phys. Soc.* **50** (1938), 70–76.

Figure 3.31 Campanile at Cattedrale di S. Giusto, Trieste. Photo ©Dave Benson.

3.12 The bell

A bell can be thought of as a very deformed plate; its vibrational modes are similar in nature, but starting with $n = 2$. But the exact shape of the bell is made so as to tune the various vibrational modes relative to each other. There are five modes with special names, which are as follows. The mode $(n, k) = (2, 0)$ is the fundamental, and is called the *hum*. The *prime* is $(2, 1)$, and is tuned to twice the frequency, putting it an octave higher. There are two different modes $(3, 1)$, one of which has the stationary circle around the waist, and the other nearer the rim. The one with the waist is called the *tierce*, and is tuned a minor third above the prime. The other mode, sometimes denoted $(3, 1^\sharp)$, with the stationary circle nearer the rim is called the *quint*. It is pitched a perfect fifth above the prime. The *nominal* mode is $(4, 1)$, tuned an octave above the prime, so that it is two octaves above the hum. The nominal mode is by far the one with the largest amplitude, so that this is the perceived pitch of the bell. Mode $(4, 1^\sharp)$ is sometimes called the *deciem*, and is usually tuned a *major* third above the nominal. It can be imagined that a great deal of skill goes into the tuning of the vibrational modes of a bell. It is an art which has developed over many centuries. Particular attention is given to the construction of the thick ring near the rim. The information described above is summarized in Figure 3.32.

The singing bowl

Our discussion of the bell applies equally well to other objects such at the Tibetan *singing bowl*, which is used mainly for ritual purposes. The singing bowl may be

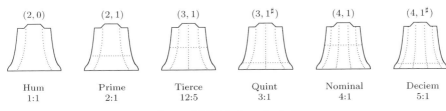

(2, 0)	(2, 1)	(3, 1)	(3, 1$^\sharp$)	(4, 1)	(4, 1$^\sharp$)
Hum	Prime	Tierce	Quint	Nominal	Deciem
1:1	2:1	12:5	3:1	4:1	5:1

Figure 3.32 Modes of a bell.

Figure 3.33 Singing bowl from Tibet. Photo ©Dave Benson.

struck with the wooden mallet on the inside of the rim to set it vibrating, or it may be stroked around the outside of the rim with the mallet to sustain a vibration in the manner of stroking a wineglass. After the mallet is removed, the tone can often last in excess of a minute before it is inaudible.

I have a Tibetan singing bowl in my living room, which is approximately 19 cm (or 7 inches) in diameter. It has two clearly audible partials, and some others too high to hear the pitch very precisely. The fundamental sounds at about 196 Hz, and the second partial sounds at about 549 Hz, giving a ratio of about 2.8:1.

Chinese bells

In 1977, an extraordinary discovery was made in the Hubei province of China. A huge burial pit was found, containing over four thousand bronze items. This was the tomb of Marquis Yi of the state of Zeng, and inscriptions date it very precisely at 433 BCE. The tomb contains many musical instruments, but the most extraordinary is a set of sixty-five bronze bells. These are able to play all twelve notes of the chromatic scale over a range of three octaves, and further bells fill this out to a five octave range.

Figure 3.34 Bell from the tomb of Marquis Yi (middle tier, height 75 cm, weight 32.2 kg) Picture from *Music in the age of Confucius*, p. 43 (see Further reading, below).

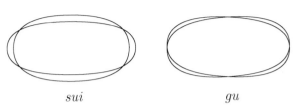

sui gu

Figure 3.35 Modes of Chinese bell.

Each bell is roughly elliptical in cross-section. There are two separate strike points, and the bell is designed so that the normal modes excited at the strike points have essentially nothing in common. So the perceived pitches are quite different. The strike point for the lower pitch is called the *sui*, and the one for the higher pitch is the *gu*. The bells are tuned so that this difference is either a major third or a minor third. The separation of the modes is achieved through the use of an elaborate set of nipples on the outer surface of the bell. See Figure 3.34. The values of n and k are the same for the *sui* and *gu* version of a vibrational mode, but the orientation is different. This may be illustrated as shown in Figure 3.35 for the modes with $n = 2$, where the diagram represents the movement of the lower rim.

It seems very hard to understand how these two tone bells were cast. The inscriptions naming the two tones were cast with the bell, so they must have been predetermined. What's more, the design does not just scale up proportionally, and there is no easy formula for how to produce a larger bell with the same musical interval. Modern physics does not lead to any understanding of the design procedures that were used to produce this set of bells.

Further reading

Lothar von Falkenhausen, *Suspended Music: Chime-bells in the Culture of Bronze Age China*, University of California Press, 1993.

Neville H. Fletcher and Thomas D. Rossing (1991), *The Physics of Musical Instruments*, Section 21.

M. Jing, A theoretical study of the vibration and acoustics of ancient Chinese bells, *J. Acoust. Soc. Amer.* **114** (3) (2003), 1622–1628.

Yuan-Yuan Lee and Sin-Yan Shen, *Chinese Musical Instruments*, Chinese Music Society of North America, 1999.

N. McLachlan, B. K. Nikjeh and A. Hasell, The design of bells with harmonic overtones, *J. Acoust. Soc. Amer.* **114** (1) (2003), 505–511.

J. Pan, X. Li, J. Tian and T. Lin, Short sound decay of ancient Chinese music bells, *J. Acoust. Soc. Amer.* **112** (6) (2002), 3042–3045.

R. Perrin and T. Chanley, The normal modes of the modern English church bell, *J. Sound Vib.* **90** (1983), 29–49.

Thomas D. Rossing, *The Acoustics of Bells*, Van Nostrand Reinhold, 1984.

Thomas D. Rossing (1990), *The Science of Sound*, Section 13.4.

Thomas D. Rossing (2000), *Science of Percussion Instruments*, Chapters 11–13.

Jenny So, *Eastern Zhou Ritual Bronzes from the Arthur M. Sackler Collections*,
    Smithsonian Institution, 1995. This is a large format book with photographs and
    descriptions of Chinese bronzes from the Eastern Zhou. Pages 357–397 describe the
    two tone bells from the collection. There is also an extensive appendix (pages
    431–484) titled 'Acoustical and musical studies on the Sackler bells', by Lothar von
    Falkenhausen and Thomas D. Rossing. This appendix gives a great many technical
    details of the acoustics and tuning of two tone bells.
Jenny So (ed.), *Music in the Age of Confucius*, Sackler Gallery, Washington, 2000. This
    beautifully produced book contains an extensive set of photographs of the set of bells
    from the tomb of Marquis Yi.

3.13 Acoustics

The basic equation of acoustics is the three dimensional wave equation, which
describes the movement of air to form sound. The discussion is similar to the one
dimensional discussion in Section 3.5. Recall that acoustic pressure p is measured
by subtracting the (constant) ambient air pressure ρ from the absolute pressure P. In
three dimensions, p is a function of x, y, z and t. This is related to the displacement
vector field $\xi(x, y, z, t)$ by two equations. The first is *Hooke's law*, which in this
situation can be written as

$$p = -B \, \nabla . \xi,$$

where B is the bulk modulus of air. Newton's second law of motion implies that

$$\nabla p = -\rho \frac{\partial^2 \xi}{\partial t^2}.$$

Putting these two equations together gives

$$\nabla^2 p = \frac{1}{c^2} \frac{\partial^2 p}{\partial t^2}, \qquad (3.13.1)$$

where $c = \sqrt{B/\rho}$. So p satisfies the three dimensional wave equation.

In an enclosed space, the boundary conditions are given by $\nabla p = 0$ on the walls
of the enclosure for all t. Looking for separable solutions leads to the theory of
Dirichlet and Neumann eigenvalues, just as in the two dimensional case when we
discussed the drum in Section 3.6. So there is a certain set of *resonant frequencies*
for the enclosure, determined by the eigenvalues of ∇^2 in the region. The same
reasoning as in Section 3.6 leads to the conclusion that the relationship between
frequency and eigenvalue is $\nu = c\sqrt{\lambda}/2\pi$, see equation (3.7.4). For an enclosure of
small total volume, the eigenvalues are widely spaced. But as the volume increases,
the eigenvalues get closer together. So, for example, a concert hall has a large
total volume, and the eigenvalues are typically at intervals of a few hertz, and
the spacing is somewhat erratic. Fortunately, the ear is performing a windowed

Fourier analysis with a relatively short time window, so that in accordance with Heisenberg's uncertainty principle, fluctuations on a fine frequency scale are not noticed.[11]

There is one useful situation where we can explicitly solve the three dimensional wave equation, namely where there is complete spherical symmetry. This corresponds to a physical situation where sound waves are generated at the origin in an anisotropic fashion. In this case, we convert into spherical coordinates, and ignore derivatives with respect to the angles. Denoting radial distance from the origin by r, the equation becomes

$$\frac{\partial^2(rp)}{\partial r^2} = \frac{1}{c^2}\frac{\partial^2(rp)}{\partial t^2}.$$

Regarding rp as the dependent variable, this is really just the one dimensional wave equation. So d'Alembert's theorem 3.2.1 shows that the general solution is given by

$$p = (f(r+ct) + g(r-ct))/r.$$

The functions f and g represent waves travelling towards and away from the origin, respectively. Notice that the sound source needs to have finite size, so that we do not run into problems at $r = 0$.

Exercises

1. Show that if **u** is a unit vector in some direction in three dimensions, then the function

$$p(\mathbf{x}, t) = e^{i\omega(ct - \mathbf{u}.\mathbf{x})}$$

satisfies the three dimensional wave equation (3.13.1). This (or rather its real part) represents a sound wave travelling in the direction of **u** with speed c and angular velocity ω.

2. Find the solutions to the three dimensional wave equation for an enclosed region in the shape of a cuboid. Use separation of variables for all four variables, and place the origin at a corner of the region to make the calculations easier.

---

[11] See pages 72–73 of Manfred Schroeder, *Fractals, Chaos and Power Laws*, Springer-Verlag, 1991.

4

Consonance and dissonance

In this chapter, we investigate the relationship between consonance and dissonance, and simple integer ratios of frequencies.

4.1 Harmonics

We saw in Sections 3.2 and 3.5 that when a note on a stringed instrument or a wind instrument sounds at a certain pitch, say with frequency v, sound is essentially periodic with that frequency. The theory of Fourier series shows that such a sound can be decomposed as a sum of sine waves with various phases, at integer multiples of the frequency v, as in Bernoulli's solution (3.2.7) to the wave equation. The component of the sound with frequency v is called the *fundamental*. The component with frequency mv is called the mth *harmonic*, or the $(m-1)$st *overtone*. So, for example, if $m = 3$ we obtain the third harmonic, or the second overtone.[1]

Figure 4.1 represents the series of harmonics based on a fundamental at the C below middle C. The seventh harmonic is actually somewhat flatter than the B♭ above the treble clef. In the modern equally tempered scale, even the third and fifth harmonics are very slightly different from the notes G and E shown above – this is more extensively discussed in Chapter 5.

There is another word which we have been using in this context: the mth *partial* of a sound is the mth frequency component, counted from the bottom. So, for example, on a clarinet, where only the odd harmonics are present, the first partial is the fundamental, or first harmonic, and the second partial is the third harmonic. This term is very useful when discussing sounds where the partials are not simple multiples of the fundamental, such as for example the drum, the gong, or the various instruments of the gamelan.

---

[1] I find that the numbering of overtones is confusing, and I shall not use this numbering.

144

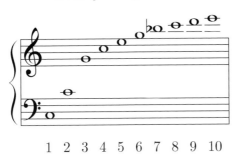

1 2 3 4 5 6 7 8 9 10

Figure 4.1 Harmonics.

440 Hz

220 Hz

Figure 4.2 The As above and below middle C.

Exercise

Define the following terms, making the distinctions between them clear:
(a) the mth harmonic, (b) the mth overtone, (c) the mth partial.

4.2 Simple integer ratios

Why is it that two notes an octave apart sound consonant, while two notes a little more or a little less than an octave apart sound dissonant? An interval of one octave corresponds to doubling the frequency of the vibration. So, for example, the A above middle C corresponds to a frequency of 440 Hz, while the A below middle C corresponds to a frequency of 220 Hz as shown in Figure 4.2.

We have seen in Chapter 3 that if we play these notes on conventional stringed or wind (but not percussive) instruments, each note will contain not only a component at the given frequency, but also partials corresponding to multiples of that frequency. So for these two notes we have partials at:

440 Hz, 880 Hz, 1320 Hz, 1760 Hz, ...
220 Hz, 440 Hz, 660 Hz, 880 Hz, 1100 Hz, 1320 Hz, ...

On the other hand, if we play two notes with frequencies 445 Hz and 220 Hz, then

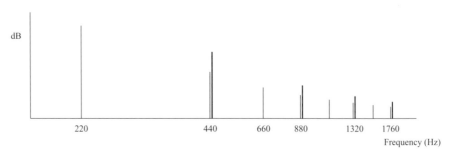

Figure 4.3 Partials of two notes witt frequencies 445 Hz and 220 Hz.

the partials occur at:

445 Hz, 890 Hz, 1335 Hz, 1780 Hz, . . .
220 Hz, 440 Hz, 660 Hz, 880 Hz, 1100 Hz, 1320 Hz, . . .

See Figure 4.3.

The presence of components at 440 Hz and 445 Hz, and at 880 Hz and 890 Hz, and so on, causes a sensation of roughness which is interpreted by the ear as dissonance. We shall discuss at length, later in this chapter, the history of different explanations of consonance and dissonance, and why this should be taken to be the correct one.

Because of the extreme consonance of an interval of an octave, and its role in the series of partials of a note, the human brain often perceives two notes an octave apart as being "really" the same note but higher. This is so heavily reinforced by musical usage in almost every genre that we have difficulty imagining that it could be otherwise. When choirs sing "in unison", this usually means that the men and women are singing an octave apart.[2] The idea that notes differing by a whole number of octaves should be considered as equivalent is often referred to as *octave equivalence*.

The musical interval of a perfect fifth[3] corresponds to a frequency ratio of 3:2. If two notes are played with a frequency ratio of 3:2, then the third partial of the lower note will coincide with the second partial of the upper note, and the notes will have a number of higher partials in common. If, on the other hand, the ratio is slightly different from 3:2, then there will be a sensation of roughness between the third partial of the lower note and the second partial of the upper note, and the notes will sound dissonant.

---

[2] It is interesting to speculate what effect it would have on the theory of colour if visible light had a span greater than an octave; in other words, if there were to exist two visible colours, one of which had exactly twice the frequency of the other. In fact, the span of human vision is just shy of an octave. One might be tempted to suppose that this explains why the colours of the rainbow seem to join up into a circle, although the analysis of this chapter suggests that this explanation is probably wrong, as light sources usually don't contain harmonics.
[3] We shall see in the next chapter that the fifth from C to G in the modern Western scale is not precisely a perfect fifth.

Figure 4.4 The Experiences of Pythagoras (Gaffurius, 1492).

In this manner, small integer ratios of frequencies are picked out as more consonant than other intervals. We stress that this discussion only works for notes whose partials are at multiples of the fundamental frequency. Pythagoras essentially discovered this in the sixth century BCE; he discovered that when two similar strings under the same tension are sounded together, they give a pleasant sound if the lengths of the strings are in the ratio of two small integers. This was the first known example of a law of nature ruled by the arithmetic of integers, and greatly influenced the intellectual development of his followers, the Pythagoreans. They considered that a liberal education consisted of the "quadrivium", or four divisions: numbers in the abstract, numbers applied to music, geometry, and astronomy. They expected that the motions of the planets would be governed by the arithmetic of ratios of small integers in a similar way. This belief has become encoded in the phrase "the music of the spheres",[4] literally denoting the inaudible sound produced

---

[4] Plato, *Republic*, 10.617, ca. 380 BCE.

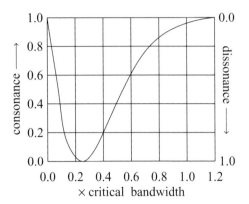

Figure 4.5 Plomp and Levelt's results.

defined below. This means that the actual scale in hertz on the horizontal axis of the graph varies according to the pitch of the notes, but the shape of the graph remains constant; the scaling factor was shown by Plomp and Levelt to be proportional to critical bandwidth.

The salient features of the above graph are that the maximum dissonance occurs at roughly one quarter of a critical bandwidth, and consonance levels off at roughly one critical bandwidth.

It should be stressed that this curve is for pure sine waves, with no harmonics; also that consonance and dissonance are different from recognition of intervals. Anyone with any musical training can recognize an interval of an octave or a fifth, but for pure sine waves, these intervals sound no more nor less consonant than nearby frequency ratios.

Exercise

Show that the function $f(t) = A\sin(at) + B\sin(bt)$ is periodic when the ratio of a to b is a rational number, and nonperiodic if the ratio is irrational. (Hint: differentiate twice and take linear combinations of the result and the original function to get a single sine wave; use this to get information about possible periods.)

Further reading

Galileo Galilei, *Discorsi e dimonstrazioni matematiche interno à due nuove scienze attenenti alla mecanica & i movimenti locali*, Elsevier, 1638. Translated by H. Crew and A. de Salvio as *Dialogues Concerning Two New Sciences*, McGraw-Hill, 1963.
D. D. Greenwood, Critical bandwidth and the frequency coordinates of the basilar membrane, *J. Acoust. Soc. Amer.* **33** (10) (1961), 1344–1356.

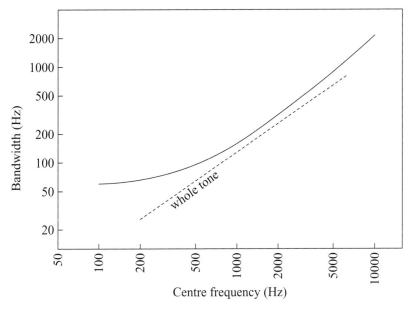

Figure 4.6 Critical bandwidth as a function of centre frequency.

R. Plomp and W. J. M. Levelt, Tonal consonance and critical bandwidth, *J. Acoust. Soc. Amer.* **38** (4) (1965), 548–560.

R. Plomp and H. J. M. Steeneken, Interference between two simple tones, *J. Acoust. Soc. Amer.* **43** (4) (1968), 883–884.

J. Tenney, *A History of 'Consonance' and 'Dissonance'*, Excelsior, 1988.

4.4 Critical bandwidth

To introduce the notion of *critical bandwidth*, each point of the basilar membrane in the cochlea is thought of as a band pass filter, which lets through frequencies in a certain band, and blocks out frequencies outside that band. The actual shape of the filter is certainly more complicated than this simplified model, in which the left, top and right edges of the envelope of the filter are straight vertical and horizontal lines. This is exactly analogous to the definition of bandwidth given in Section 1.11, and introducing a smoother shape for the filter does not significantly alter the discussion. The width of the filter in this model is called the critical bandwidth. Experimental data for the critical bandwidth as a function of centre frequency is available from a number of sources, listed at the end of this section. Figure 4.6 shows a sketch of the results.

A rough calculation based on this graph shows that the size of the critical bandwidth is somewhere between a whole tone and a minor third throughout most of the audible range, and increasing to a major third for small frequencies.

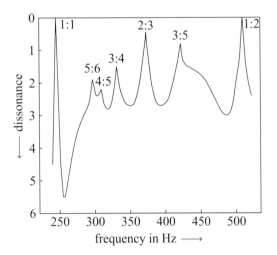

Figure 4.7 Plomp-Levelt curves.

Further reading

B. R. Glasberg and B. C. J. Moore, Derivation of auditory filter shapes from notched-noise
    data, *Hear. Res.* **47** (1990), 103–138.
E. Zwicker, Subdivision of the audible frequency range into critical bands
    (Frequenzgruppen), *J. Acoust. Soc. Amer.* **33** (2) (1961), 248.
E. Zwicker, G. Flottorp and S. S. Stevens, Critical band width in loudness summation,
    J. Acoust. Soc. Amer. **29** (5) (1957), 548–557.
E. Zwicker and E. Terhardt, Analytical expressions for critical-band rate and critical
    bandwidth as a function of frequency, *J. Acoust. Soc. Amer.* **68** (5) (1980),
    1523–1525.

4.5 Complex tones

Plomp and Levelt took the analysis one stage further, and examined what would
happen for tones with a more complicated harmonic content. They worked under
the simplifying assumption that the total dissonance is the sum of the dissonances
caused by each pair of adjacent partials, and used the graph in Figure 4.6 for the
individual dissonances. They did a sample calculation in which a note has partials
at the fundamental and its multiples up to the sixth harmonic. The graph they
obtained is shown in Figure 4.7. Notice the sharp peaks at the fundamental (1:1),
the octave (1:2) and the perfect fifth (2:3), and the smaller peaks at ratios 5:6 (just
minor third), 4:5 (just major third), 3:4 (perfect fourth) and 3:5 (just major sixth).
If higher harmonics are taken into account, the graph acquires more peaks.

In order to be able to draw such Plomp–Levelt curves more systematically, we
choose a formula which gives a reasonable approximation to the curve displayed
in Figure 4.5. Writing *x* for the frequency difference in multiples of the critical

bandwidth, we choose the dissonance function to be[7]

$$f(x) = 4|x|e^{1-4|x|}.$$

This takes its maximum value $f(x) = 1$ when $x = \frac{1}{4}$, as can easily be seen by differentiating. It satisfies $f(0) = 0$, and $f(1.2)$ is small (about 0.1), but not zero. This last feature does not quite match the graph given by Plomp and Levelt, but a closer examination of their data shows that the value $f(1.2) = 0$ is not quite justified.

Further reading

R. Plomp and W. J. M. Levelt, Tonal consonance and critical bandwidth, *J. Acoust. Soc. Amer.* **38** (4) (1965), 548–560.

4.6 Artificial spectra

So what would happen if we artificially manufacture a note having partials which are not exact multiples of the fundamental? It is easy to perform such experiments using a digital synthesizer. We make a note whose partials are at

440 Hz, 860 Hz, 1203 Hz, 1683 Hz, ...

and another with partials at

225 Hz, 440 Hz, 615 Hz, 860 Hz, ...

to represent slightly squeezed harmonics. These notes sound consonant, despite the fact that they are slightly less than an octave apart, whereas scaling the second down to

220 Hz, 430 Hz, 602 Hz, 841 Hz, ...

causes a distinctly dissonant sounding exact octave.

If we are allowed to change the harmonic content of a note in this way, we can make almost any set of intervals seem consonant. This idea was put forward by Pierce (1966, reference below), who designed a spectrum suitable for an equal temperament scale with eight notes to the octave. Namely, he used the following partials, given as multiples of the fundamental frequency:

$$1:1, \quad 2^{\frac{5}{4}}:1, \quad 4:1, \quad 2^{\frac{5}{2}}:1, \quad 2^{\frac{11}{4}}:1, \quad 8:1.$$

This may be thought of as a stretched version of the ordinary series of harmonics of the fundamental. When two notes of the eight tone equal tempered scale are played

---

[7] Sethares (1998) takes for the dissonance function $f(x) = e^{-b_1x} - e^{-b_2x}$ where $b_1 = 3.5$ and $b_2 = 5.75$. This needs normalizing by multiplication by about 5.5, and then gives a graph very similar to the one I have chosen. The particular choice of function is somewhat arbitrary, because of a lack of precision in the data as well as in the subjective definition of dissonance. The main point is to mimic the visible features of the graph.

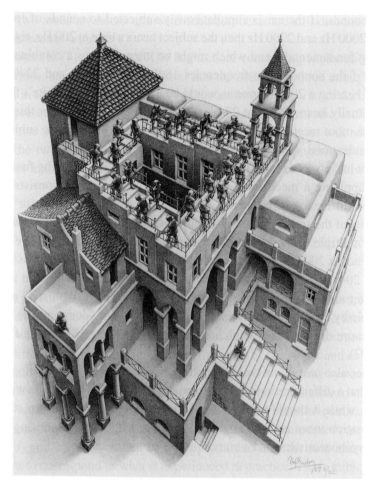

Figure 4.8 M. C. Escher, *Ascending and Descending* © 2006 The M. C. Escher Company-Holland. All rights reserved.

J. F. Schouten, The residue and the mechanism of hearing, *Proceedings of the Koningklijke Nederlandse Akademie van Wetenschappen* **43** (1940), 991–999.

K. Walliser, Über ein Funktionsschema für die Bildung der Periodentonhöhe aus dem Schallreiz, *Kybernetik* **6** (1969), 65–72.

4.8 Musical paradoxes

One of the most famous paradoxes of musical perception was discovered by R. N. Shepard, and goes under the name of the Shepard scale. Listening to the Shepard scale, one has the impression of an ever-ascending scale where the end joins up with the beginning, just like Escher's famous ever ascending staircase in his picture, *Ascending and Descending* (Figure 4.8). This effect is achieved by

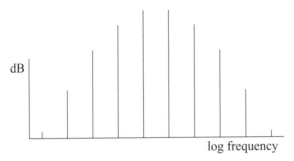

Figure 4.9 Frequencies of a note in the Shepard scale.

building up each note out of a complex tone consisting of ten partials spaced at one octave intervals. These are passed through a filter so that the middle partials are the loudest, and they tail off at both the bottom and the top. See Figure 4.9. The same filter is applied for all notes of the scale, so that after ascending through one octave, the dominant part of the sound has shifted downwards by one partial.

The partials present in this sound are of the form $2^n.f$, where f is the lowest audible frequency component.

A related paradox, discovered by Diana Deutsch (1975), is called the *tritone paradox*. If two Shepard tones are separated by exactly half an octave (a tritone in the equal tempered scale), or a frequency factor of $\sqrt{2}$, then it might be expected that the listener would be confused as to whether the interval is ascending or descending. In fact, only some listeners experience confusion. Others are quite definite as to whether the interval is ascending or descending, and consistently judge half the possible cases as ascending and the complementary half as descending.

Diana Deutsch is also responsible for discovering a number of other paradoxes. For example, when tones of 400 Hz and 800 Hz are presented to the two ears with opposite phase, about 99% of subjects experience the lower tone in one ear and the higher tone in the other ear. When the headphones are reversed, the lower tone stays in the same ear as before. See her 1974 article in *Nature* for further details.

Further reading

E. M. Burns, Circularity in relative pitch judgments: the Shepard demonstration revisited, again, *Perception and Psychophys.* **21** (1977), 563–568.

Diana Deutsch, An auditory illusion, *Nature* **251** (1974), 307–309.

Diana Deutsch, Musical illusions, *Scientific American* **233** (1975), 92–104.

Diana Deutsch, A musical paradox, *Music Percept.* **3** (1986), 275–280.

Diana Deutsch, The tritone paradox: an influence of language on music perception, *Music Percept.* **8** (1990), 335–347.

R. N. Shepard, Circularity in judgments of relative pitch, *J. Acoust. Soc. Amer.* **36** (12) (1964), 2346–2353.

Further listening (See Appendix G)

Auditory Demonstrations CD (Houtsma, Rossing and Wagenaars), track 52 is a demonstration of Shepard's scale, followed by an analogous continuously varying tone devised by Jean-Claude Risset.

5

Scales and temperaments: the fivefold way

A perfect fourth? cries Tom. Whoe'er gave birth
To such a riddle, should stick or fiddle
On his numbskull ring until he sing
A scale of perfect fourths from end to end.
Was ever such a noddy? Why, almost everybody
Knows that not e'en one thing perfect is on earth—
How then can we expect to find a perfect fourth?
Musical World, 1863. Quoted in Nicolas
Slonimsky's Book of Musical Anecdotes,
reprinted by Schirmer, 1998, p. 299.

Figure 5.1 From J. Frazer, A new visual illusion of direction, *British Journal of Psychology*, 1908. And yes, check it out, they *are* concentric circles, not a spiral.

Figure 5.2 Pythagoras.

5.1 Introduction

We saw in the last chapter that for notes played on conventional instruments, where partials occur at integer multiples of the fundamental frequency, intervals corresponding to frequency ratios expressable as a ratio of small integers are favoured as consonant. In this chapter, we investigate how this gives rise to the scales and temperaments found in the history of western music.

Scales based around the octave are categorized by Barbour (1951) into five broad groups: Pythagorean, just, meantone, equal, and irregular systems. The title of this chapter refers to this fivefold classification, as well as to the use of the first five harmonics as the starting point for the development of scales. We shall try to indicate where these five types of scales come from.

In Chapter 6 we shall discuss further developments in the theory of scales and temperaments, and in particular, we shall study some scales which are not based around the interval of an octave. These are the Bohlen–Pierce scale, and the scales of Wendy Carlos.

5.2 Pythagorean scale

As we saw in Section 4.2, Pythagoras (Figure 5.2) discovered that the interval of a perfect fifth, corresponding to a frequency ratio of 3:2, is particularly consonant. He concluded from this that a convincing scale could be constructed just by using the ratios 2:1 and 3:2. Greek music scales of the Pythagorean school are built using only these intervals, although other ratios of small integers played a role in classical Greek scales.

So, for example, if we use the ratio 3:2 twice, we obtain an interval with ratio 9:4, which is a little over an octave. Reducing by an octave means halving this ratio to give 9:8. Using the ratio 3:2 again will then bring us to 27:16, and so on.

What we now refer to as the Pythagorean scale is the one obtained by tuning a sequence of fifths

$$\text{fa–do–so–re–la–mi–ti.}$$

This gives the following table of frequency ratios for a major scale:[1]

note	do	re	mi	fa	so	la	ti	do
ratio	1:1	9:8	81:64	4:3	3:2	27:16	243:128	2:1

In this system, the two intervals between successive notes are a *major tone* of 9:8 and a *minor semitone* of 256:243 or $2^8:3^5$. The semitone is not quite half of a tone in this system: two minor semitones give a frequency ratio of $2^{16}:3^{10}$ rather than 9:8. The Pythagoreans noticed that these were almost equal:

$$2^{16}/3^{10} = 1.109\,857\,15\ldots,$$
$$9/8 = 1.125.$$

In other words, the Pythagorean system is based on the fact that

$$2^{19} \approx 3^{12}, \quad \text{or} \quad 524\,288 \approx 531\,441,$$

so that going up 12 fifths and then down 7 octaves brings you back to almost exactly where you started. The fact that this is not quite so gives rise to the *Pythagorean comma* or *ditonic comma*, namely the frequency ratio

$$3^{12}/2^{19} = 1.013\,643\,265\ldots$$

or just slightly more than one ninth of a whole tone.[2]

It seems likely that the Pythagoreans thought of musical intervals as involving the process of continued subtraction or *antanairesis*, which later formed the basis of Euclid's algorithm for finding the greatest common divisor of two integers (if you don't remember how Euclid's algorithm goes, it is described in Lemma 9.7.1). A 2:1 octave minus a 3:2 perfect fifth is a 4:3 perfect fourth. A perfect fifth minus a perfect fourth is a 9:8 Pythagorean whole tone. A perfect fourth minus two whole tones is a 256:243 Pythagorean minor semitone. It was called a *diesis* (difference), and was later referred to as a *limma* (remnant). A tone minus a diesis is a 2187:2048

---

[1] A Pythagorean minor scale can be constructed using ratios 32:27 for the minor third, 128:81 for the minor sixth and 16:9 for the minor seventh.

[2] Musical intervals are measured logarithmically, so dividing a whole tone by nine really means taking the ninth root of the ratio, see Section 5.4.

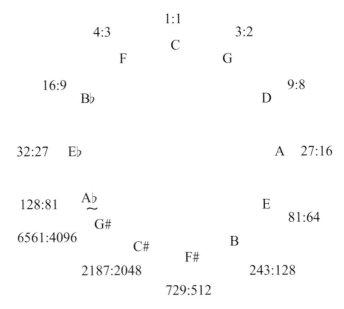

Figure 5.3 Cycle of fifths in the Pythagorean tuning system.

Pythagorean major semitone, called an *apotomē*. An apotomē minus a diesis is a 531 441:524 288 Pythagorean comma.

5.3 The cycle of fifths

The Pythagorean tuning system can be extended to a twelve tone scale by tuning perfect fifths and octaves, at ratios of 3:2 and 2:1. This corresponds to tuning a 'cycle of fifths' as in Figure 5.3.

In this picture, the Pythagorean comma appears as the difference between the notes A♭ and G♯, or indeed any other *enharmonic* pair of notes:

$$\frac{6561/4096}{128/81} = \frac{3^{12}}{2^{19}} = \frac{531\,441}{524\,288}.$$

In these days of equal temperament (see Section 5.14), we think of A♭ and G♯ as just two different names for the same note, so that there is really a *circle* of fifths. Other notes also have several names, for example the notes C and B♯, or the notes E♭♭, D and C×.[3] In each case, the notes are said to be enharmonic, and in the Pythagorean system that means a difference of exactly one Pythagorean comma. So the Pythagorean system does not so much have a circle of fifths, more a sort of *spiral of fifths* as in Figure 5.4.

---

[3] The symbol × is used in music instead of ♯♯ to denote a double sharp.

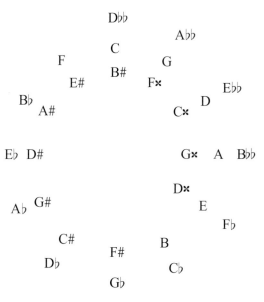

Figure 5.4 Spiral of fifths.

So, for example, going clockwise one complete revolution takes us from the note C to B♯, one Pythagorean comma higher. Going round the other way would take us to D♭♭, one Pythagorean comma lower. We shall see in Section 6.2 that the Pythagorean spiral never joins up. In other words, no two notes of this spiral are equal. The twelfth note is reasonably close, the 53rd is closer, and the 665th is very close indeed.

Exercise

What is the name of the note
(a) one Pythagorean comma lower than F,
(b) two Pythagorean commas higher than B,
(c) two Pythagorean commas lower than B?

Further listening (See Appendix G)

Guillaume de Machaut, *Messe de Notre Dame*, Hilliard Ensemble, sung in Pythagorean intonation.

5.4 Cents

Adding musical intervals corresponds to *multiplying* frequency ratios. So, for example, if an interval of an octave corresponds to a ratio of 2:1 then an interval of two octaves corresponds to a ratio of 4:1, three octaves to 8:1, and so on. In other words, our perception of musical distance between two notes is *logarithmic* in frequency, as logarithms turn products into sums.

We now explain the system of *cents*, first introduced by Alexander Ellis around 1875, for measuring frequency ratios. This is the system most often employed in the modern literature. This is a logarithmic scale in which there are 1200 cents to the octave. Each whole tone on the modern equal tempered scale (described below) is 200 cents, and each semitone is 100 cents. To convert from a frequency ratio of r:1 to cents, the value in cents is

$$1200 \log_2(r) = 1200 \ln(r)/\ln(2).$$

To convert an interval of n cents to a frequency ratio, the formula is

$$2^{\frac{n}{1200}} : 1.$$

For example, the interval from C to D in the Pythagorean scale represents a frequency ratio of 9:8, so in cents this comes out as

$$1200 \log_2(9/8) = 1200 \ln(9/8)/\ln(2)$$

or approximately 203.910 cents. The Pythagorean scale in the key of C major comes out as follows:

note	C	D	E	F	G	A	B	C
ratio	1:1	9:8	81:64	4:3	3:2	27:16	243:128	2:1
cents	0.000	203.910	407.820	498.045	701.955	905.865	1109.775	1200.000

We shall usually give our scales in the key of C, and assign the note C a value of 0 cents. Everything else is measured in cents above the note C.

In France, rather than measuring intervals in cents, they use as their basic unit the *savart*, named after its proponent, the French physicist Félix Savart (1791–1841). In this system, a ratio of 10:1 is assigned a value of 1000 savarts. So a 2:1 octave is

$$1000 \log_{10}(2) \approx 301.030 \text{ savarts.}$$

One savart corresponds to a frequency ratio of $10^{\frac{1}{1000}}$:1, and is equal to

$$\frac{1200}{1000 \log_{10}(2)} = \frac{6}{5 \log_{10}(2)} \approx 3.98631 \text{ cents.}$$

Exercises

1. Show that, to three decimal places, the Pythagorean comma is equal to 23.460 cents. What is it in savarts?

2. Convert the frequency ratios for the vibrational modes of a drum, given in Section 3.6, into cents above the fundamental.

3. Assigning C the value of 0 cents, what is the value of the note E♭♭ in the Pythagorean scale?

Further reading

Parry Moon, A scale for specifying frequency levels in octaves and semitones, *J. Acoust. Soc. Amer.* **25** (3) (1953), 506–515.

5.5 **Just intonation**

> Just intonation refers to any tuning system that uses small, whole numbered ratios between the frequencies in a scale. This is the natural way for the ear to hear harmony, and it's the foundation of classical music theory. The dominant Western tuning system – equal temperament – is merely a 200 year old compromise that made it easier to build mechanical keyboards. Equal temperament is a lot easier to use than JI, but I find it lacks expressiveness. It sounds dead and lifeless to me. As soon as I began working microtonally, I felt like I moved from black & white into color. I found that certain combinations of intervals moved me in a deep physical way. Everything became clearer for me, more visceral and expressive. The trade-off is that I had to be a lot more careful with my compositions, for while I had many more interesting consonant intervals to choose from, I also had new kinds of dissonances to avoid. Just intonation also opened me up to a greater appreciation of non-Western music, which has clearly had a large impact on my music.
>
> *Robert Rich (synthesist)*

After the octave and the fifth, the next most interesting ratio is 4:3. If we follow a perfect fifth (ratio 3:2) by the ratio 4:3, we obtain a ratio of 4:2 or 2:1, which is an octave. So 4:3 is an octave minus a perfect fifth, or a perfect fourth. So this gives us nothing new. The next new interval is given by the ratio 5:4, which is the fifth harmonic brought down two octaves.

If we continue this way, we find that the series of harmonics of a note can be used to construct scales consisting of notes that are for the most part related by small integer ratios. Given the fundamental role of the octave, it is natural to take the harmonics of a note and move them down a number of octaves to place them all in the same octave. In this case, the ratios we obtain are:

1:1 for the first, second, fourth, eighth, etc. harmonic,

3:2 for the third, sixth, twelfth, etc. harmonic,

5:4 for the fifth, tenth, etc. harmonic,

7:4 for the seventh, fourteenth, etc. harmonic,

and so on.

As we have already indicated, the ratio of 3:2 (or 6:4) is a perfect fifth. The ratio of 5:4 is a more consonant major third than the Pythagorean one, since it is a ratio of smaller integers. So we now have a *just major triad* (do–mi–so) with frequency ratios 4:5:6. Most scales in the world incorporate the major triad in some form. In western music it is regarded as the fundamental building block on which chords and scales are built. Scales in which the frequency ratio 5:4 are included were first developed by Didymus in the first century BCE and Ptolemy in the second century CE. The difference between the Pythagorean major third 81:64 and the Ptolemy–Didymus major third 5:4 is a ratio of 81:80. This interval is variously called the *syntonic comma*, *comma of Didymus*, *Ptolemaic comma*, or *ordinary comma*. When we use the word *comma* without further qualification, we shall always be referring to the syntonic comma.

Just intonation in its most limited sense usually refers to the scales in which each of the major triads I, IV and V (i.e., C–E–G, F–A–C and G–B–D) is taken to have frequency ratios 4:5:6. Thus we obtain the following table of ratios for a just major scale:

note	do	re	mi	fa	so	la	ti	do
ratio	1:1	9:8	5:4	4:3	3:2	5:3	15:8	2:1
cents	0.000	203.910	386.314	498.045	701.955	884.359	1088.269	1200.000

The *just major third* is therefore the name for the interval (do–mi) with ratio 5:4, and the *just major sixth* is the name for the interval (do–la) with ratio 5:3. The complementary intervals (mi–do) of 8:5 and (la–do) of 6:5 are called the *just minor sixth* and the *just minor third*.

The differences between various versions of just intonation mostly involve how to fill in the remaining notes of a twelve tone scale. In order to qualify as just intonation, each of these notes must differ by a whole number of commas from the Pythagorean value. In this context, the comma may be thought of as the result of going up four perfect fifths and then down two octaves and a just major third. In some versions of just intonation, a few of the notes of the above basic scale have also been altered by a comma.

5.6 Major and minor

In the last section, we saw that the basic building block of western music is the major triad, which in just intonation is built up out of the fourth, fifth and sixth notes in the harmonic series. See Figure 5.6.

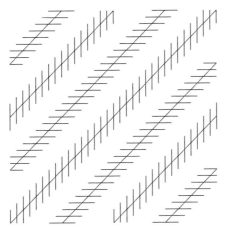

Figure 5.5 Parallel lines.

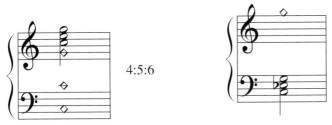

4:5:6

Figure 5.6 Major triad.

10:12:15

Figure 5.7 Minor triad.

The minor triad is built by reversing the order of the two intervals, to obtain a chord of the form C–E♭–G. The ratios are 5:6 for the C–E♭ and 4:5 for the E♭–G. It seems futile to try to understand these as the harmonics of a common fundamental, because we would have to express the ratios as 10:12:15, making the fundamental 1/10 of the frequency of the C. It makes more sense to look at the harmonics of the notes in the triad, and to notice that all three notes have a common harmonic. Namely,

$$6 \times C = 5 \times E♭ = 4 \times G.$$

So if we play a minor triad, if we listen carefully we can pick out this common harmonic, which is a G two octaves higher. For some subtle psychoacoustic reason, it sometimes sounds as though it's just one octave higher. It is probably

4:5:6:7

Figure 5.8 Chord with seventh harmonic.

G^7 C

Figure 5.9 Dominant seventh.

the high common harmonic which causes us to associate minor chords with sadness.

Another point of view regarding the minor triad is to view it as a *modification* of a major triad by slightly lowering the middle note to change the flavour. Music theory is full of modified chords, usually meaning that one of the notes in the chord has been raised or lowered by a semitone.

Further reading

P. Hindemith, *Craft of Musical Composition, I. Theory.* Schott, 1937, Section III.5, *The minor triad.*

5.7 The dominant seventh

If we go as far as the seventh harmonic, we obtain a chord with ratios 4:5:6:7. This can be thought of as C–E–G–B♭, with a 7:4 B♭. See Figure 5.8.

There is a closely related chord called the *dominant seventh* chord, in which the B♭ is the Pythagorean minor seventh, 16:9 higher than the C instead of 7:4. If we start this chord on G (3:2 above C) instead of C, we will obtain a chord G–B–D–F, and the F will be 4:3 above C. This chord has a strong tendency to resolve to C major, whereas the 4:5:6:7 version feels a lot more stable. See Figure 5.9.

We shall have more to say about the seventh harmonic in Section 6.9.

Further reading

Martin Vogel (1991), *Die Naturseptime*.

5.8 Commas and schismas

Recall from Section 5.2 that the Pythagorean comma is defined to be the difference
between twelve perfect fifths and seven octaves, which gives a frequency ratio of
531 441:524 288, or a difference of about 23.460 cents. Recall also from Section
5.5 that the word *comma*, used without qualification, refers to the syntonic comma,
which is a frequency ratio of 81:80. This is a difference of about 21.506 cents.

So the syntonic comma is very close in value to the Pythagorean comma, and
the difference is called the *schisma*. This represents a frequency ratio of

$$\frac{531\,441/524\,288}{81/80} = \frac{32805}{32768},$$

or about 1.953 cents.

The *diaschisma*[4] is defined to be one schisma less than the comma, or a frequency
ratio of 2048:2025. This may be viewed as the result of going up three octaves, and
then down four perfect fifths and two just major thirds.

The *great diesis*[5] is one octave minus three just major thirds, or three syntonic
commas minus a Pythagorean comma. This represents a frequency ratio of 128:125
or a difference of 41.059 cents.

The *septimal comma* is the amount by which the seventh harmonic 7:4 is flatter
than the Pythagorean minor seventh 16:9. So it represents a ratio of $(16/9)(4/7) =$
64/63 or a difference of 27.264 cents.

Exercises

1. Show that, to three decimal places, the (syntonic) comma is equal to 21.506 cents and
   the schisma is equal to 1.953 cents.
2. (G. B. Benedetti)[6] Show that if all the major thirds and sixths and the perfect fourths
   and fifths are taken to be just in the harmonic progression shown in Figure 5.10, then
   the pitch will drift upwards by exactly one comma from the initial G to the final G.

---

[4] Historically, the Roman theorist Boethius (ca. 480–524 CE) attributes to Philolaus of Pythagoras' school a
definition of schisma as one half of the Pythagorean comma and the diaschisma for one half of the diesis, but
this does not correspond to the common modern usage of the terms.

[5] The word *diesis* in Greek means 'leak' or 'escape', and is based on the technique for playing the aulos, an
ancient Greek wind instrument. To raise the pitch of a note on the aulos by a small amount, the finger on the
lowest closed hole is raised slightly to allow a small amount of air to escape.

[6] G. B. Benedetti, *Diversarum Speculationum*, Turin, 1585, page 282. The example is borrowed from Lindley
and Turner-Smith (1993), page 16.

$$\left(\tfrac{3}{4} \times \tfrac{3}{2} \times \tfrac{3}{5} \times \tfrac{3}{2} = \tfrac{81}{80}\right)$$

Figure 5.10 Upward drift.

This example was given by Benedetti in 1585 as an argument against Zarlino's[7] assertion (1558) that unaccompanied singers will tend to sing in just intonation. For a further discussion of the syntonic comma in the context of classical harmony, see Section 5.11.

3. Here is a quote from Karlheinz Stockhausen[8] (*Lectures and Interviews*, compiled by Robert Maconie, Marion Boyars, 1989, pages 110–111):

> With the purest tones you can make the most subtle melodic gestures, much, much more refined than what the textbooks say is the smallest interval we can hear, namely the Pythagorean comma 80:81. That's not true at all. If I use sine waves, and make little glissandi instead of stepwise changes, then I can really feel that little change, going far beyond what people say about Chinese music, or in textbooks of physics or perception.
>
> But it all depends on the tone: you cannot just use any tone in an interval relationship. We have discovered a new law of relationship between the nature of the sound and the scale on which it may be composed. Harmony and melody are no longer abstract systems to be filled with any given sounds we may choose as material. There is a very subtle relationship nowadays between form and material.

(a) Find the error in this quote, and explain why it does not really matter.

(b) What is the new law of relationship to which Stockhausen is referring?

5.9 Eitz's notation

Eitz[9] devised a system of notation, used in Barbour (1951), which is convenient for describing scales based around the octave. His method is to start with the Pythagorean definitions of the notes and then put a superscript describing how many commas to adjust by. Each comma multiplies the frequency by a factor of 81/80.

---

[7] G. Zarlino, *Istitutione Harmoniche*, Venice, 1558.

[8] Karlheinz Stockhausen was much maligned in the German press in the months following September 2001. I urge anyone with a brain to go to his home page at www.stockhausen.org and find out what he really said, and what the context was. The full text of the original interview is there.

[9] Carl A. Eitz, *Das mathematisch-reine Tonsystem*, Leipzig, 1891. A similar notation was used earlier by Hauptmann and modified by Helmholtz (1877).

As an example, the Pythagorean E, notated E^0 in this system, is 81:64 of C, while E^{-1} is decreased by a factor of $81/80$ from this value, to give the just ratio of 80:64 or 5:4.

In this notation, the basic scale for just intonation is given by

$$C^0 - D^0 - E^{-1} - F^0 - G^0 - A^{-1} - B^{-1} - C^0.$$

A common variant of this notation is to use subscripts rather than superscripts, so that the just major third in the key of C is E_{-1} instead of E^{-1}.

An often used graphical device for denoting just scales, which we use here in combination with Eitz's notation, is as follows. The idea is to place notes in a triangular array in such a way that moving to the right increases the note by a 3:2 perfect fifth, moving up and a little to the right increases a note by a 5:4 just major third, and moving down and a little to the right increases a note by a 6:5 just minor third. So a just major 4:5:6 triad is denoted

$$
\begin{array}{ccc}
 & E^{-1} & \\
C^0 & & G^0.
\end{array}
$$

A *just minor triad* has these intervals reversed:

$$
\begin{array}{ccc}
C^0 & & G^0 \\
 & E\flat^{+1} &
\end{array}
$$

and the notes of the just major scale form the following array:

$$
\begin{array}{cccc}
 & A^{-1} & E^{-1} & B^{-1} \\
F^0 & C^0 & G^0 & D^0
\end{array}
$$

This method of forming an array is usually ascribed to Hugo Riemann,[10] although such arrays have been common in German music theory since the eighteenth century to denote key relationships and functional interpretation rather than frequency relationships.

It is sometimes useful to extend Eitz's notation to include other commas. Several different notations appear in the literature, and we choose to use p to denote the Pythagorean comma and z to denote the septimal comma. So, for example, $G\sharp^{-p}$ is the same note as $A\flat^0$, and the interval from C^0 to $B\flat^{-z}$ is a ratio of $\frac{16}{9} \times \frac{63}{64} = \frac{7}{4}$, namely the seventh harmonic.

---

[10] Hugo Riemann, *Ideen zu einer 'Lehre von den Tonvorstellungen'*, Jahrbuch der Musikbibliothek Peters, 1914–1915, page 20; *Grosse Kompositionslehre*, W. Spemann, 1902, Volume 1, page 479.

Exercises

1. Show that in Eitz's notation, the example of Section 5.8, Exercise 2 looks like:

$$G^0 \quad D^0 \quad A^0 \qquad E^0$$
$$C^{+1} \qquad G^{+1}.$$

2. (a) Show that the schisma is equal to the interval between Dbb^{+1} and C^0, and the interval between C^0 and $B\sharp^{-1}$.

(b) Show that the diaschisma is equal to the interval between C^0 and Dbb^{+2}.

(c) Give an example to show that a sequence of six overlapping chords in just intonation can result in a drift of one diaschisma.

(d) How many overlapping chords in just intonation are needed in order to achieve a drift of one schisma?

5.10 Examples of just scales

Using Eitz's notation, we list the examples of just intonation given in Barbour (1951) for comparison. The dates and references have also been taken from that work.

Ramis' Monochord
(Bartolomeus Ramis de Pareja, *Musica Practica*, Bologna, 1482)

$$\qquad\qquad D^{-1} \quad A^{-1} \quad E^{-1} \quad B^{-1} \quad F\sharp^{-1} \quad C\sharp^{-1}$$
$$Ab^0 \quad Eb^0 \quad Bb^0 \quad\quad F^0 \quad C^0 \quad G^0$$

Erlangen Monochord
(anonymous German manuscript, second half of fifteenth century)

$$\qquad\qquad\qquad\qquad\qquad\qquad\qquad E^{-1} \quad B^{-1}$$
$$Gb^0 \qquad Db^0 \quad Ab^0 \quad Eb^0 \quad Bb^0 \quad F^0 \quad C^0 \quad G^0$$
$$Ebb^{+1} \qquad Bbb^{+1}$$

Erlangen Monochord, revised
The deviations of Ebb^{+1} from D^0, and of Bbb^{+1} from A^0 are equal to the schisma, as are the deviations of Gb^0 from $F\sharp^{-1}$, of Db^0 from $C\sharp^{-1}$, and of Ab^0 from $G\sharp^{-1}$. So Barbour conjectures that the Erlangen monochord was really intended as

$$\qquad\qquad\qquad\qquad E^{-1} \quad B^{-1} \quad F\sharp^{-1} \quad C\sharp^{-1} \quad G\sharp^{-1}$$
$$Eb^0 \quad Bb^0 \quad F^0 \quad C^0 \quad G^0 \quad D^0 \quad A^0$$

Fogliano's Monochord No. 1

(Lodovico Fogliano, *Musica Theorica*, Venice, 1529)

$$F\sharp^{-2} \qquad C\sharp^{-2} \qquad G\sharp^{-2}$$
$$D^{-1} \qquad A^{-1} \qquad E^{-1} \qquad B^{-1}$$
$$B\flat^{0} \qquad F^{0} \qquad C^{0} \qquad G^{0}$$
$$E\flat^{+1}$$

Fogliano's Monochord No. 2

$$F\sharp^{-2} \qquad C\sharp^{-2} \qquad G\sharp^{-2}$$
$$A^{-1} \qquad E^{-1} \qquad B^{-1}$$
$$F^{0} \qquad C^{0} \qquad G^{0} \qquad D^{0}$$
$$E\flat^{+1} \qquad B\flat^{+1}$$

Agricola's Monochord

(Martin Agricola, *De monochordi dimensione*, in *Rudimenta Musices*, Wittemberg, 1539)

$$F\sharp^{-1} \qquad C\sharp^{-1} \qquad G\sharp^{-1} \qquad D\sharp^{-1}$$
$$B\flat^{0} \quad F^{0} \quad C^{0} \quad G^{0} \quad D^{0} \qquad A^{0} \qquad E^{0} \qquad B^{0}$$

De Caus's Monochord

(Salomon de Caus, *Les raisons des forces mouvantes avec diverses machines*, Francfort, 1615, Book 3, Problem III)

$$F\sharp^{-2} \qquad C\sharp^{-2} \qquad G\sharp^{-2} \qquad D\sharp^{-2}$$
$$D^{-1} \qquad A^{-1} \qquad E^{-1} \qquad B^{-1}$$
$$B\flat^{0} \qquad F^{0} \qquad C^{0} \qquad G^{0}$$

Kepler's Monochord No. 1

(Johannes Kepler, *Harmonices Mundi*, Augsburg, 1619)

$$E^{-1} \qquad B^{-1} \qquad F\sharp^{-1} \qquad C\sharp^{-1} \quad G\sharp^{-1}$$
$$F^{0} \quad C^{0} \qquad G^{0} \qquad D^{0} \qquad A^{0}$$
$$E\flat^{+1} \qquad B\flat^{+1}$$

(Note: the $G\sharp^{-1}$ is incorrectly labelled $G\sharp^{+1}$ in Barbour, but his numerical value in cents is correct.)

Kepler's Monochord No. 2

$$E^{-1} \qquad B^{-1} \qquad F\sharp^{-1} \qquad C\sharp^{-1}$$
$$F^{0} \qquad C^{0} \qquad G^{0} \qquad D^{0} \qquad A^{0}$$
$$A\flat^{+1} \qquad E\flat^{+1} \qquad B\flat^{+1}$$

Figure 5.14 Colin Brown's voice harmonium. Reproduced with permission of Science Museum/Science and Society Picture Library.

We shall return to the discussion of just intonation in Section 6.1, where we consider scales built using primes higher than 5. In Section 6.8, we look at a way of systematizing the discussion by using lattices, and we interpret the above scales as periodicity blocks.

Exercises

1. Choose several of the just scales described in this section, and write down the values of the notes
  (i) in cents, and
  (ii) as frequencies, giving the answers as multiples of the frequency for C.

2. Show that the Pythagorean scale with perfect fifths

$$G\flat^0 - D\flat^0 - A\flat^0 - E\flat^0 - B\flat^0 - F^0 - C^0 - G^0 - D^0 - A^0 - E^0 - B^0$$

gives good approximations to just major triads on D, A and E, in the form $D^0 - G\flat^0 - A^0$, $A^0 - D\flat^0 - E^0$ and $E^0 - A\flat^0 - B^0$. How far from just are the thirds of these chords (in cents)?

3. The voice harmonium of Colin Brown (1875) is shown in Figure 5.14. A plan of a little more than one octave of the keyboard is shown in Figure 5.15. Diagonal rows of black keys and white keys alternate, and each black key has a red peg sticking out of its upper left corner, represented by a small circle in the plan. The purpose of this keyboard is to be able to play in a number of different keys in just intonation. Locate examples of the following on this keyboard:
  (i) a just major triad,

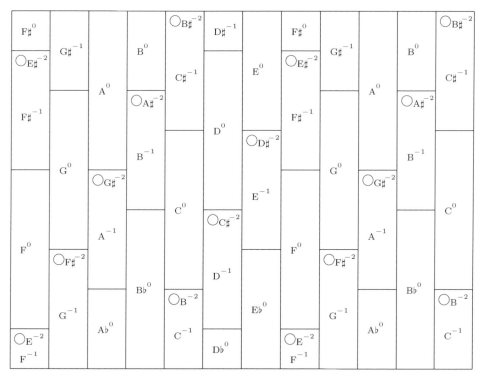

Figure 5.15 Keyboard diagram for Colin Brown's voice harmonium.

  (ii) a just minor triad,
 (iii) a just major scale,
 (iv) two notes differing by a syntonic comma,
  (v) two notes differing by a schisma,
 (vi) two notes differing by a diesis,
(vii) two notes differing by an apotomē.

5.11 Classical harmony

The main problem with the just major scale introduced in Section 5.5 is that certain harmonic progressions which form the basis of classical harmony don't quite work. This is because certain notes in the major scale are being given two different just interpretations, and switching from one to the other is a part of the progression. In this section, we discuss the progressions which form the basis for classical harmony,[12] and find where the problems lie.

---

[12] The phrase 'classical harmony' here is used in its widest sense, to include not only classical, romantic and baroque music, but also most of the rock, jazz and folk music of western culture.

We begin with the names of the triads. An upper case roman numeral denotes a major chord based on the given scale degree, whereas a lower case roman numeral denotes a minor chord. So, for example, the major chords I, IV and V form the basis for the just major scale in Section 5.5, namely $C^0 - E^{-1} - G^0$, $F^0 - A^{-1} - C^0$ and $G^0 - B^{-1} - D^0$ in the key of C major. The triads $A^{-1} - C^0 - E^{-1}$ and $E^{-1} - G^0 - B^{-1}$ are the minor triads vi and iii. The problem comes from the triad on the second note of the scale, $D^0 - F^0 - A^{-1}$. If we alter the D^0 to a D^{-1}, this is a just minor triad, which we would then call ii.

Classical harmony makes use of ii as a minor triad, so maybe we should have used D^{-1} instead of D^0 in our just major scale. But then the triad V becomes $G^0 - B^{-1} - D^{-1}$, which doesn't quite work. We shall see that there is no choice of just major scale which makes all the required triads work. To understand this, we discuss classical harmonic practice.

We begin at the end. Most music in the western world imparts a sense of finality through the sequence V–I, or variations of it (V^7–I, vii^0–I).[13] It is not fully understood why V–I imparts such a feeling of finality, but it cannot be denied that it does. A great deal of music just consists of alternate triads V and I.

The progression V–I can stand on its own, or it can be approached in a number of ways. A sequence of fifths forms the basis for the commonest method, so that we can extend to ii–V–I, then to vi–ii–V–I, and even further to iii–vi–ii–V–I, each of these being less common than the previous ones. Here is a chart of the most common harmonic progressions in the music of the western world, in the major mode:

$$[\text{iii}] \rightarrow [\text{vi}] \rightarrow \begin{bmatrix} \text{IV} \\ \downarrow \\ \text{ii} \end{bmatrix} \searrow \begin{bmatrix} \text{vii}^0 \\ \\ \text{V} \end{bmatrix} \searrow \text{I}$$

and then either end the piece, or go back from I to any previous triad. Common exceptions are to jump from iii to IV, from IV to I and from V to vi.

Now take a typical progression from the above chart, such as

$$\text{I–vi–ii–V–I,}$$

and let us try to interpret this in just intonation. Let us stipulate one simple rule, namely that if a note on the diatonic scale appears in two adjacent triads, it should be given the same just interpretation. So if I is $C^0 - E^{-1} - G^0$ then vi must be interpreted as $A^{-1} - C^0 - E^{-1}$, since the C and E are in common between the two triads. This means that the ii should be interpreted as $D^{-1} - F^0 - A^{-1}$, with the A in common with

---

[13] The superscript zero in the notation vii^0 denotes a *diminished triad* with two minor thirds as the intervals. It has nothing to do with the Eitz comma notation.

Figure 5.16 W. A. Mozart, *Sonata* (K. 333), third movement, beginning.

Figure 5.17 J. S. Bach, *Partita no. 5, Gigue,* bars 23–24.

vi. Then V needs to be interpreted as $G^{-1} - B^{-2} - D^{-1}$ because it has D in common with ii. Finally, the I at the end is forced to be interpreted as $C^{-1} - E^{-2} - G^{-1}$ since it has G in common with V. We are now one syntonic comma lower than where we started.

To put the same problem in terms of ratios, in the second triad the A is $\frac{5}{3}$ of the frequency of the C, then in the third triad, D is $\frac{2}{3}$ of the frequency of A. In the fourth triad, G is $\frac{4}{3}$ of the frequency of D, and finally in the last triad, C is $\frac{2}{3}$ of the frequency of G. This means that the final C is

$$\frac{5}{3} \times \frac{2}{3} \times \frac{4}{3} \times \frac{2}{3} = \frac{80}{81}$$

of the frequency of the initial one.

A similar drift downward through a syntonic comma occurs in the sequences

$$\text{I–IV–ii–V–I,}$$

$$\text{I–iii–vi–ii–V–I,}$$

and so on. Figures 5.16 to 5.20 show some actual musical examples, chosen pretty much at random.

The meantone scale, which we shall discuss in the next section, solves the problem of the syntonic comma by deviating slightly from the just values of notes in such

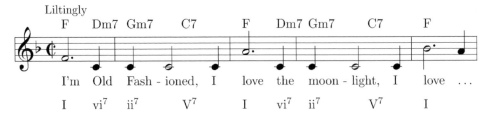

Figure 5.18 *I'm Old Fashioned* (1942). Music by Jerome Kern, words by Johnny Mercer.

Figure 5.19 W. A. Mozart, *Fantasie* (K. 397), bars 55–59.

Figure 5.20 A minor example: J. S. Bach, *Jesu, der du meine Seele.*

a way that the comma is spread equally between the four perfect fifths involved, shaving one quarter of a comma from each of them.

Harry Partch discusses the issue of the syntonic comma at length, towards the end of Chapter 11 of Partch (1974). He arrives at a different conclusion from the one adopted historically, namely that the progressions above sound fine, played in just intonation in such a way that the second note on the scale is played (in C major) as D^{-1} in ii, and as D^0 in V. This means that these two versions of the 'same' note are played in consecutive triads, but the sense of the harmonic progression is not lost.

5.12 Meantone scale

A *tempered scale* is a scale in which adjustments are made to the Pythagorean or just scale in order to spread around the problem caused by wishing to regard two notes differing by various commas as the same note, as in the example of Section 5.8, Exercise 2, and the discussion in Section 5.11.

The meantone scales are the tempered scales formed by making adjustments of a fraction of a (syntonic) comma to the fifths in order to make the major thirds better.

The commonest variant of the meantone scale, sometimes referred to as the classical meantone scale, or quarter-comma meantone scale, is the one in which the major thirds are made in the ratio 5:4 and then the remaining notes are interpolated as equally as possible. So C–D–E are in the ratios $1 : \sqrt{5}/2 : 5/4$, as are F–G–A and G–A–B. This leaves two semitones to decide, and they are made equal. Five tones of ratio $\sqrt{5}/2 : 1$ and two semitones make an octave 2:1, so the ratio for the semitone is

$$\sqrt{2/\left(\sqrt{5}/2\right)^5} : 1 = 8 : 5^{\frac{5}{4}}.$$

The table of ratios is therefore as follows:

note	do	re	mi	fa	so	la	ti	do
ratio	1:1	$\sqrt{5}$:2	5:4	$2:5^{\frac{1}{4}}$	$5^{\frac{1}{4}}$:1	$5^{\frac{3}{4}}$:2	$5^{\frac{5}{4}}$:4	2:1
cents	0.000	193.157	386.314	503.422	696.579	889.735	1082.892	1200.000

The fifths in this scale are no longer perfect.

Another, more enlightening way to describe the classical meantone scale is to temper each fifth by making it narrower than the Pythagorean value by exactly one quarter of a comma, in order for the major thirds to come out right. So, working from C, the G is one quarter comma flat from its Pythagorean value, the D is one half comma flat, the A is three quarters of a comma flat, and finally, E is one comma flat from a Pythagorean major third, which makes it exactly equal to the just major third. Continuing in the same direction, this makes the B five quarters of a comma flatter than its Pythagorean value. Correspondingly, the F should be made one quarter comma sharper than the Pythagorean fourth.

Thus, in Eitz's notation, the classical meantone scale can be written as

$$C^0 - D^{-\frac{1}{2}} - E^{-1} - F^{+\frac{1}{4}} - G^{-\frac{1}{4}} - A^{-\frac{3}{4}} - B^{-\frac{5}{4}} - C^0.$$

Writing these notes in the usual array notation, we obtain

$$
\begin{array}{cccccc}
 & E^{-1} & & B^{-\frac{5}{4}} & & \\
C^0 & & G^{-\frac{1}{4}} & & D^{-\frac{1}{2}} & A^{-\frac{3}{4}} \quad E^{-1} \\
 & & & F^{+\frac{1}{4}} & & C^0
\end{array}
$$

The meantone scale can be completed by filling in the remaining notes of a twelve (or more) tone scale according to the same principles. The only question is how far to go in each direction with the quarter comma tempered fifths. Some examples, again taken from Barbour (1951), follow.

Aaron's Meantone Temperament
(Pietro Aaron, *Toscanello in Musica*, Venice, 1523)

$$C^0 \; C\sharp^{-\frac{7}{4}} \; D^{-\frac{1}{2}} \; E\flat^{+\frac{3}{4}} \; E^{-1} \; F^{+\frac{1}{4}} \; F\sharp^{-\frac{3}{2}} \; G^{-\frac{1}{4}} \; A\flat^{+1} \; A^{-\frac{3}{4}} \; B\flat^{+\frac{1}{2}} \; B^{-\frac{5}{4}} \; C^0$$

Gibelius' Monochord for Meantone Temperament
(Otto Gibelius, *Propositiones mathematico-musicæ*, Münden, 1666)
is the same, but with two extra notes

$$C^0 \; C\sharp^{-\frac{7}{4}} \; D^{-\frac{1}{2}} \; D\sharp^{-\frac{9}{4}} \; E\flat^{+\frac{3}{4}} \; E^{-1} \; F^{+\frac{1}{4}} \; F\sharp^{-\frac{3}{2}} \; G^{-\frac{1}{4}} \; G\sharp^{-2} \; A\flat^{+1} \; A^{-\frac{3}{4}} \; B\flat^{+\frac{1}{2}} \; B^{-\frac{5}{4}} \; C^0$$

These meantone scales are represented in array notation as follows:

$$
\begin{array}{cccccccc}
 & (G\sharp^{-2}) & & (D\sharp^{-\frac{9}{4}}) & & & & \\
 & E^{-1} & & B^{-\frac{5}{4}} & & F\sharp^{-\frac{3}{2}} & C\sharp^{-\frac{7}{4}} & (G\sharp^{-2}) \\
C^0 & & G^{-\frac{1}{4}} & & D^{-\frac{1}{2}} & A^{-\frac{3}{4}} & E^{-1} & \\
A\flat^{+1} & E\flat^{+\frac{3}{4}} & & B\flat^{+\frac{1}{2}} & & F^{+\frac{1}{4}} & C^0 &
\end{array}
$$

where the right hand edge is thought of as equal to the left hand edge. Thus the notes can be thought of as lying on a cylinder, with four quarter-comma adjustments taking us once round the cylinder, as shown in Figure 5.21.

So the syntonic comma has been taken care of, and modulations can be made to a reasonable number of keys. The Pythagorean comma has not been taken care of, so that modulation around an entire circle of fifths is still not feasible. Indeed, the difference between the enharmonic notes $A\flat^{+1}$ and $G\sharp^{-2}$ is three syntonic commas minus a Pythagorean comma, which is a ratio of 128:125, or a difference of 41.059 cents. This interval, called the great diesis, is nearly half a semitone, and is very noticeable to the ear. The imperfect fifth between C♯ and A♭ (or wherever else it may happen to be placed) in the meantone scale is sometimes referred to as the *wolf*[14] interval of the scale. We shall see in Section 6.5 that one way of dealing with the wolf fifth is to use thirty-one tones to an octave instead of twelve.

---

[14] This has nothing to do with the 'wolf' notes on a stringed instrument such as the cello, which has to do with the sympathetic resonance of the body of the instrument.

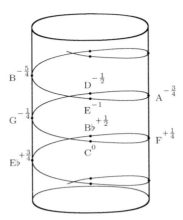

Figure 5.21 Meantone scale on a cylinder.

Although what we have described is the commonest form of meantone scale, there are others formed by taking different divisions of the comma. In general, the α-comma meantone temperament refers to the following temperament:

$$
\begin{array}{ccccccccc}
& E^{-4\alpha} & & B^{-5\alpha} & & F\sharp^{-6\alpha} & & C\sharp^{-7\alpha} & & G\sharp^{-8\alpha} \\
C^0 & & G^{-\alpha} & & D^{-2\alpha} & & A^{-3\alpha} & & E^{-4\alpha} & \\
& Eb^{+3\alpha} & & Bb^{+2\alpha} & & F^{+\alpha} & & C^0 &
\end{array}
$$

Without any qualification, the phrase 'meantone temperament' refers to the case $\alpha = \frac{1}{4}$. The following names are associated with various values of α:

0	Pythagoras	
$\frac{1}{7}$	Romieu	(1755); *Mémoire théorique et pratique sur les systèmes tempérés de musique*, Paris, 1758
$\frac{1}{6}$	Silbermann	Sorge, *Gespräch zwischen einem Musico theoretico und einem Studioso musices*, Lobenstein, 1748, p. 20
$\frac{1}{5}$	Abraham Verheijen, Lemme Rossi	Simon Stevin, *Van de Spiegeling der Singconst*, c. 1600 *Sistema musico*, Perugia, 1666, p. 58
$\frac{2}{9}$	Lemme Rossi	*Sistema musico*, Perugia, 1666, p. 64
$\frac{1}{4}$	Aaron/Gibelius/Zarlino/...	Aaron, 1523...
$\frac{2}{7}$	Gioseffo Zarlino	*Istitutioni armoniche*, Venice, 1558
$\frac{1}{3}$	Francisco de Salinas	*De musica libri VII*, Salamanca, 1577

So, for example, Zarlino's $\frac{2}{7}$ comma meantone temperament is as follows:

$$
\begin{array}{ccccccccc}
& E^{-\frac{8}{7}} & & B^{-\frac{10}{7}} & & F\sharp^{-\frac{12}{7}} & & C\sharp^{-2} & & G\sharp^{-\frac{16}{7}} \\
C^0 & & G^{-\frac{2}{7}} & & D^{-\frac{4}{7}} & & A^{-\frac{6}{7}} & & E^{-\frac{8}{7}} & \\
& Eb^{+\frac{6}{7}} & & Bb^{+\frac{4}{7}} & & F^{+\frac{2}{7}} & & C^0 &
\end{array}
$$

more irregular temperament in which all keys are more or less satisfactorily in tune.[15]

A typical example of such a temperament is Werckmeister's most frequently used temperament. This is usually referred to as Werckmeister III (although Barbour (1951) refers to it as Werckmeister's Correct Temperament No. 1),[16] which is as follows.

Werckmeister III (Correct Temperament No. 1)

(Andreas Werckmeister, *Musicalische Temperatur* Frankfort and Leipzig, 1691; reprinted by Diapason Press, 1986, with commentary by Rudolph Rasch)

$$\begin{array}{ccccc} \text{E}^{-\frac{3}{4}p} & \text{B}^{-\frac{3}{4}p} & \text{F}\sharp^{-1p} & \text{C}\sharp^{-1p} & \text{G}\sharp^{-1p} \\ \text{C}^{0} \quad \text{G}^{-\frac{1}{4}p} & \text{D}^{-\frac{1}{2}p} & \text{A}^{-\frac{3}{4}p} & \text{E}^{-\frac{3}{4}p} \\ \text{E}\flat^{0} & \text{B}\flat^{0} & \text{F}^{0} & \text{C}^{0} \end{array}$$

In this temperament, the Pythagorean comma (*not* the syntonic comma) is distributed equally on the fifths from C–G–D–A and B–F♯. We use a modified version of Eitz's notation to denote this, in which 'p' is used to denote the usage of the Pythagorean comma rather than the syntonic comma. A good way to think of this is to use the approximation discussed in Section 5.14 which says '$p = \frac{12}{11}$', so that, for example, $\text{E}^{-\frac{3}{4}p}$ is essentially the same as $\text{E}^{-\frac{9}{11}}$. Note that $\text{A}\flat^{0}$ is equal to $\text{G}\sharp^{-1p}$, so the circle of fifths does join up properly in this temperament. In fact, this was the first temperament to be widely adopted which has this property.

In this and other irregular temperaments, different key signatures have different characteristic sounds, with some keys sounding direct and others more remote. This may account for the modern myth that the same holds in equal temperament.[17]

---

[15] It is a common misconception that Bach intended the *Well Tempered Clavier* to be played in equal temperament. He certainly knew of equal temperament, but did not use it by preference, and it is historically much more likely that the 48 Preludes and Fugues were intended for an irregular temperament of the kind discussed in this section. (It should be mentioned that there is also evidence that Bach did intend equal temperament, see Rudolf A. Rasch, *Does 'Well-tempered' mean 'Equal-tempered'?*, in Williams (ed.), *Bach, Händel, Scarlatti Tercentenary Essays*, Cambridge University Press, 1985, pp. 293–310.)

[16] Werckmeister I usually refers to just intonation, and Werckmeister II to classical meantone temperament. Werckmeister IV and V are described later in this section. There is also a temperament known as Werckmeister VI, or 'septenarius', which is based on a division of a string into 196 equal parts. This scale gives the ratios 1:1, 196:186, 196:176, 196:165, 196:156, 4:3, 196:139, 196:131, 196:124, 196:117, 196:110, 196:104, 2:1.

[17] If this were really true, then the shift of nearly a semitone in pitch between Mozart's time and our own would have resulted in a permutation of the resulting moods, which seems to be nonsense. Actually, this argument really only applies to keyboard instruments. It is still possible in equal temperament for string and wind instruments to give different characters to different keys. For example, a note on an open string on a violin sounds different in character from a stopped string. Mozart and others have made use of this difference with a technique called *scordatura* (Italian *scordare*, to mistune), which involves unconventional retuning of stringed instruments. A well known example is his *Sinfonia Concertante*, in which all the strings of the solo viola are tuned a semitone sharp. The orchestra plays in E♭ for a softer sound, and the solo viola plays in D for a more brilliant sound.

A more shocking example (communicated to me by Markus Linckelmann) is Schubert's *Impromptu No. 3* for piano in G♭ major. The same piece played in G major on a modern piano has a very different feel to it. It is possible that, in this case, the mechanics of the fingering are responsible.

C major	Completely pure. Its character is: innocence, simplicity, naivety, children's talk.
C minor	Declaration of love and at the same time the lament of unhappy love. All languishing, longing, sighing of the lovesick soul lies in this key.
Db major	A leering key, degenerating into grief and rapture. It cannot laugh, but it can smile; it cannot howl, but it can at least grimace its crying. Consequently only unusual characters and feelings can be brought out in this key.
C♯ minor	Penitential lamentation, intimate conversation with God, the friend and help-meet of life; sighs of disappointed friendship and love lie in its radius.
D major	The key of triumph, of Halelujahs, of war-cries, of victory-rejoicing. Thus, the inviting symphonies, the marches, holiday songs and heaven-rejoicing choruses are set in this key.
D minor	Melancholy womanliness, the spleen and humours brood.
Eb minor	Feelings of the anxiety of the soul's deepest distress, of brooding despair, of blackest depression, of the most gloomy condition of the soul. Every fear, every hesitation of the shuddering heart, breathes out of horrible Eb minor. If ghosts could speak, their speech would approximate this key.
Eb major	The key of love, of devotion, of intimate conversation with God; through its three flats [1789: according to Euler] expressing the holy trinity.
E major	Noisy shouts of joy, laughing pleasure and not yet complete, full delight lies in E major.
E minor	Naive, womanly, innocent declaration of love, lament without grumbling; sighs accompanied by few tears; this key speaks of the imminent hope of resolving in the pure happiness of C major. Since by nature it has only one colour, it can be compared to a maiden, dressed in white, with a rose-red bow at her breast. From this key one steps with inexpressible charm back again to the fundamental key of C major, where heart and ear find the most complete satisfaction.
F major	Complaisance and calm.
F minor	Deep depression, funereal lament, groans of misery and longing for the grave.
Gb major	Triumph over difficulty, free sigh of relief uttered when hurdles are surmounted; echo of a soul which has fiercely struggled and finally conquered lies in all uses of this key.
F♯ minor	A gloomy key: it tugs at passion as a dog biting a dress. Resentment and discontent are its language. It really does not seem to like its own position: therefore it languishes ever for the calm of A major or for the triumphant happiness of D major.
G major	Everything rustic, idyllic and lyrical, every calm and satisfied passion, every tender gratitude for true friendship and faithful love, – in a word, every gentle and peaceful emotion of the heart is correctly expressed by this key. What a pity that because of its seeming lightness it is so greatly neglected nowadays…
G minor	Discontent, uneasiness, worry about a failed scheme; bad-tempered gnashing of teeth; in a word: resentment and dislike.
Ab major	The key of the grave. Death, grave, putrefaction, judgement, eternity lie in its radius.
G♯ minor	Grumbler, heart squeezed until it suffocates; wailing lament which sighs in double sharps, difficult struggle; in a word, the colour of this key is everything struggling with difficulty.
A major	This key includes declarations of innocent love, satisfaction with one's state of affairs; hope of seeing one's beloved again when parting; youthful cheerfulness and trust in God.
A minor	Pious womanliness and tenderness of character.
Bb major	Cheerful love, clear conscience, hope, aspiration for a better world.
Bb minor	A quaint creature, often dressed in the garment of night. It is somewhat surly and very seldom takes on a pleasant countenance. Mocking God and the world; discontented with itself and with everything; preparation for suicide sounds in this key.
B major	Strongly coloured, announcing wild passions, composed from the most glaring colours. Anger, rage, jealousy, fury, despair and every emotion of the heart lies in its sphere.
B minor	This is as it were the key of patience, of calm awaiting one's fate and of submission to divine dispensation. For that reason its lament is so mild, without ever breaking out into offensive murmuring or whimpering. The use of this key is rather difficult for all instruments; therefore so few pieces are found which are expressly set in this key.

Key characteristics, from Christian Schubart, *Ideen zu einer Aesthetik der Tonkunst*, written 1784, published 1806, translated by Rita Steblin.

Berlin, 1764)

$$
\begin{array}{ccccccc}
 & E^{-1} & & B^{-1} & & F\sharp^{-1} & & \\
C^0 & & G^0 & & D^0 & & A^{-\frac{1}{2}} & E^{-1} \\
Ab^0 & Eb^0 & & Bb^0 & & F^0 & & C^0 \\
 & & & Db^0 & & Ab^0 & &
\end{array}
$$

Kirnberger III

(Johann Phillip Kirnberger, *Die Kunst des reinen Satzes in der Musik*, 2nd part, 3rd division, Berlin, 1779)

$$
\begin{array}{ccccccc}
 & E^{-1} & & B^{-1} & & F\sharp^{-1} & & \\
C^0 & & G^{-\frac{1}{4}} & & D^{-\frac{1}{2}} & & A^{-\frac{3}{4}} & E^{-1} \\
Ab^0 & Eb^0 & & Bb^0 & & F^0 & & C^0 \\
 & & & Db^0 & & Ab^0 & &
\end{array}
$$

Lambert's $\frac{1}{7}$-comma temperament

(Johann Heinrich Lambert, *Remarques sur le tempérament en musique*, Nouveaux mémoires de l'Académie Royale, 1774)

$$
\begin{array}{ccccc}
E^{-\frac{4}{7}p} & B^{-\frac{5}{7}p} & F\sharp^{-\frac{6}{7}p} & C\sharp^{-\frac{6}{7}p} & G\sharp^{-\frac{6}{7}p} \\
C^0 \quad G^{-\frac{1}{7}p} & D^{-\frac{2}{7}p} & A^{-\frac{3}{7}p} & E^{-\frac{4}{7}p} & \\
Eb^{+\frac{1}{7}p} & Bb^{+\frac{1}{7}p} & F^{+\frac{1}{7}p} & C^0 &
\end{array}
$$

Marpurg's Temperament I

(Friedrich Wilhelm Marpurg, *Versuch über die musikalische Temperatur*, Breslau, 1776)

$$
\begin{array}{ccccc}
E^{-\frac{1}{3}p} & B^{-\frac{1}{3}p} & F\sharp^{-\frac{1}{3}p} & C\sharp^{-\frac{1}{3}p} & G\sharp^{-\frac{2}{3}p} \\
C^0 \quad G^0 & D^0 & A^0 & E^{-\frac{1}{3}p} & \\
Eb^{+\frac{1}{3}p} & Bb^{+\frac{1}{3}p} & F^{+\frac{1}{3}p} & C^0 &
\end{array}
$$

Barca's $\frac{1}{6}$-comma temperament

(Alessandro Barca, *Introduzione a una nuova teoria di musica, memoria prima*, Accademia di scienze, lettere ed arti in Padova. Saggi scientifici e lettari (Padova, 1786), 365–418)

$$
\begin{array}{ccccc}
E^{-\frac{2}{3}} & B^{-\frac{5}{6}} & F\sharp^{-1} & C\sharp^{-1} & G\sharp^{-1} \\
C^0 \quad G^{-\frac{1}{6}} & D^{-\frac{1}{3}} & A^{-\frac{1}{2}} & E^{-\frac{2}{3}} & \\
Eb^0 & Bb^0 & F^0 & C^0 &
\end{array}
$$

Young's Temperament, No. 1

(Thomas Young, Outlines of experiments and inquiries respecting sound and light,

Figure 5.22 Francescantonio Vallotti (1697–1780).

Philosophical Transactions, **XC** (1800), 106–150)

$$C^0 \quad E^{-\frac{3}{4}} \quad G^{-\frac{3}{16}} \quad B^{-\frac{5}{6}} \quad D^{-\frac{3}{8}} \quad F\sharp^{-\frac{11}{12}} \quad A^{-\frac{9}{16}} \quad C\sharp^{-\frac{11}{12}} \quad E^{-\frac{3}{4}} \quad G\sharp^{-\frac{11}{12}}$$

$$E\flat^{+\frac{1}{6}} \quad B\flat^{+\frac{1}{6}} \quad F^{+\frac{1}{12}} \quad C^0$$

Vallotti and Young $\frac{1}{6}$-comma temperament (Young's Temperament, No. 2)

(Francescantonio Vallotti, *Trattato della scienza teoretica e pratica della moderna musica*, 1780; Thomas Young, Outlines of experiments and inquiries respecting sound and light, *Philosophical Transactions*, **XC** (1800), 106–150. Below is Young's version of this temperament. In Vallotti's version, the fifths which are narrow by $\frac{1}{6}$ Pythagorean commas are F–C–G–D–A–E–B instead of C–G–D–A–E–B–F♯)

$$E^{-\frac{2}{3}p} \quad B^{-\frac{5}{6}p} \quad F\sharp^{-1p} \quad C\sharp^{-1p} \quad G\sharp^{-1p}$$
$$C^0 \quad G^{-\frac{1}{6}p} \quad D^{-\frac{1}{3}p} \quad A^{-\frac{1}{2}p} \quad E^{-\frac{2}{3}p}$$
$$E\flat^0 \quad B\flat^0 \quad F^0 \quad C^0$$

The temperament of Vallotti and Young is probably closest to the intentions of J. S. Bach for his *Well-Tempered Clavier*. According to the researches of Barnes, it is possible that Bach preferred the F♯ to be one sixth of a Pythagorean comma sharper than in this temperament, so that the fifth from B to F♯ is pure. Barnes based his work on a statistical study of prominence of the different major thirds, and the mathematical procedure of Donald Hall for evaluating suitability of temperaments. Other authors, such as Kelletat and Kellner, have come to slightly different conclusions, and we will probably never find out who is right. Here are these reconstructions for comparison.

Kelletat's Bach reconstruction (1966),

$$E^{-\frac{5}{6}p} \quad B^{-1p} \quad F\sharp^{-1p} \quad C\sharp^{-1p} \quad G\sharp^{-1p}$$
$$C^0 \quad G^{-\frac{1}{12}p} \quad D^{-\frac{1}{3}p} \quad A^{-\frac{7}{12}p} \quad E^{-\frac{5}{6}p}$$
$$E\flat^0 \quad B\flat^0 \quad F^0 \quad C^0$$

Kellner's Bach reconstruction (1975),

$$E^{-\frac{4}{5}p} \quad B^{-\frac{4}{5}p} \quad F\sharp^{-1p} \quad C\sharp^{-1p} \quad G\sharp^{-1p}$$
$$C^0 \quad G^{-\frac{1}{5}p} \quad D^{-\frac{2}{5}p} \quad A^{-\frac{3}{5}p} \quad E^{-\frac{4}{5}p}$$
$$E\flat^0 \quad B\flat^0 \quad F^0 \quad C^0$$

Barnes' Bach reconstruction (1979)

$$E^{-\frac{2}{3}p} \quad B^{-\frac{5}{6}p} \quad F\sharp^{-\frac{5}{6}p} \quad C\sharp^{-1p} \quad G\sharp^{-1p}$$
$$C^0 \quad G^{-\frac{1}{6}p} \quad D^{-\frac{1}{3}p} \quad A^{-\frac{1}{2}p} \quad E^{-\frac{2}{3}p}$$
$$E\flat^0 \quad B\flat^0 \quad F^0 \quad C^0$$

More recently, in the late 1990s, Andreas Sparschuh[19] and Michael Zapf came up with the interesting idea that the series of squiggles at the top of the title page of the *Well-Tempered Clavier* (see Figure 5.23) encode instructions for laying the temperament.

Each loop in this squiggle has zero, one or two twists, giving the following sequence:

$$1\text{–}1\text{–}1\text{–}0\text{–}0\text{–}0\text{–}2\text{–}2\text{–}2\text{–}2\text{–}2,$$

---

[19] An announcement appears in Andreas Sparschuh, *Stimm-Arithmetik des wohltemperierten Klaviers*, Deutsche Mathematiker Vereinigung Jahrestagung 1999, Mainz S.154–155. There seems to be no full article by either Sparschuh or Zapf.

Figure 5.23 Title page of Bach's *Well-Tempered Clavier*.

to be interpreted as telling the tuner by how much to make the eleven fifths narrow from a perfect fifth. The twelfth fifth, completing the circle, does not have to be specified.

In 2005, Bradley Lehman described a modified version of this idea in which the top stroke of the C of 'Clavier' is interpreted as giving the position of C in the circle. He chooses to orient the cycle so that going to the left ascends through the cycle of fifths, and he interprets the numbers as numbers of twelfths of a Pythagorean comma. Here is the result of this interpretation.

Lehman's Bach reconstruction (2005)

$$E^{-\frac{2}{3}p} \qquad B^{-\frac{2}{3}p} \qquad F\sharp^{-\frac{2}{3}p} \qquad C\sharp^{-\frac{2}{3}p} \qquad G\sharp^{-\frac{3}{4}p}$$
$$C^0 \qquad G^{-\frac{1}{6}p} \qquad D^{-\frac{1}{3}p} \qquad A^{-\frac{1}{2}p} \qquad E^{-\frac{2}{3}p}$$
$$E\flat^{\frac{1}{6}p} \qquad B\flat^{\frac{1}{12}p} \qquad F^{\frac{1}{6}p} \qquad C^0$$

Exercises

1. Take the information on various temperaments given in this section, and work out a table of values in cents for the notes of the scale.
2. If you have a synthesizer where each note of the scale can be retuned separately, retune it to some of the temperaments given in this section, using your answers to Exercise 1. Sequence some harpsichord music and play it through your synthesizer using these temperaments, and compare the results.
3. In a well tempered scale, take three major thirds adding up to an octave. The total amount by which these are sharp from the just major third does not depend on the temperament. Show that this amount is equal to a great diesis (~ 41.059 cents).

Further reading

Pierre-Yves Asselin (1997), *Musique et tempérament*.
Murray Barbour (1951), *Tuning and Temperament, a Historical Survey*.

Murray Barbour, Bach and 'The art of temperament', *Musical Quarterly* **33** (1) (1947), 64–89.
John Barnes, Bach's Keyboard Temperament, *Early Music* **7** (2) (1979), 236–249.
Dominique Devie (1990), *Le tempérament musical*.
D. E. Hall, The objective measurement of goodness-of-fit for tuning and temperaments, *J. Music Theory* **17** (2) (1973), 274–290.
D. E. Hall, Quantitative evaluation of musical scale tuning, *Amer. J. Phys.* **42** (1974), 543–552.
Owen Jorgensen (1991), *Tuning*.
Herbert Kelletat, *Zur musikalischen Temperatur insbesondere bei J. S. Bach*. Onkel Verlag, 1960 and 1980.
Herbert Anton Kellner, Eine Rekonstruktion der wohltemperierten Stimmung von Johann Sebastian Bach. *Das Musikinstrument* **26** (1977), 34–35.
Herbert Anton Kellner, Was Bach a mathematician? *English Harpsichord Magazine* **2/2** April 1978, 32–36.
Herbert Anton Kellner, Comment Bach accordait-il son clavecin? *Flûte à bec et instruments anciens* **13–14**, SDIA, Paris 1985.
Bradley Lehman, Bach's extraordinary temperament: our Rosetta Stone, *Early Music* **33** (2005), 3–24; 211–232; 545–548 (correspondence).
Rita Steblin, *A History of Key Characteristics in the 18th and Early 19th Centuries*, UMI Research Press, 1983. Second edition, University of Rochester Press, 2002.

Further listening (See Appendix G)

Johann Sebastian Bach, *The Complete Organ Music*, Volumes 6 and 8, recorded by Hans Fagius, using Neidhardt's Circulating Temperament No. 3 'für eine grosse Stadt' (for a large town).
Johann Sebastian Bach, *Italian Concerto*, etc., recorded by Christophe Rousset, Editions de l'Oiseau-Lyre 433 054-2, Decca 1992. These works were recorded using Werckmeister III.
Lou Harrison, *Complete Harpsichord Works*, New Albion, 2002. These works were recorded using Werckmeister III and other temperaments.
The Katahn/Foote recording, *Six Degrees of Tonality*, contains tracks comparing Mozart's *Fantasie* K. 397 in equal temperament, meantone, and an irregular temperament of Prelleur.
Johann Gottfried Walther, *Organ Works*, Volume 1, played by Craig Cramer on the organ of St. Bonifacius, Tröchtelborn, Germany. This organ was restored in Kellner's reconstruction of Bach's temperament, shown above.
Aldert Winkelman, *Works by Mattheson, Couperin, and Others*. The pieces by Johann Mattheson, François Couperin, Johann Jakob Froberger, Joannes de Gruytters and Jacques Duphly are played on a harpsichord tuned to Werckmeister III.

5.14 Equal temperament

Music is a science which should have definite rules; these rules should be drawn from an evident principle; and this principle cannot really be known to us without the aid of mathematics. Notwithstanding all the experience I may have acquired in music from being associated with it

for so long, I must confess that only with the aid of mathematics did my
ideas become clear and did light replace a certain obscurity of which I
was unaware before.

Rameau (1722).[20]

Each of the scales described in the previous sections has its advantages and disad-
vantages, but the one disadvantage of most of them is that they are designed to make
one particular key signature or a few adjacent key signatures as good as possible,
and leave the remaining ones to look after themselves.

Twelve tone equal temperament is a natural endpoint of these compromises. This
is the scale that results when all twelve semitones are taken to have equal ratios.
Since an octave is a ratio of 2:1, the ratios for the equal tempered scale give all
semitones a ratio of $2^{\frac{1}{12}}:1$ and all tones a ratio of $2^{\frac{1}{6}}:1$. So the ratios come out as
follows:

note	do	re	mi	fa	so	la	ti	do
ratio	1:1	$2^{\frac{1}{6}}:1$	$2^{\frac{1}{3}}:1$	$2^{\frac{5}{12}}:1$	$2^{\frac{7}{12}}:1$	$2^{\frac{3}{4}}:1$	$2^{\frac{11}{12}}:1$	2:1
cents	0.000	200.000	400.000	500.000	700.000	900.000	1100.000	1200.000

Equal tempered thirds are about 14 cents sharper than perfect thirds, and sound
nervous and agitated. As a consequence, the just and meantone scales are more
calm temperaments. To my ear, tonal polyphonic music played in meantone temperament has a clarity and sparkle that I do not hear on equal tempered instruments.
The irregular temperaments described in the previous section have the property
that each key retains its own characteristics and colour; keys with few sharps
and flats sound similar to meantone, while the ones with more sharps and flats
have a more remote feel to them. Equal temperament makes all keys essentially
equivalent.

Twelve elevenths of a syntonic comma, or a factor of

$$\left(\frac{81}{80}\right)^{\frac{12}{11}} \approx 1.013\,644\,082,$$

or 23.461 4068 cents is an extremely good approximation to the Pythagorean
comma of

$$\frac{531\,441}{524\,288} \approx 1.013\,643\,265,$$

or 23.460 0104 cents. It follows that equal temperament is almost exactly equal to the $\frac{1}{11}$-comma meantone scale:

$$E^{-\frac{4}{11}} \qquad B^{-\frac{5}{11}} \qquad F\sharp^{-\frac{6}{11}} \qquad C\sharp^{-\frac{7}{11}} \qquad G\sharp^{-\frac{8}{11}}$$
$$C^{0} \qquad G^{-\frac{1}{11}} \qquad D^{-\frac{2}{11}} \qquad A^{-\frac{3}{11}} \qquad E^{-\frac{4}{11}}$$
$$A\flat^{+\frac{4}{11}} \quad E\flat^{+\frac{3}{11}} \quad B\flat^{+\frac{2}{11}} \qquad F^{+\frac{1}{11}} \qquad C^{0}$$

where the difference between $A\flat^{+\frac{4}{11}}$ and $G\sharp^{-\frac{8}{11}}$ is 0.001 3964 cents.

This observation was first made by Kirnberger[21] who used it as the basis for a recipe for tuning keyboard instruments in equal temperament. His recipe was to obtain an interval of an equal tempered fourth by tuning up three perfect fifths and one major third, and then down four perfect fourths. This corresponds to equating the equal tempered F with $E\sharp^{-1}$. The disadvantage of this method is clear: in order to obtain one equal tempered interval, one must tune eight intervals by eliminating beats. The fifths and fourths are not so hard, but tuning a major third by eliminating beats is considered difficult. This method of tuning equal temperament was discovered independently by John Farey[22] nearly twenty years later.

Alexander Ellis, in his Appendix XX (Section G, Article 11) to Helmholtz (1877), gives an easier practical rule for tuning in equal temperament. Namely, tune the notes in the octave above middle C by tuning fifths upwards and fourths downwards. Make the fifths perfect and then flatten them (make them more narrow) by one beat per second (see Section 1.8). Make the fourths perfect and then flatten them (make them wider) by three beats every two seconds. The result will be accurate to within two cents on every note. Having tuned one octave using this rule, tuning out beats for octaves allows the entire piano to be tuned.

It is desirable to apply spot checks throughout the piano to ensure that the fifths remain slightly narrow and the fourths slightly wide. Ellis states at the end of Article 11 that there is no way of distinguishing slightly narrow fourths or fifths from slightly wide ones using beats. In fact, there is a method, which was not yet conceived in 1885, as follows (Jorgensen (1991), Section 227, see Figures 5.24 and 5.25).

For the fifth, say C3–G3, compare the intervals C3–E♭3 and E♭3–G3. If the fifth is narrow, as desired, the first interval will beat more frequently than the second. If perfect, the beat frequencies will be equal. If wide, the second interval will beat more frequently than the first.

For the fourth, say G3–C4, compare the intervals C4–E♭4 and G3–E♭4, or compare E♭3–C4 and E♭3–G3. If the fourth is wide, as desired, the first interval will

[21] Johann Philipp Kirnberger, *Die Kunst des reinen Satzes in der Musik*, 2nd part, 3rd division (Berlin, 1779), pp. 197f.

[22] John Farey, On a new mode of equally tempering the musical scale, *Philosophical Magazine*, **XXVII** (1807), 65–66.

Figure 5.24 Jorgensen's tuning method for fifths.

Figure 5.25 Jorgensen's tuning method for fourths.

beat more frequently than the second. If perfect, the beat frequencies will be equal. If narrow, the second interval will beat more frequently than the first. This method is based on the observation that, in equal temperament, the major third is enough wider than a just major third that gross errors would have to be made in order for it to have ended up narrower and spoil the test.

Exercises

1. Show that taking eleventh powers of the approximation of Kirnberger and Farey described in this section gives the approximation

$$2^{161} \approx 3^{84} 5^{12}.$$

   The ratio of these two numbers is roughly 1.000 008 873, and the eleventh root of this is roughly 1.000 000 8066.

2. Use the ideas of Section 4.6 to construct a spectrum which is close to the usual harmonic spectrum, but in such a way that the twelve tone equal tempered scale has consonant major thirds and fifths, as well as consonant seventh harmonics.

3. Calculate the accuracy of the method of Alexander Ellis for tuning equal temperament, described in this section.

4. Draw up a table of scale degrees in cents for the twelve notes in the Pythagorean, just, meantone and equal scales.

5. (Serge Cordier's equal temperament for piano with perfect fifths) Serge Cordier formalized a technique for piano tuning in the tradition of Pleyel (France). Cordier's recipe is as follows.[23] Make the interval F–C a perfect fifth, and divide it into seven equal semitones. Then use perfect fifths to tune from these eight notes to the entire piano.

   Show that this results in octaves which are stretched by one seventh of a Pythagorean comma. This is of the same order of magnitude as the natural stretching of the octaves

---

[23] Serge Cordier, *L'accordage des instruments à claviers. Bulletin du Groupe Acoustique Musicale* (G. A. M.) **75** (1974), Paris VII; *Piano bien tempéré et justesse orchestrale*, Buchet-Chastel, 1982.

due to the inharmonicity of physical piano strings. Draw a diagram in Eitz's notation to demonstrate this temperament. This should consist of a horizontal strip with the top and bottom edges identified. Calculate the deviation of major and minor thirds from just in this temperament.

Further reading

Ian Stewart, *Another Fine Math You've Got Me Into...*, W. H. Freeman & Co., 1992. Chapter 15 of this book, The well tempered calculator, contains a description of some of the history of practical approximations to equal temperament. Particularly interesting is his description of Strähle's method of 1743.

5.15 Historical remarks

Ancient Greek music

The word *music* ($\mu o \upsilon \sigma \iota \kappa \acute{\eta}$) in ancient Greece had a wider meaning than it does for us, embracing the idea of ratios of integers as the key to understanding both the visible physical universe and the invisible spiritual universe.

It should not be supposed that the Pythagorean scale discussed in Section 5.2 was the main one used in ancient Greece in the form described there. Rather, this scale is the result of applying the Pythagorean ideal of using only the ratios 2:1 and 3:2 to build the intervals. The Pythagorean scale as we have presented it first occurs in Plato's *Timaeus*, and was used in medieval European music from about the eighth to the fourteenth century CE.

The *diatonic syntonon* of Ptolemy is the same as the major scale of just intonation, with the exception that the classical Greek octave was usually taken to be made up of two Dorian[24] tetrachords, E–F–G–A and B–C–D–E, as described below, so that C was not the tonal centre. It should be pointed out that Ptolemy recorded a long list of Greek diatonic tunings, and there is no reason to believe that he preferred the diatonic syntonic scale to any of the others he recorded.

The point of the Greek tunings was the construction of tetrachords, or sequences of four consecutive notes encompassing a perfect fourth; the ratio of 5:4 seems to have been an incidental consequence rather than representing a recognized consonant major third.

A Greek scale consisted of two tetrachords, either in *conjunction*, which means overlapping (for example two Dorian tetrachords B–C–D–E and E–F–G–A) or in *disjunction*, which means non-overlapping (for example E–F–G–A and B–C–D–E) with a whole tone as the gap. The tetrachords came in three types, called *genera*

---

[24] Dorian tetrachords should not be confused with the Dorian mode of medieval church music, which is D–E–F–G–A–B–C–D. See Appendix F.

(plural of *genus*), and the two tetrachords in a scale belong to the same genus. The first genus is the *diatonic genus* in which the lowest interval is a semitone and the two upper ones are tones. The second is the *chromatic genus* in which the lowest two intervals are semitones and the upper one is a tone and a half. The third is the *enharmonic genus* in which the lowest two intervals are quarter tones and the upper one is two tones. The exact values of these intervals varied somewhat according to usage.[25] The interval between the lowest note and the higher of the two movable notes of a chromatic or enharmonic tetrachord is called the *pyknon*, and is always smaller than the remaining interval at the top of the tetrachord.

Medieval to modern music

Little is known about the harmonic content, if any, of European music prior to the decline of the Roman Empire. The music of ancient Greece, for example, survives in a small handful of fragments, and is mostly melodic in nature. There is little evidence of continuity of musical practice from ancient Greece to medieval European music, although the theoretical writings had a great deal of impact.

Harmony, in a primitive form, seems to have first appeared in liturgical plainchant around 800 CE, in the form of *parallel organum*, or melody in parallel fourths and fifths. Major thirds were not regarded as consonant, and a Pythagorean tuning system with its perfect fourths and fifths works well for such music.

Polyphonic music started developing around the eleventh century CE. Pythagorean intonation continued to be used for several centuries, and so the consonances in this system were the perfect fourths, fifths and octaves. The major third was still not regarded as a consonant interval, and it was something to be used only in passing.

The earliest known advocates of the 5:4 ratio as a consonant interval are the Englishmen Theinred of Dover (twelfth century) and Walter Odington (fl. 1298–1316),[26] in the context of early English polyphonic music. One of the earliest recorded uses of the major third in harmony is the four part vocal canon *sumer is icumen in*, of English origin, dating from around 1250. But for keyboard music, the question of tuning delayed its acceptance.

British folk music from the fourteenth and fifteenth centuries involved harmonizing around a melodic line by adding major thirds under it and perfect fourths over it to give parallel $\frac{6}{3}$ chords. The consonant major third travelled from England to the European continent in the early fifteenth century. But when the French imitated

---

[25] For example, Archytas described tetrachords using the ratios 1:1, 28:27, 32:27, 4:3 (diatonic), 1:1, 28:27, 9:8, 4:3 (chromatic) and 1:1, 28:27, 16:15, 4:3 (enharmonic), in which the primes 2, 3, 5 and 7 appear. Plato, his contemporary, does not allow primes other than 2 and 3, in better keeping with the Pythagorean tradition.

[26] The 'fl.' indicates that these are the years in which he is known to have flourished.

the sound of the parallel $\frac{6}{3}$ chords, they use the top line rather than the middle line as the melody, giving what is referred to as *Faux Bourdon*.

In more formal music, Dunstable was one of the most well known British composers of the early fifteenth century to use the consonant major third. The story goes that the Duke of Bedford, who was Dunstable's employer, inherited land in the north of France and moved there some time in the 1420s or 1430s. The French heard Dunstable's consonant major thirds and latched onto the idea. Guillaume Dufay was the first major French composer to use it extensively. The accompanying transition from modality to tonality can be traced from Dufay through Ockeghem, Josquin, Palestrina and Monteverdi during the fifteenth and sixteenth centuries.

The method for obtaining consonant major thirds in fourteenth and fifteenth century keyboard music is interesting. Starting with a series of Pythagorean fifths

$$G\flat^0 - D\flat^0 - A\flat^0 - E\flat^0 - B\flat^0 - F^0 - C^0 - G^0 - D^0 - A^0 - E^0 - B^0,$$

the triad $D^0 - G\flat^0 - A^0$ is used as a major triad. A just major triad would be $D^0 - F\sharp^{-1} - A^0$, and the difference between $F\sharp^{-1}$ and $G\flat^0$ is one schisma, or 1.953 cents. This is much more consonant than the modern equal temperament, in which the major thirds are impure by 13.686 cents. Other major triads available in this system are $A^0 - D\flat^0 - E^0$ and $E^0 - A\flat^0 - B^0$, but the system does not include a consonant $C - E - G$ triad.

By the mid- to late fifteenth century, especially in Italy, many aspects of the arts were reaching a new level of technical and mathematical precision. Leonardo da Vinci was integrating the visual arts with the sciences in revolutionary ways. In music, the meantone temperament was developed around this time, allowing the use of major and minor triads in a wide range of keys, and allowing harmonic progressions and modulations which had previously not been possible.

Many keyboard instruments from the sixteenth century have split keys for one or both of G$\sharp$/A$\flat$ and D$\sharp$/E$\flat$ to extend the range of usable key signatures. This was achieved by splitting the key across the middle, with the back part higher than the front part. Figure 5.26 shows the split keys of the Malamini organ in San Petronio, Bologna, Italy. Figure 5.27 shows a clavecin with split keys.

Meantone tuning has lasted for a long time. It is still common today for organs to be tuned in quarter comma meantone. The English concertinas made by Wheatstone & Co. in the nineteenth century, of which many are still in circulation, are tuned to quarter comma meantone, with separate keys for D$\sharp$/E$\flat$ and for G$\sharp$/A$\flat$.

The practice for music of the sixteenth and seventeenth century was to choose a tonal centre and gradually move further away. The furthest reaches were sparsely used, before gradually moving back to the tonal centre.

Figure 5.26 Malamini organ (split keys).

Exact meantone tuning was not achieved in practice before the twentieth century, for lack of accurate prescriptions for tuning intervals. Keyboard instrument tuners tended to colour the temperament, so that different keys had slightly different sounds to them. The irregular temperaments of Section 5.13 took this process further, and to some extent formalized it.

An early advocate for equal temperament for keyboard instruments was Rameau (1730). This helped it gain in popularity, until by the early nineteenth century it was fairly widely used, at least in theory. However, much of Beethoven's piano music is best played with an irregular temperament (see Section 5.13), and Chopin was reluctant to compose in certain keys (notably D minor) because their characteristics did not suit him. In practice, equal temperament did not really take full hold until the end of the nineteenth century. Nineteenth century piano tuning practice often involved slight deviation from equal temperament in order to preserve, at least to some extent, the individual characteristics of the different keys. In the twentieth century, the dominance of chromaticism and the advent of twelve tone music have pretty much forced the abandonment of unequal temperaments, and piano tuning practice has reflected this.

Twelve tone music

Equal temperament is an essential ingredient in twentieth century twelve tone music, where combinatorics and chromaticism seem to supersede harmony. Some interesting evidence that harmonic content is irrelevant in Schoenberg's music is

Figure 5.27 Italian clavecin with split keys (1619) Musical Instruments Museum, Brussels (inv. 464).

that the performance version of one of his most popular works, *Pierrot Lunaire*, contained many transcription errors confusing sharps, naturals and flats, until it was re-edited for his collected works in the 1980s.

The mathematics involved in twelve tone music of the twentieth century is different in nature to most of the mathematics we have described so far. It is more combinatorial in nature, and involves discussions of subsets and permutations of the twelve tones of the chromatic scale. We shall have more to say on this subject in Chapter 9.

The role of the synthesizer

Before the days of digital synthesizers, we had a choice of several different versions of the tuning compromise. The just scales have perfect intervals, but do not allow us to modulate from the original key, and have problems with the triad on ii, and with syntonic commas interfering in fairly short harmonic sequences. Meantone scales sacrifice a little perfection in the fifths in order to remove the problem of the syntonic comma, but still have a problem with keys far removed from the original key, and with enharmonic modulations. Equal temperament works in all keys equally well, or rather, one might say equally badly. In particular, the equal tempered major third is nervous and agitated.

In these days of digitally synthesized and controlled music, there is very little reason to make do with the equal tempered compromise, because we can retune any note by any amount as we go along. It may still make sense to prefer a meantone scale to a just one on the grounds of interference of the syntonic comma, but it may also make sense to turn the situation around and use the syntonic comma for effect.

It seems that for most users of synthesizers the extra freedom has not had much effect, in the sense that most music involving synthesizers is written using the equal tempered twelve tone scale. A notable exception is Wendy Carlos, who has composed a great deal of music for synthesizers using many different scales. I particularly recommend *Beauty in the Beast*, which has been released on compact disc (SYNCD 200, Audion, 1986, Passport Records, Inc.). For example, the fourth track, called Just Imaginings, uses a version of just intonation with harmonics all the way up to the nineteenth, and includes some deft modulations. Other tracks use other scales, including Carlos' *alpha* and *beta* scales and the Balinese gamelan *pelog* and *slendro*.

Wendy Carlos' earlier recordings, *Switched on Bach* and *The Well Tempered Synthesizer*, were recorded on a Moog synthesizer fixed in equal temperament. But when *Switched on Bach 2000* came out in 1992, twenty-five years after the original, it made use of a variety of meantone and unequal temperaments. It is not hard to hear from this recording the difference in clarity between these and equal temperament.

Further reading

(1) History of music theory

Thomas Christensen (ed.)(2002), *The Cambridge History of Music Theory*, 2002.

Leo Treitler, *Strunk's Source Readings in Music History*, revised edition, Norton & Co., 1998. This 1552 page book, originally by Strunk but revised extensively by Treitler, contains translations of historical documents from ancient Greece to the twentieth century. It comes in seven sections, which are available in separate paperbacks.

(2) Ancient Greek music

W. D. Anderson, *Music and Musicians in Ancient Greece*, Cornell University Press, 1994; paperback edition 1997.

Andrew Barker, *Greek Musical Writings, Vol. 2: Harmonic and Acoustic Theory*, Cambridge University Press, 1989. This 581 page book contains translations and commentaries on many of the most important ancient Greek sources, including Aristoxenus' *Elementa Harmonica*, the Euclidean *Sectio Canonis*, Nicomachus' *Enchiridion*, Ptolemy's *Harmonics*, and Aristides Quintilianus' *De Musica*.

Giovanni Comotti, *Music in Greek and Roman culture*, Johns Hopkins University Press, 1989; paperback edition 1991.

John G. Landels, *Music in Ancient Greece and Rome*, Routledge, 1999; paperback edition 2001.

Thomas J. Mathiesen, *Apollo's Lyre: Greek Music and Music Theory in Antiquity and the Middle Ages*, University of Nebraska Press, 1999.

M. L. West, *Ancient Greek Music*, Oxford University Press, 1992; paperback edition 1994. Chapter 10 of this book reproduces all 51 known fragments of ancient Greek music.

R. P. Winnington-Ingram, *Mode in Ancient Greek Music*, Cambridge University Press, 1936. Reprinted by Hakkert, Amsterdam, 1968.

(3) Medieval to modern music

Gustave Reese, *Music in the Middle Ages*, Norton, 1940, reprinted 1968. Despite the age of this text, it is still regarded as an invaluable source because of the quality of the scholarship. But the reader should bear in mind that much information has come to light since it appeared.

D. J. Grout and C. V. Palisca, *A History of Western Music*, fifth edition, Norton, 1996. Originally written in the 1950s by Grout, and updated a number of times by Palisca. This is a standard text used in many music history departments.

Owen H. Jorgensen (1991), *Tuning* contains an excellent discussion of the development of temperament, and argues that equal temperament was not commonplace in practice until the twentieth century.

(4) Twelve tone music

Allen Forte (1973), *The Structure of Atonal Music*.

George Perle (1977), *Twelve Tone Tonality*.

(5) The role of the synthesizer

Easley Blackwood, Discovering the microtonal resources of the synthesizer, *Keyboard*, May 1982, 26–38.

Benjamin Frederick Denckla, *Dynamic Intonation in Synthesizer Performance*, M.Sc. Thesis, MIT, 1997 (61 pp).

Henry Lowengard, Computers, digital synthesizers and microtonality, *Pitch* **1** (1) (1986), 6–7.

Robert Rich, Just intonation for MIDI synthesizers, *Electronic Musician*, Nov 1986, 32–45.

M. Yunik and G. W. Swift, Tempered music scales for sound synthesis, *Computer Music Journal* **4** (4) (1980), 60–65.

6

More scales and temperaments

6.1 Harry Partch's 43 tone and other just scales

In Section 5.5, we talked about just intonation in its narrowest sense. This involved building up a scale using ratios only involving the primes 2, 3 and 5, to obtain a twelve tone scale. Just intonation can be extended far beyond this limitation. The

Figure 6.1 Harry Partch playing the bamboo marimba (Boo I).

phrase *super just* is sometimes used to denote a scale formed with exact rational multiples for the intervals, but using primes other than the 2, 3 and 5. Most of these come from the twentieth century.

Harry Partch (see Figure 6.1) developed a just scale of 43 notes which he used in a number of his compositions. The tonic for his scale is G^0. The scale is symmetric, in the sense that every interval upwards from G^0 is also an interval downwards from G^0.

The primes involved in Partch's scale are 2, 3, 5, 7 and 11. The terminology used by Partch to describe this is that his scale is based on the 11-limit, while the Pythagorean scale is based on the 3-limit and the just scales of Sections 5.5 and 5.10 are based on the 5-limit. More generally, if p is a prime, then a p-limit scale only uses rational numbers whose denominators and numerators factor as products of prime numbers less than or equal to p (repetitions are allowed).

Harry Partch's 43 tone scale					
G^0	1:1	0.000		10:7	617.488
G^{+1}	81:80	21.506		16:11	648.682
	33:32	53.273	D^{-1}	40:27	680.449
	21:20	84.467	D^0	3:2	701.955
Ab^{+1}	16:15	111.713		32:21	729.219
	12:11	150.637		14:9	764.916
	11:10	165.004		11:7	782.492
A^{-1}	10:9	182.404	Eb^{+1}	8:5	813.686
A^0	9:8	203.910		18:11	852.592
	8:7	231.174	E^{-1}	5:3	884.359
	7:6	266.871	E^0	27:16	905.865
Bb^0	32:27	294.135		12:7	933.129
Bb^{+1}	6:5	315.641		7:4	968.826
	11:9	347.408	F^0	16:9	996.090
B^{-1}	5:4	386.314	F^{+1}	9:5	1017.596
	14:11	417.508		20:11	1034.996
	9:7	435.084		11:6	1049.363
	21:16	470.781	$F\sharp^{-1}$	15:8	1088.269
C^0	4:3	498.045		40:21	1115.533
C^{+1}	27:20	519.551		64:33	1146.727
	11:8	551.318	G^{-1}	160:81	1178.494
	7:5	582.512	G^0	2:1	1200.000

Here are some other just scales. The Chinese Lü scale by Huai-nan-dsi of the Han dynasty is the twelve tone just scale with ratios

1:1, 18:17, 9:8, 6:5, 54:43, 4:3, 27:19, 3:2, 27:17, 27:16, 9:5, 36:19, (2:1).

The Great Highland Bagpipe of Scotland is tuned to a ten tone 7-limit just scale based on a drone pitched at A (slightly sharper than modern concert pitch), and with ratios

$$(7:8), (8:9), 1:1 \text{ (A)}, 9:8, 5:4, 4:3, 27:20, 3:2, 5:3, 7:4, 16:9, 9:5, (2:1).$$

Wendy Carlos has developed several just scales. The 'Wendy Carlos super just intonation' is the twelve tone scale with ratios

$$1:1, 17:16, 9:8, 6:5, 5:4, 4:3, 11:8, 3:2, 13:8, 5:3, 7:4, 15:8, (2:1).$$

The 'Wendy Carlos harmonic scale' also has twelve tones, with ratios

$$1:1, 17:16, 9:8, 19:16, 5:4, 21:16, 11:8, 3:2, 13:8, 27:16, 7:4, 15:8, (2:1).$$

A better way of writing this might be to multiply all the entries by 16:

$$16, \ 17, \ 18, \ 19, \ 20, \ 21, \ 22, \ 24, \ 26, \ 27, \ 28, \ 30, \ (32).$$

Lou Harrison has a 16 tone just scale with ratios

$$1:1, 16:15, 10:9, 8:7, 7:6, 6:5, 5:4, 4:3, 17:12,$$
$$3:2, 8:5, 5:3, 12:7, 7:4, 9:5, 15:8, (2:1).$$

Wilfrid Perret[1] has a 19 tone 7-limit just scale with ratios

$$1:1, 21:20, 35:32, 9:8, 7:6, 6:5, 5:4, 21:16, 4:3, 7:5, 35:24,$$
$$3:2, 63:40, 8:5, 5:3, 7:4, 9:5, 15:8, 63:32, (2:1).$$

John Chalmers also has a 19 tone 7-limit just scale, differing from this in just two places. The ratios are

$$1:1, 21:20, 16:15, 9:8, 7:6, 6:5, 5:4, 21:16, 4:3, 7:5, 35:24,$$
$$3:2, 63:40, 8:5, 5:3, 7:4, 9:5, 28:15, 63:32, (2:1).$$

Michael Harrison has a 24 tone 7-limit just scale with ratios

$$1:1, 28:27, 135:128, 16:15, 243:224, 9:8, 8:7, 7:6, 32:27, 6:5, 135:112, 5:4,$$
$$81:64, 9:7, 21:16, 4:3, 112:81, 45:32, 64:45, 81:56, 3:2, 32:21, 14:9, 128:81,$$
$$8:5, 224:135, 5:3, 27:16, 12:7, 7:4, 16:9, 15:8, 243:128, 27:14, (2:1).$$

Harrison writes:

Beginning in 1986, I spent two years extensively modifying a seven-foot Schimmel grand piano to create the *Harmonic Piano*. It is the first piano tuned in Just Intonation with the

---

[1] W. Perret, *Some Questions of Musical Theory*, W. Heffer & Sons Ltd., 1926.

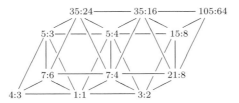

Figure 6.2 Twelve tone 7-limit just scale.

flexibility to modulate to multiple key centers at the press of a pedal. With its unique pedal mechanism, the Harmonic Piano can differentiate between notes usually shared by the same piano key (for example, C-sharp and D-flat). As a result, the Harmonic Piano is capable of playing 24 notes per octave. In contrast to the three unison strings per note of the standard piano, the Harmonic Piano uses only single strings, giving it a 'harp-like' timbre. Special muting systems are employed to dampen unwanted resonances and to enhance the instrument's clarity of sound.[2]

The Indian Sruti scale,[3] commonly used to play ragas, is a 5-limit just scale with 22 tones, but has some large numerators and denominators:

1:1, 256:243, 16:15, 10:9, 9:8, 32:27, 6:5, 5:4, 81:64, 4:3, 27:20, 45:32,

729:512, 3:2, 128:81, 8:5, 5:3, 27:16, 16:9, 9:5, 15:8, 243:128, (2:1).

Various notations have been designed for describing just scales. For example, for 7-limit scales, a three-dimensional lattice of tetrahedra and octahedra can just about be drawn on paper. Figure 6.2 is an example of a twelve tone 7-limit just scale drawn three-dimensionally in this way.[4]

The lines indicate major and minor thirds, perfect fifths, and three different septimal consonances 7:4, 7:5 and 7:6 (notes have been normalized to lie inside the octave 1:1 to 2:1). We return to the discussion of just intonation in Section 6.8, where we discuss unison vectors and periodicity blocks. We put the above diagram into context in Section 6.9.

Exercise

Taking 1:1 to be C^0, write the Indian Sruti scale described in this section as an array using Eitz's comma notation (like the scales in Section 5.10).

---

[2] From the liner notes to Harrison's CD *From Ancient Worlds, for Harmonic Piano*, see Appendix G.
[3] Taken from B. Chaitanya Deva, *The Music of India* (1981), Table 9.2. Note that the fractional value of note 5 given in this table should be 32/27, not 64/45, to match the other information given in this table. This also matches the value given in Tables 9.4 and 9.8 of the same work. Beware that the exact values of the intervals in Indian scales is a subject of much debate and historical controversy.
[4] This way of drawing the scale comes from Paul Erlich. According to Paul, the scale was probably first written down by Erv Wilson in the 1960s.

Further reading

David B. Doty, *The Just Intonation Primer* (1993), privately published and available from
  the Just Intonation Network at www.justintonation.net.
Harry Partch (1974), *Genesis of a Music*.
Joseph Yasser (1932), *A Theory of Evolving Tonality*.

Further listening (See Appendix G)

Bill Alves, *Terrain of Possibilities*.
Wendy Carlos, *Beauty in the Beast*.
Michael Harrison, *From Ancient Worlds*.
Harry Partch, *Bewitched*.
Robert Rich, *Rainforest, Gaudi*.

6.2 Continued fractions

$$e^{2\pi/5}\left(\sqrt{\frac{5+\sqrt{5}}{2}} - \frac{\sqrt{5}+1}{2}\right) = \frac{1}{1+}\frac{e^{-2\pi}}{1+}\frac{e^{-4\pi}}{1+}\frac{e^{-6\pi}}{1+}\cdots$$

(Srinivasa Ramanujan)

The modern twelve tone equal tempered scale is based around the fact that

$$7/12 = 0.58333\ldots$$

is a good approximation to

$$\log_2(3/2) = 0.584\,962\,5007\ldots,$$

so that if we divide the octave into twelve equal semitones, then seven semitones is
a good approximation to a perfect fifth. This suggests the following question. Can
$\log_2(3/2)$ be expressed as a ratio of two integers, m/n? In other words, is $\log_2(3/2)$
a rational number? Since $\log_2(3/2)$ and $\log_2(3)$ differ by one, this is the same as
asking whether $\log_2(3)$ is rational.

Lemma 6.2.1 *The number* $\log_2(3)$ *is irrational.*

Proof Suppose that $\log_2(3) = m/n$ with m and n positive integers. Then $3 = 2^{m/n}$,
or $3^n = 2^m$. Now 3^n is always odd while 2^m is always even (since $m > 0$). So this
is not possible.   □

So the best we can expect to do is to approximate $\log_2(3/2)$ by rational numbers
such as $7/12$. There is a systematic theory of such rational approximations to

irrational numbers, which is the theory of continued fractions.[5] A continued fraction is an expression of the form

$$a_0 + \cfrac{1}{a_1 + \cfrac{1}{a_2 + \cfrac{1}{a_3 + \cdots}}}$$

where $a_0, a_1, \ldots$ are integers, and a_i is usually taken to be positive for $i \geq 1$. The expression is allowed to stop at some finite stage, or it may go on for ever. If it stops, the last a_n is usually not allowed to equal 1, because if it does, it can just be absorbed into a_{n-1} to make it finish sooner (for example $1 + \frac{1}{2+\frac{1}{1}}$ can be rewritten as $1 + \frac{1}{3}$). For typographic convenience, we write the continued fraction in the form

$$a_0 + \frac{1}{a_1+} \frac{1}{a_2+} \frac{1}{a_3+} \cdots$$

For even greater compression of notation, this is sometimes written as

$$[a_0; a_1, a_2, a_3, \ldots].$$

Every real number has a unique continued fraction expansion, and it stops precisely when the number is rational. The easiest way to see this is as follows. If x is a real number, then the largest integer less than or equal to x (the *integer part* of x) is written $\lfloor x \rfloor$.[6] So $\lfloor x \rfloor$ is what we take for a_0. The remainder $x - \lfloor x \rfloor$ satisfies $0 \leq x - \lfloor x \rfloor < 1$, so if it is nonzero, we now invert it to obtain a number $1/(x - \lfloor x \rfloor)$ which is strictly larger than one.

Writing $x_0 = x$, $a_0 = \lfloor x_0 \rfloor$ and $x_1 = 1/(x_0 - \lfloor x_0 \rfloor)$, we have

$$x = a_0 + \frac{1}{x_1}.$$

Now just carry on going. Let $a_1 = \lfloor x_1 \rfloor$, and $x_2 = 1/(x_1 - \lfloor x_1 \rfloor)$, so that

$$x = a_0 + \frac{1}{a_1+} \frac{1}{x_2}.$$

Inductively, we set $a_n = \lfloor x_n \rfloor$ and $x_{n+1} = 1/(x_n - \lfloor x_n \rfloor)$ so that

$$x = a_0 + \frac{1}{a_1+} \frac{1}{a_2+} \frac{1}{a_3+} \cdots$$

---

[5] The first mathematician known to have made use of continued fractions was Rafael Bombelli in 1572. The modern notation for them was introduced by P. A. Cataldi in 1613.
[6] In some books, $[x]$ is used instead.

This algorithm continues provided each $x_n \neq 0$, which happens exactly when x is irrational. Otherwise, if x is rational, the algorithm terminates to give a finite continued fraction. For irrational numbers the continued fraction expansion is unique. For rational numbers, we have uniqueness provided we stipulate that the last a_n is larger than one.

As an example, let us compute the continued fraction expansion of

$$\pi = 3.141\,592\,653\,589\,793\,238\,462\,643\,383\,279\,502\,884\,197\,169\,399\,375$$
$$105\,820\,974\,944\,592\,307\,8164\ldots$$

In this case, we have $a_0 = 3$ and

$$x_1 = 1/(\pi - 3) = 7.062\,513\,086\ldots$$

So $a_1 = 7$, and

$$x_2 = 1/(x_1 - 7) = 15.996\,65\ldots$$

Continuing this way, we obtain

$$\pi = 3 + \cfrac{1}{7+}\ \cfrac{1}{15+}\ \cfrac{1}{1+}\ \cfrac{1}{292+}\ \cfrac{1}{1+}\ \cfrac{1}{1+}\ \cfrac{1}{1+}\ \cfrac{1}{2+}\ \cfrac{1}{1+}\ \cfrac{1}{3+}\ \cfrac{1}{1+}\ \cfrac{1}{14+}\ \cdots$$

In the more compressed notation, here are more terms:[7]

$\pi = [3; 7, 15, 1, 292, 1, 1, 1, 2, 1, 3, 1, 14, 2, 1, 1, 2, 2, 2, 2, 1, 84, 2, 1, 1, 15, 3, 13,$
$\quad 1, 4, 2, 6, 6, 99, 1, 2, 2, 6, 3, 5, 1, 1, 6, 8, 1, 7, 1, 2, 3, 7, 1, 2, 1, 1, 12, 1, 1, 1, 3,$
$\quad 1, 1, 8, 1, 1, 2, 1, 6, 1, 1, 5, 2, 2, 3, 1, 2, 4, 4, 16, 1, 161, 45, 1, 22, 1, 2, 2, 1, 4,$
$\quad 1, 2, 24, 1, 2, 1, 3, 1, 2, 1, 1, 10, 2, 5, 4, 1, 2, 2, 8, 1, 5, 2, 2, 26, 1, 4, 1, 1, 8, 2,$
$\quad 42, 2, 1, 7, 3, 3, 1, 1, 7, 2, 4, 9, 7, 2, 3, 1, 57, 1, 18, 1, 9, 19, 1, 2, 18, 1, 3, 7, 30,$
$\quad 1, 1, 1, 3, 3, 3, 1, 2, 8, 1, 1, 2, 1, 15, 1, 2, 13, 1, 2, 1, 4, 1, 12, 1, 1, 3, 3, 28, 1, 10,$
$\quad 3, 2, 20, 1, 1, 1, 1, 4, 1, 1, 1, 5, 3, 2, 1, 6, 1, 4, 1, 120, 2, 1, 1, 3, 1, 23, 1, 15, 1, 3,$
$\quad 7, 1, 16, 1, 2, 1, 21, 2, 1, 1, 2, 9, 1, 6, 4, 127, 14, 5, 1, 3, 13, 7, 9, 1, 1, 1, 1, 1, 5,$
$\quad 4, 1, 1, 3, 1, 1, 29, 3, 1, 1, 2, 2, 1, 3, 1, 1, 1, 3, 1, 1, 10, 3, 1, 3, 1, 2, 1, 12, 1, 4, 1,$
$\quad 1, 1, 1, 7, 1, 1, 2, 1, 11, 3, 1, 7, 1, 4, 1, 48, 16, 1, 4, 5, 2, 1, 1, 4, 3, 1, 2, 3, 1, 2, 2,$
$\quad 1, 2, 5, 20, 1, 1, 5, 4, 1, 436, 8, 1, 2, 2, 1, 1, 1, 1, 1, 5, 1, 2, 1, 3, 6, 11, 4, 3, 1, 1, 1,$
$\quad 2, 5, 4, 6, 9, 1, 5, 1, 5, 15, 1, 11, 24, 4, 4, 5, 2, 1, 4, 1, 6, 1, 1, 1, 4, 3, 2, 2, 1, 1, 2,$
$\quad 1, 58, 5, 1, 2, 1, 2, 1, 1, 2, 2, 7, 1, 15, 1, 4, 8, 1, 1, 4, 2, 1, 1, 1, 3, 1, 1, 1, 2, 1, 1, 1,$
$\quad 1, 1, 9, 1, 4, 3, 15, 1, 2, 1, 13, 1, 1, 1, 3, 24, 1, 2, 4, 10, 5, 12, 3, 3, 21, 1, 2, 1, 34,$
$\quad 1, 1, 1, 4, 15, 1, 4, 44, 1, 4, 20776, 1, 1, 1, 1, 1, 1, 1, 23, 1, 7, 2, 1, 94, 55, 1, 1, 2, \ldots].$

To get good rational approximations, we stop just before a large value of a_n. So for example, stopping just before the 15, we obtain the well-known

---

[7] Note that the values given in Hua (1982), page 252, are errenious. The correct values for the first 20 000 000 terms in the continued fraction expansion of π can be downloaded from www.lacim.uqam.ca/piDATA.

approximation $\pi \approx 22/7$. Stopping just before the 292 gives us the extremely good approximation

$$\pi \approx 355/113 = 3.141\,5929\ldots$$

which was known to the Chinese mathematician Chao Jung-Tze (or Tsu Ch'ung-Chi, depending on how you transliterate the name) in 500 AD.

The rational approximations obtained by truncating the continued fraction expansion of a number are called the *convergents*. So the convergents for π are

$$\frac{3}{1},\frac{22}{7},\frac{333}{106},\frac{355}{113},\frac{103\,993}{33\,102},\frac{104\,348}{33\,215},\ldots$$

There is an extremely efficient way to calculate the convergents from the continued fraction.

Theorem 6.2.2 *Define numbers p_n and q_n inductively as follows:*

$$p_0 = a_0, \qquad p_1 = a_1 a_0 + 1, \qquad p_n = a_n p_{n-1} + p_{n-2} \quad (n \geq 2), \qquad (6.2.1)$$
$$q_0 = 1, \qquad q_1 = a_1, \qquad q_n = a_n q_{n-1} + q_{n-2} \quad (n \geq 2). \qquad (6.2.2)$$

Then we have

$$a_0 + \frac{1}{a_1+}\frac{1}{a_2+}\cdots\frac{1}{a_n} = \frac{p_n}{q_n}.$$

Proof (see Hardy and Wright (1980), Theorem 149, or Hua (1982), Theorem 10.1.1).

The proof goes by induction on n. It is easy enough to check the cases $n = 0$ and $n = 1$, so we assume that $n \geq 2$ and that the theorem holds for smaller values of n. Then we have

$$a_0 + \frac{1}{a_1+}\frac{1}{a_2+}\cdots\frac{1}{a_{n-1}+}\frac{1}{a_n} = a_0 + \frac{1}{a_1+}\frac{1}{a_2+}\cdots\frac{1}{a_{n-1}+\frac{1}{a_n}}.$$

So we can use the formula given by the theorem with $n - 1$ in place of n to write this as

$$\frac{\left(a_{n-1} + \frac{1}{a_n}\right)p_{n-2} + p_{n-3}}{\left(a_{n-1} + \frac{1}{a_n}\right)q_{n-2} + q_{n-3}} = \frac{a_n(a_{n-1}p_{n-2} + p_{n-3}) + p_{n-2}}{a_n(a_{n-1}q_{n-2} + q_{n-3}) + q_{n-2}}$$
$$= \frac{a_n p_{n-1} + p_{n-2}}{a_n q_{n-1} + q_{n-2}} = \frac{p_n}{q_n}.$$

So the theorem is true for n, and the induction is complete. $\qquad\square$

So, in the above example for π, we have $p_0 = a_0 = 3, q_0 = 1, p_1 = a_1 a_0 + 1 = 22, q_1 = a_1 = 7$, and we get

$$\frac{p_2}{q_2} = \frac{p_0 + 15 p_1}{q_0 + 15 q_1} = \frac{333}{106},$$

so that $p_2 = 333, q_2 = 106$,

$$\frac{p_3}{q_3} = \frac{p_1 + p_2}{q_1 + q_2} = \frac{355}{113},$$

so that $p_3 = 355, q_3 = 113$, and so on.

Examining the value of x_2 in the case $x = \pi$ above, it may look as though it would be of advantage to allow negative as well as positive values for a_n. However, this doesn't really help, because if x_n is very slightly less than $a_n + 1$ then a_{n+1} will be equal to one, and from there on the sequence is as it would have been. In other words, the rational approximations obtained this way are no better. A related observation is that if $a_{n+1} = 2$ then it is worth examining the approximation given by replacing a_n by $a_n + 1$ and stopping there.

The continued fraction expansion for the base of natural logarithms

$$e = 2.718\,281\,828\,459\,045\,235\,360\,287\,471\,352\,662\,497\,757\,247\,093 \ldots$$

$$= 2 + \frac{1}{1+} \frac{1}{2+} \frac{1}{1+} \frac{1}{1+} \frac{1}{4+} \frac{1}{1+} \frac{1}{1+} \frac{1}{6+} \frac{1}{1+} \frac{1}{1+} \frac{1}{8+} \frac{1}{1+} \frac{1}{1+} \ldots$$

follows an easily described pattern, as was discovered by Leonhard Euler. The continued fraction expansion of the golden ratio is even easier to describe:

$$\tau = \tfrac{1}{2}(1 + \sqrt{5}) = 1 + \frac{1}{1+} \frac{1}{1+} \frac{1}{1+} \frac{1}{1+} \frac{1}{1+} \ldots$$

Although the continued fraction expansion of π is not regular in this way, there is a closely related formula (Brouncker)

$$\frac{\pi}{4} = \frac{1}{1+} \frac{1}{3+} \frac{4}{5+} \frac{9}{7+} \frac{16}{9+} \ldots$$

which is a special case of the arctan formula

$$\tan^{-1} z = \frac{z}{1+} \frac{z^2}{3+} \frac{4z^2}{5+} \frac{9z^2}{7+} \frac{16z^2}{9+} \ldots$$

The tan formula

$$\tan z = \frac{z}{1+} \frac{-z^2}{3+} \frac{-z^2}{5+} \frac{-z^2}{7+} \ldots$$

can be used to show that π is irrational (Pringsheim).

How good are the rational approximations obtained from continued fractions? This is answered by the following theorems. Recall that $x_n = p_n/q_n$ denotes the nth convergent. In other words,

$$\frac{p_n}{q_n} = a_0 + \frac{1}{a_1+} \frac{1}{a_2+} \cdots \frac{1}{a_{n-1}+} \frac{1}{a_n}.$$

Theorem 6.2.3 *The error in the nth convergent of the continued fraction expansion of a real number x is bounded by*

$$\left| \frac{p_n}{q_n} - x \right| < \frac{1}{q_n^2}.$$

Proof (see Hardy and Wright (1980), Theorem 171, or Hua (1982), Theorem 10.2.6).

First, we notice that $p_{n-1}q_n - p_n q_{n-1} = (-1)^n$. This is easiest to see by induction. For $n = 1$, we have $p_0 = a_0$, $q_0 = 1$, $p_1 = a_0 a_1 + 1$, $q_1 = a_1$, so $p_0 a_1 - p_1 a_0 = -1$. For $n > 1$, using Equations (6.2.1) and (6.2.2), we have

$$p_{n-1}q_n - p_n q_{n-1} = p_{n-1}(q_{n-2} + a_n q_{n-1}) - (p_{n-2} + a_n p_{n-1})q_{n-1}$$
$$= p_{n-1}q_{n-2} - p_{n-2}q_{n-1}$$
$$= -(p_{n-2}q_{n-1} - p_{n-1}q_{n-2})$$
$$= -(-1)^{n-1} = (-1)^n.$$

Now we use the fact that x lies between

$$\frac{p_{n-2} + a_n p_{n-1}}{q_{n-2} + a_n q_{n-1}} \quad \text{and} \quad \frac{p_{n-2} + (a_n + 1)p_{n-1}}{q_{n-2} + (a_n + 1)q_{n-1}},$$

or in other words between $\dfrac{p_n}{q_n}$ and $\dfrac{p_n + p_{n-1}}{q_n + q_{n-1}}$. The distance between these two numbers is

$$\left| \frac{p_n + p_{n-1}}{q_n + q_{n-1}} - \frac{p_n}{q_n} \right| = \left| \frac{(p_n + p_{n-1})q_n - p_n(q_n + q_{n-1})}{(q_n + q_{n+1})q_n} \right|$$
$$= \left| \frac{p_{n-1}q_n - p_n q_{n-1}}{q_n^2 + q_n q_{n-1}} \right| = \left| \frac{(-1)^n}{q_n^2 + q_n q_{n-1}} \right| < \frac{1}{q_n^2}. \qquad \square$$

Notice that if we choose a denominator q at random, then the intervals between the rational numbers of the form p/q are of size $1/q$. So by choosing p to minimize the error, we get $|p/q - x| \le 1/2q$. So the point of the above theorem is that the convergents in the continued fraction expansion are considerably better than random denominators. In fact, more is true.

Theorem 6.2.4 *Among the fractions p/q with $q \le q_n$, the closest to x is p_n/q_n.*

Proof See Hardy and Wright (1980), Theorem 181. ☐

It is not true that if p/q is a rational number satisfying $|p/q - x| < 1/q^2$ then p/q is a convergent in the continued fraction expansion of x. However, a theorem of Hurwitz (see Hua (1982), Theorem 10.4.1) says that of any two consecutive convergents to x, at least one of them satisfies $|p/q - x| < 1/2q^2$. Moreover, if a rational number p/q satisfies this inequality then it is a convergent in the continued fraction expansion of x (see Hua (1982), Theorem 10.7.2).

Distribution of the a_n

If we perform continued fractions on a transcendental number x, given an integer k, how likely is it that $a_n = k$? It seems plausible that $a_n = 1$ is the most likely, and that the probabilities decrease rapidly as k increases, but what is the exact distribution of probabilities?

Gauss answered this question in a letter addressed to Laplace, although he never published a proof.[8] Writing $\mu\{-\}$ for the measure of a set $\{-\}$, what he proved is the following. Given any t in the range $(0, 1)$, in the limit the measure of the set of numbers x in the interval $(0, 1)$ for which $x_n - \lfloor x_n \rfloor$ is at most t is given by[9]

$$\lim_{n \to \infty} \mu\{x \in (0, 1) \mid x_n - \lfloor x_n \rfloor \le t\} = \log_2(1 + t).$$

The continued fraction process says that we should then invert $x_n - \lfloor x_n \rfloor$. Writing u for $1/t$, we obtain

$$\lim_{n \to \infty} \mu\left\{x \in (0, 1) \mid \frac{1}{x_n - \lfloor x_n \rfloor} \ge u\right\} = \log_2(1 + 1/u).$$

Now we need to take the integer part of $1/(x_n - \lfloor x_n \rfloor)$ to obtain a_{n+1}. So, if k is an integer with $k \ge 1$, then

$$\lim_{n \to \infty} \mu\{x \in (0, 1) \mid a_n = k\} = \log_2\left(1 + \frac{1}{k}\right) - \log_2\left(1 + \frac{1}{k+1}\right)$$

$$= \log_2\left(\frac{(k+1)^2}{k(k+2)}\right) = \log_2\left(1 + \frac{1}{k(k+2)}\right).$$

We now tabulate the probabilities given by this formula.

---

[8] According to A. Ya. Khinchin, *Continued Fractions*, Dover 1964, page 72, the first published proof was by Kuz'min in 1928.

[9] If you don't know what measure means in this context, think of this as giving the probability that a randomly chosen number in the given interval satisfies the hypothesis.

Value of k	Limiting probability that $a_n = k$ as $n \to \infty$
1	0.4150375
2	0.2223924
3	0.0931094
4	0.0588937
5	0.0406420
6	0.0297473
7	0.0227201
8	0.0179219
9	0.0144996
10	0.0119726

For large k, this decreases like $1/k^2$.

Multiple continued fractions

It is sometimes necessary to make simultaneous rational approximations for more than one irrational number. For example, in the equal tempered scale, not only do seven semitones approximate a perfect fifth with ratio 3:2, but also four semitones approximate a major third with ratio 5:4. So we have

$$\log_2(3/2) \approx 7/12; \qquad \log_2(5/4) \approx 4/12.$$

A theorem of Dirichlet tells us how closely we should expect to be able to approximate a set of k real numbers simultaneously.

Theorem 6.2.5 *If $\alpha_1, \alpha_2, \ldots, \alpha_k$ are real numbers, and at least one of them is irrational, then the there exist an infinite number of ways of choosing a denominator q and numerators $p_1, p_2, \ldots, p_k$ in such a way that the approximations*

$$p_1/q \approx \alpha_1; \quad p_2/q \approx \alpha_2; \quad \ldots \quad p_k/q \approx \alpha_k$$

have the property that the errors are all less than $1/q^{1+\frac{1}{k}}$.

Proof See Hardy and Wright (1980), Theorem 200.   □

The case $k = 1$ of this theorem is just Theorem 6.2.3. There is no known method when $k \geq 2$ analogous to the method of continued fractions for obtaining the approximations whose existence is guaranteed by this theorem. Of course, we can just work through the possibilities for q one at a time, but this is much more tedious than one would like.

The power of q in the denominator in the above theorem (i.e., $1 + \frac{1}{k}$) is known to be the best possible. Notice that the error term remains better than the error term

$1/2q$ which would result by choosing q randomly. But the extent to which it is better diminishes to insignificant as k grows large.

Exercises

1. Investigate the convergents for the continued fraction expansion of the golden ratio $\tau = (1 + \sqrt{5})/2$. What do these convergents have to do with the Fibonacci series?

   Coupled oscillators have a tendency to seek frequency ratios which can be expressed as rational numbers with small numerators and denominators. For example, Mercury rotates on its axis exactly three times for every two rotations around the Sun, so that one Mercurial day lasts two Mercurial years. In a similar way, the orbital times of Jupiter and the minor planet Pallas around the Sun are locked in a ratio of 18 to 7 (Gauss calculated in 1812 that this would be true, and observation has confirmed it). This is also why the Moon rotates once around its axis for each rotation around the Earth, so that it always shows us the same face.

   Among small frequency ratios for coupled oscillators, the golden ratio is the least likely to lock in to a nearby rational number. Why?

2. Find the continued fraction expansion of $\sqrt{2}$. Show that if a number has a periodic continued fraction expansion then it satisfies a quadratic equation with integer coefficients. In fact, the converse is also true: if a number satisfies a quadratic equation with integer coefficients then it has a periodic continued fraction expansion. See, for example, Hardy and Wright (1980), Section 10.12.

3. (Hua, 1982) The synodic month is the period of time between two new moons, and is 29.5306 days. When projected onto the star sphere, the path of the Moon intersects the ecliptic (the path of the Sun) at the ascending and the descending nodes. A draconic month is the period of time for the Moon to return to the same node, and is 27.2123 days. Show that the solar and lunar eclipses occur in cycles with a period of 18 years 10 days.

4. In this problem, you will prove that π is not equal to $\frac{22}{7}$. This problem is not really relevant to the text, but it is interesting anyway.

   Use partial fractions (actually, just the long division part of the algorithm) to prove that
   $$\int_0^1 \frac{x^4(1-x)^4 \, dx}{1+x^2} = \tfrac{22}{7} - \pi.$$

Deduce that $\pi < \frac{22}{7}$. Show that
$$\int_0^1 x^4(1-x)^4 \, dx = \tfrac{1}{630},$$

and use this to deduce that
$$\tfrac{1}{1260} < \tfrac{22}{7} - \pi < \tfrac{1}{630}.$$

What would this sentence be like if π were 3?

> If π were equal to 3, this sentence would look something like this.

(Scott Kim/Harold Cooper, quoted from Douglas Hofstadter's *Metamagical Themas*, Basic Books, 1985).

5. Show that if a and b have no common factor then $\log_a(b)$ is irrational. Show that if no pair among a, b and c has a common factor then $\log_a(b)$ and $\log_a(c)$ are rationally independent. In other words, there cannot exist nonzero integers n_1, n_2 and n_3 such that $n_1 \log_a(b) + n_2 \log_a(c) = n_3$.

6. Find the continued fraction expansion for the rational number $531\,441/524\,288$, which represents the frequency ratio for the Pythagorean comma. Explain in terms of this example the relationship between the continued fraction expansion of a rational number and Euclid's algorithm for finding highest common factors (if you don't remember how Euclid's algorithm goes, it is described in Lemma 9.7.1).

7. The *Gaussian integers* are the complex numbers of the form $a + bi$ where a and b are in the rational integers $\mathbb{Z}$. Develop a theory of continued fractions for simultaneously approximating two real numbers α and β, by considering the complex number $\alpha + \beta i$. Explain why this method favours denominators which can be expressed as a sum of two squares, so that it does not always find the best approximations.

8. A certain number is known to be a ratio of two 3-digit integers. Its decimal expansion, to nine significant figures, is $0.137\,637\,028$. What are the integers?

Further reading

G. H. Hardy and E. M. Wright (1980), *Number Theory*, chapter X.

Hua (1982), *Introduction to Number Theory*, Chapter 10.

Hubert Stanley Wall, *Analytic Theory of Continued Fractions*. Chelsea, 1948.

J. Murray Barbour, Music and ternary continued fractions, *Amer. Math. Mon.* **55** (9) (1948), 545–555.

Viggo Brun, Music and ternary continued fractions, *Norske Vid. Selsk. Forh., Trondheim* **23** (1950), 38–40. This article is a response to the above article of Murray Barbour.

Viggo Brun, Music and Euclidean algorithms, *Nordisk Mat. Tidskr.* **9** (1961), 29–36, 95.

Edward Dunne and Mark McConnell, Pianos and continued fractions, *Math. Mag.*, **72** (2) (1999), 104–115.

J. B. Rosser, Generalized ternary continued fractions, *Amer. Math. Mon.* **57** (8) (1950), 528–535. This is another response to Murray Barbour's article.

Murray Schechter, Tempered scales and continued fractions, *Amer. Math. Mon.* **87** (1) (1980), 40–42.

6.3 Fifty-three tone scale

The first continued fraction expansion of interest to us is the one for $\log_2(3/2)$. The first few terms are

$$\log_2(3/2) = \cfrac{1}{1+}\cfrac{1}{1+}\cfrac{1}{2+}\cfrac{1}{2+}\cfrac{1}{3+}\cfrac{1}{1+}\cfrac{1}{5+}\cfrac{1}{2+}\cfrac{1}{23+}\cfrac{1}{2+}\cfrac{1}{2+}\cfrac{1}{1+}\cdots$$

Figure 6.3 Bosanquet's harmonium. Reproduced with permission of Science Museum/Science and Society Picture Library.

The sequence of convergents for the continued fraction expansion of $\log_2(3/2)$ is

$$1, \frac{1}{2}, \frac{3}{5}, \frac{7}{12}, \frac{24}{41}, \frac{31}{53}, \frac{179}{306}, \frac{389}{665}, \frac{9126}{15\,601}, \ldots$$

The denominators of these fractions tell us how many equal notes to divide an octave into, and the numerators tell us how many of these notes make up one approximate fifth. The fourth of the above approximations give us our western scale. The next obvious places to stop are at 31/53 and 389/665, just before large denominators.

The fifty-three tone equally tempered scale is interesting enough to warrant some discussion. In 1876, Robert Bosanquet made a 'generalized keyboard harmonium' with fifty-three notes to an octave.[10] Figure 6.3 is a photograph of this instrument. A discussion of this harmonium can be found in the translator's Appendix XX.F.8 (pages 479–481) in Helmholtz (1877). One way of thinking of the fifty-three note scale is that it is based around the approximation which makes the Pythagorean

---

[10] Described in Bosanquet, *Musical Intervals and Temperaments*, Macmillan and Co., 1876. Reprinted with commentary by Rudolph Rasch, Diapason Press, 1986.

comma equal to one fifty-third of an octave, or $1200/53 = 22.642$ cents, rather than the true value of 23.460 cents. So if we go around a complete circle of fifths, we get from C to a note which we may call B♯ 22.642 cents higher. This corresponds to the equation

$$12 \times 31 - 7 \times 53 = 1,$$

which can be interpreted as saying that twelve 53-tone equal temperament fifths minus seven octaves equals one step in the 53-tone scale.

The following table shows the fifty-three tone equivalents of the notes on the Pythagorean scale:

note	C	B♯	D♭	C♯	D	E♭	D♯	E	F	G♭
degree	0	1	4	5	9	13	14	18	22	26

note	F♯	G	A♭	G♯	A	B♭	A♯	C♭	B	C
degree	27	31	35	36	40	44	45	48	49	53

Thus the fifty-three tone scale is made up of five whole tones each of nine scale degrees and two semitones each of four scale degrees, $5 \times 9 + 2 \times 4 = 53$. Flattening or sharpening a note changes it by five scale degrees. The perfect fifth is extremely closely approximated in this scale by the thirty-first degree, which is

$$\frac{31}{53} \times 1200 = 701.887$$

cents rather than the true value of 701.955.

The just major third is also closely approximated in this scale by the seventeenth degree, which is

$$\frac{17}{53} \times 1200 = 384.906$$

cents rather than the true value of 386.314 cents. In effect, what is happening is that we are approximating both the Pythagorean comma and the syntonic comma by a single scale degree in the 53 note scale, which is roughly half way between them. So, in Eitz's notation, we are identifying the note G♯^{0} with A♭$^{+1}$, whose difference is one schisma. Similarly, we are identifying the note B^{-1} with C♭^{0}, B♯$^{-1}$ with C^0, and so on. We are also identifying the note G^{+2} with A♭$^{-2}$, whose difference is a diesis minus four commas, or

$$\frac{256}{243} \left(\frac{80}{81}\right)^4 = \frac{2^{24}5^4}{3^{21}} = \frac{10\,485\,760\,000}{10\,460\,353\,203},$$

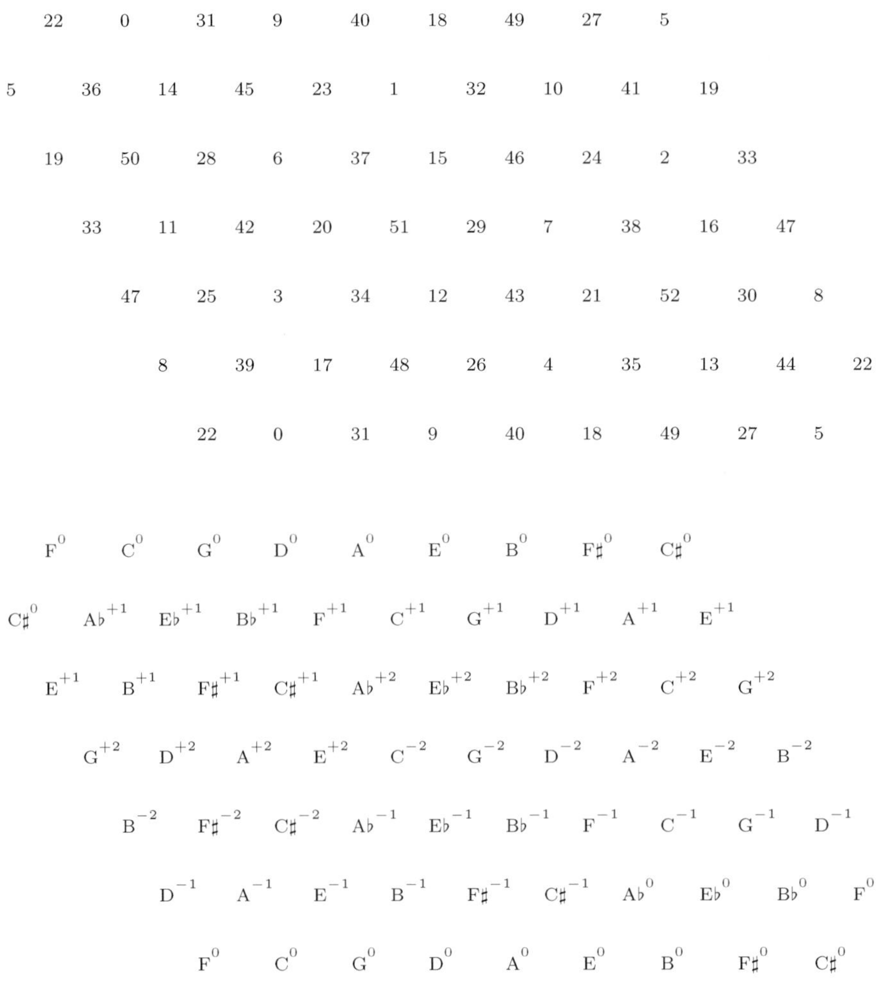

22	0	31	9	40	18	49	27	5		
5	36	14	45	23	1	32	10	41	19	
19	50	28	6	37	15	46	24	2	33	
33	11	42	20	51	29	7	38	16	47	
47	25	3	34	12	43	21	52	30	8	
8	39	17	48	26	4	35	13	44	22	
22	0	31	9	40	18	49	27	5		

$F^0 \quad C^0 \quad G^0 \quad D^0 \quad A^0 \quad E^0 \quad B^0 \quad F\sharp^0 \quad C\sharp^0$

$C\sharp^0 \quad A\flat^{+1} \quad E\flat^{+1} \quad B\flat^{+1} \quad F^{+1} \quad C^{+1} \quad G^{+1} \quad D^{+1} \quad A^{+1} \quad E^{+1}$

$E^{+1} \quad B^{+1} \quad F\sharp^{+1} \quad C\sharp^{+1} \quad A\flat^{+2} \quad E\flat^{+2} \quad B\flat^{+2} \quad F^{+2} \quad C^{+2} \quad G^{+2}$

$G^{+2} \quad D^{+2} \quad A^{+2} \quad E^{+2} \quad C^{-2} \quad G^{-2} \quad D^{-2} \quad A^{-2} \quad E^{-2} \quad B^{-2}$

$B^{-2} \quad F\sharp^{-2} \quad C\sharp^{-2} \quad A\flat^{-1} \quad E\flat^{-1} \quad B\flat^{-1} \quad F^{-1} \quad C^{-1} \quad G^{-1} \quad D^{-1}$

$D^{-1} \quad A^{-1} \quad E^{-1} \quad B^{-1} \quad F\sharp^{-1} \quad C\sharp^{-1} \quad A\flat^{0} \quad E\flat^{0} \quad B\flat^{0} \quad F^{0}$

$F^0 \quad C^0 \quad G^0 \quad D^0 \quad A^0 \quad E^0 \quad B^0 \quad F\sharp^0 \quad C\sharp^0$

Figure 6.4 Torus of thirds and fifths in 53 tone equal temperament.

or about 4.200 cents. The effect of this is that the array notation introduced in Section 5.9 becomes periodic in both directions, so that we obtain the diagram in Figure 6.4. In this diagram, the top and bottom row are identified with each other, and the left and right walls are identified with each other. The resulting geometric figure is called a torus, and it looks like a bagel, or a tyre.

It appears that the Pythagoreans were aware of the 53 tone equally tempered scale. Philolaus, a disciple of Pythagoras, thought of the tone as being divided into two minor semitones and a Pythagorean comma, and took each minor semitone to be four commas. This makes nine commas to the whole tone and four commas to the minor semitone, for a total of 53 commas to the octave. The Chinese theorist

King Fâng of the third century BCE also seems to have been aware that the 54th note in the Pythagorean system is almost identical to the first.

After 53, the next good denominator in the continued fraction expansion of $\log_2(3/2)$ is 665. The extra advantages obtained by going to an equally tempered 665 tone scale, which is gives a remarkably good approximation to the perfect fifth, are far outweighed by the fact that adjacent tones are so close together (1.805 cents) as to be almost indistinguishable. If 53 tone equal temperament is thought of as the scale of commas, then 665 tone equal temperament might be thought of as a scale of schismas.

6.4 Other equal tempered scales

Other divisions of the octave into equal intervals which have been used for experimental tunings have included 19, 24, 31 and 43. The 19 tone scale has the advantage of excellent approximations to the 6:5 minor third and the 5:3 major sixth as well as reasonable approximations to the 5:4 major third and the 8:5 minor sixth. The eleventh degree gives an approximation to the 3:2 perfect fifth which is somewhat worse than in twelve tone equal temperament, but still acceptable.

name	ratio	cents	19-tone degree	cents
fundamental	1:1	0.000	0	0.000
minor third	6:5	315.641	5	315.789
major third	5:4	386.314	6	378.947
perfect fifth	3:2	701.955	11	694.737
minor sixth	8:5	813.687	13	821.053
major sixth	5:3	884.359	14	884.211
octave	2:1	1200.000	19	1200.000

Christiaan Huygens, in the late 17th century, seems to have been the first to use the equally tempered 19 tone scale as a way of approximating just intonation in a way that allowed for modulation into other keys. Yasser[11] was an important twentieth century proponent. The properties of 19 tone equal temperament with respect to formation of a diatonic scale are very similar to those for 12 tones. But accidentals and chromatic scales behave very differently.

The main purpose I can see for the equally tempered 24 tone scale, usually referred to as the *quarter-tone scale*, is that it increases the number of tones available without throwing out the familiar twelve tones. It contains no better approximations to the ratios 3:2 and 5:4 than the twelve tone scale, but has a marginally better approximation to 7:4 and a significantly better approximation to 11:8. The two sets

---

[11] Joseph Yasser, *A Theory of Evolving Tonality*, American Library of Musicology, New York, 1932.

of twelve notes formed by taking every other note from the 24 tone scale can be alternated with interesting effect, but using notes from both sets of twelve at once has a strong tendency to make discords. Examples of works using the quarter-tone scale include the German composer Richard Stein's *Zwei Konzertstücke* Op. 26, 1906 for cello and piano and Alois Hába's Suite for String Orchestra, 1917.[12] Twentieth century American composers such as Howard Hanson and Charles Ives have composed music designed for two pianos tuned a quarter tone apart.

Appendix B contains a table of various equal tempered scales, quantifying how well they approximate perfect fifths, just major thirds, and seventh harmonics. An examination of this table reveals that the 31 tone scale is unusually good at approximating all three at once. We examine this scale in the next section.

Further reading

Jim Aikin, Discover 19-tone equal temperament, *Keyboard*, March 1988, p. 74–80.
Easley Blackwood, Modes and chords progressions in equal tunings, *Perspectives in New Music* **29** (2) (1991), 166–200.
M. Yunik and G. W. Swift, Tempered music scales for sound synthesis, *Comp. Music J.* **4** (4) (1980), 60–65.

Further listening (See Appendix G)

Between the Keys, Microtonal Masterpieces of the 20th Century. This CD contains recordings of Charles Ives' *Three Quartertone Pieces*, and a piece by Ivan Vyshnegradsky in 72 tone equal temperament.
Easley Blackwood *Microtonal Compositions*. This is a recording of a set of microtonal compositions in each of the equally tempered scales from 13 tone to 24 tone.
Clarence Barlow's 'OTOdeBLU' is in 17 tone equal temperament, played on two pianos.
Neil Haverstick, *Acoustic Stick*. Played on custum built acoustic guitars tuned in 19 and 34 tone equal temperament.
William Sethares, *Xentonality*, Music in 10, 13, 17 and 19 tone equal temperament, using spectrally adjusted instruments.

6.5 Thirty-one tone scale

The 31 tone equal tempered scale was first investigated by Nicola Vicentino[13] and also later by Christiaan Huygens.[14] It gives a better approximation to the perfect fifth than the 19 tone scale, but it is still worse than the 12 tone scale.

---

[12] It is said that Hába practiced to the point where he could accurately sing five divisions to a semitone, or sixty to an octave.

[13] Nicola Vicentino, *L'antica musica ridotta alla moderna pratica*, Rome, 1555. Translated as *Ancient Music Adapted to Modern Practice*, Yale University Press, 1996.

[14] Christiaan Huygens, *Lettre touchant le cycle harmonique*, Letter to the editor of the journal *Histoire des Ouvrage de Sçavans*, Rotterdam 1691. Reprinted with English and Dutch translation (ed. Rudolph Rasch), Diapason Press, 1986.

It also contains good approximations to the major third and minor sixth, as well as the seventh harmonic.

name	ratio	cents	31-tone degree	cents
fundamental	1:1	0.000	0	0.000
minor third	6:5	315.641	8	309.677
major third	5:4	386.314	10	387.097
perfect fifth	3:2	701.955	18	696.774
minor sixth	8:5	813.687	21	812.903
seventh harmonic	7:4	968.826	25	967.742

The main reason for interest in 31-tone equal temperament is that note 18 of this scale is an unexpectedly good approximation to the meantone fifth (696.579) rather than the perfect fifth. So the entire meantone scale can be approximated as shown in the table below. Fokker[15] was an important twentieth century proponent of the 31 tone scale.

note	meantone	31-tone	
C	0.000	0	0.000
C♯	76.049	2	77.419
D	193.157	5	193.548
E♭	310.265	8	309.677
E	386.314	10	387.097
F	503.422	13	503.226
F♯	579.471	15	580.645
G	696.579	18	696.774
A♭	813.686	21	812.903
A	889.735	23	890.323
B♭	1006.843	26	1006.452
B	1082.892	28	1083.871
C	1200.000	31	1200.000

Figure 6.5 shows a 31 tone equal tempered instrument, made by Vitus Trasuntinis in 1606. Each octave has seven keys as usual where the white keys would normally go, and five sets of four keys where the five black keys would normally go. Then there are two keys each between the white keys that would normally not be separated by black keys, for a total of $7 + 4 \times 5 + 2 \times 2 = 31$.

---

[15] See, for example, A. D. Fokker, The qualities of the equal temperament by 31 fifths of a tone in the octave, *Report of the Fifth Congress of the International Society for Musical Research, Utrecht, 3–7 July 1952*, Vereniging voor Nederlandse Muziekgeschiedenis, Amsterdam (1953), 191–192; Equal temperament with 31 notes, *Organ Institute Quarterly* **5** (1955), 41; Equal temperament and the thirty-one-keyed organ, *Scientific Monthly* **81** (1955), 161–166. Also M. Joel Mandelbaum, 31-Tone Temperament: The Dutch Legacy, *Ear Magazine East*, New York, 1982/1983; Henk Badings, A. D. Fokker: new music with 31 notes, *Zeitschrift für Musiktheorie* **7** (1976), 46–48.

Figure 6.5 Trasuntius' 31 tone clavicord (1606), State Museum, Bologna, Italy.

Let us examine the relationship between the meantone scale and 31 tone equal temperament in terms of continued fractions. Since the meantone scale is generated by the meantone fifth, which represents a ratio of $\sqrt[4]{5} : 1$, we should look at the continued fraction for $\log_2(\sqrt[4]{5})$. We obtain

$$\log_2(\sqrt[4]{5}) = \tfrac{1}{4}\log_2(5) = 0.580\,482\,024\ldots$$

$$= \frac{1}{1+}\,\frac{1}{1+}\,\frac{1}{2+}\,\frac{1}{1+}\,\frac{1}{1+}\,\frac{1}{1+}\,\frac{1}{1+}\,\frac{1}{5+}\,\frac{1}{1+}\cdots$$

with convergents

$$1,\ \frac{1}{2},\ \frac{3}{5},\ \frac{4}{7},\ \frac{7}{12},\ \frac{11}{19},\ \frac{18}{31},\ \frac{101}{174},\ \frac{119}{205},\ \ldots$$

Cutting off just before the denominator 5 gives the approximation 18/31, which gives rise to the 31 tone equal tempered scale described above.

Exercises

1. Draw a torus of thirds and fifths, analogous to the one in Figure 6.4, for the 31 tone equal tempered scale, regarded as an approximation to meantone tuning.

Figure 6.6 Wendy Carlos, photo from her website.

2. In the text, the 31 tone equal tempered scale was compared with the usual (quarter comma) meantone scale, using the observation that taking multiples of the fifth generates a meantone scale, and then applying the theory of continued fractions to approximate the fifth. Carry out the same process to make the following comparisons.
   (i) Compare the 19 tone equal tempered scale with Salinas' $\frac{1}{3}$ comma meantone scale.
   (ii) Compare the 43 tone equal tempered scale with the $\frac{1}{5}$ comma meantone scale of Verheijen and Rossi.
   (iii) Compare the 50 tone equal tempered scale with Zarlino's $\frac{2}{7}$ comma meantone scale.
   (iv) Compare the 55 tone equal tempered scale with Silbermann's $\frac{1}{6}$ comma meantone scale.
   Appendix E has a diagram which is relevant to this question.

6.6 The scales of Wendy Carlos

The idea behind the alpha, beta and gamma scales of Wendy Carlos (Figure 6.6) is to ignore the requirement that there are a whole number of notes to an octave, and try to find equal tempered scales which give good approximations to the just intervals 3:2 and 5:4 (perfect fifth and major third). Since $6/5 = 3/2 \div 5/4$, this automatically gives good approximations to the 6:5 minor third. This means that we need $\log_2(3/2)$ and $\log_2(5/4)$ to be close to integer multiples of the scale degree. So we must find rational approximations to the ratio of these quantities.

We investigate the continued fraction expansion of the ratio:

$$\frac{\log_2(3/2)}{\log_2(5/4)} = \frac{\ln(3/2)}{\ln(5/4)} = 1 + \frac{1}{1+}\frac{1}{4+}\frac{1}{2+}\frac{1}{6+}\frac{1}{1+}\frac{1}{10+}\frac{1}{135+}\cdots$$

The sequence of convergents obtained by truncating this continued fraction is:

$$1, \ 2, \ \frac{9}{5}, \ \frac{20}{11}, \ \frac{129}{71}, \ \frac{149}{82}, \ \ldots$$

Carlos' α (alpha) scale arises from the approximation 9/5 for the above ratio. This means taking a value for the scale degree so that nine of them approximate a 3:2 perfect fifth, five of them approximate a 5:4 major third, and four of them approximate a 6:5 minor third. In order to make the approximation as good as possible we minimize the mean square deviation. So if x denotes the scale degree (taking the octave as unit) then we must minimize

$$(9x - \log_2(3/2))^2 + (5x - \log_2(5/4))^2 + (4x - \log_2(6/5))^2.$$

Setting the derivative with respect to x of this quantity equal to zero, we obtain the equation

$$x = \frac{9\log_2(3/2) + 5\log_2(5/4) + 4\log_2(6/5)}{9^2 + 5^2 + 4^2} \approx 0.064\,970\,824\,62.$$

Multiplying by 1200, we obtain a scale degree of 77.965 cents, and there are 15.3915 of them to the octave.[16]

Carlos also considers the scale α' obtained by doubling the number of notes in the octave. This gives the same approximations as before for the ratios 3:2, 5:4 and 6:5, but the 25th degree of the new scale (974.562 cents) is a good approximation to the seventh harmonic in the form of the ratio 7:4 (968.826 cents).

If instead we use the approximation

$$1 + \frac{1}{1+}\frac{1}{5} = \frac{11}{6},$$

which we get by rounding up at the end instead of down, we obtain Carlos' β (beta) scale. We choose a value of the scale degree so that eleven of them approximate a 3:2 perfect fifth, six of them approximate a 5:4 major third, and five of them approximate a 6:5 minor third. Proceeding as before, we see that the proportion of an octave occupied by each scale degree is

$$\frac{11\log_2(3/2) + 6\log_2(5/4) + 5\log_2(6/5)}{11^2 + 6^2 + 5^2} \approx 0.053\,194\,110\,48.$$

---

[16] This actually differs very slightly from Carlos' figure of 15.385 α-scale degrees to the octave. This is obtained by approximating the scale degree to 78.0 cents.

Multiplying by 1200, we obtain a scale degree of 63.833 cents, and there are 18.7991 of them to the octave.[17] One advantage of the beta scale over the alpha scale is that the 15th scale degree (957.494 cents) is a reasonable approximation to the seventh harmonic in the form of the ratio 7:4 (968.826 cents). Indeed, it may be preferable to include this approximation into the above least squares calculation to get a scale in which the proportion of an octave occupied by each scale degree is

$$\frac{15 \log_2(7/4) + 11 \log_2(3/2) + 6 \log_2(5/4) + 5 \log_2(6/5)}{15^2 + 11^2 + 6^2 + 5^2} \approx 0.053\,542\,142\,35.$$

This gives a scale degree of 64.251 cents, and there are 18.677 of them to the octave. The fifteenth scale degree is then 963.759 cents.

Going one stage further, and using the approximation 20/11, we obtain Carlos' γ (gamma) scale. We choose a value of the scale degree so that twenty of them approximate a 3:2 perfect fifth, nine of them approximate a 5:4 major third, and eleven of them approximate a 4:3 minor third. The proportion of an octave occupied by each scale degree is

$$\frac{20 \log_2(3/2) + 11 \log_2(5/4) + 9 \log_2(6/5)}{20^2 + 11^2 + 9^2} \approx 0.029\,248\,785\,23.$$

Multiplying by 1200, we obtain a scale degree of 35.099 cents, and there are 34.1895 of them to the octave.[18] This scale contains almost pure perfect fifths and major thirds, but it does not contain a good approximation to the ratio 7:4.

name	ratio	cents	α	cents	β	cents	γ	cents
fundamental	1:1	0.000	0	0.000	0	0.000	0	0.000
minor third	6:5	315.641	4	311.860	5	319.165	9	315.887
major third	5:4	386.314	5	389.825	6	382.998	11	386.084
perfect fifth	3:2	701.955	9	701.685	11	702.162	20	701.971
seventh harmonic	7:4	968.826	$12\frac{1}{2}$	974.562	15	957.494	28	982.759

6.7 The Bohlen–Pierce scale

Jaja, unlike Stravinsky, has never been guilty of composing harmony in all his life. Jaja is pure absolute twelve tone. Never tempted, like some of the French composers, to write with thirteen tones. Oh no. This, says Jaja, is the baker's dozen, the "Nadir of Boulanger."

(From Gerard Hoffnung's Interplanetary Music Festival,
analysis by two "distinguished teutonic musicologists" of the work of a
fictitious twelve-tone composer, Bruno Heinz Jaja. Text by John Amis.)

---

[17] Carlos has 18.809 β-scale degrees to the octave, corresponding to a scale degree of 63.8 cents.
[18] Carlos has 34.188 γ-scale degrees to the octave, corresponding to a scale degree of 35.1 cents.

The Bohlen–Pierce scale is the thirteen tone scale described in the article of Mathews and Pierce (1989), Chapter 13. Like the scales of Wendy Carlos, it is not based around the octave as the basic interval. But whereas Carlos uses 3:2 and 5:4, Bohlen and Pierce replace the octave by an octave and a perfect fifth (a ratio of 3:1). In the equal tempered version, this is divided into thirteen equal parts. This gives a good approximation to a 'major' chord with ratios 3:5:7. The idea is that only odd multiples of frequencies are used. Music written using this scale works best if played on an instrument such as the clarinet, which involves predominantly odd harmonics, or using specially created synthetic voices with the same property. We shall prefix all words associated with the Bohlen–Pierce scale with the letters BP to save confusion with the corresponding notions based around the octave.

The basic interval of an octave and a perfect fifth, which is a ratio of exactly 3:1 or an interval of 1901.955 cents, is called a *BP-tritave*. In the equal tempered 13 tone scale, each scale degree is one thirteenth of this, or 146.304 cents. It may be felt that the scale of cents is inappropriate for calculations with reference to this scale, but we shall stick with it nonetheless for comparison with intervals in scales based around the octave.

The Pythagorean approach to the division of the tritave begins with a ratio of 7:3 as the analogue of the fifth. We shall call this interval the perfect BP-tenth, since it will correspond to note ten in the BP-scale. The corresponding continued fraction is

$$\log_3(7/3) = \cfrac{1}{1+}\cfrac{1}{3+}\cfrac{1}{2+}\cfrac{1}{1+}\cfrac{1}{2+}\cfrac{1}{4+}\cfrac{1}{22+}\cfrac{1}{32+}\cdots,$$

whose convergents are

$$\frac{0}{1}, \frac{1}{1}, \frac{3}{4}, \frac{7}{9}, \frac{10}{13}, \frac{27}{35}, \frac{118}{153}, \cdots$$

If we perform the same calculation for the 5:3 ratio, we obtain the continued fraction

$$\log_3(5/3) = \cfrac{1}{2+}\cfrac{1}{6+}\cfrac{1}{1+}\cfrac{1}{1+}\cfrac{1}{1+}\cfrac{1}{3+}\cfrac{1}{7+}\cdots$$

with convergents

$$\frac{0}{1}, \frac{1}{2}, \frac{6}{13}, \frac{7}{15}, \frac{13}{28}, \frac{20}{43}, \frac{73}{157}, \cdots$$

Comparing these continued fractions, it looks like a good idea to divide the tritave into 13 equal intervals, with note 10 approximating the ratio 7:3, and note 6 approximating the ratio 5:3.

note	degree	7/3-Pythag.	just
C	0	1:1	1:1
D	2	19683:16807	25:21
E	3	9:7	9:7
F	4	343:243	7:5
G	6	81:49	5:3
H	7	49:27	9:5
J	9	729:343	15:7
A	10	7:3	7:3
B	12	6561:2401	25:9
C	13	3:1	3:1

Basing a BP-Pythagorean scale around the ratio 7:3, we obtain a scale of 13 notes in which the circle of BP-tenths has a BP 7/3-comma given by a ratio of

$$\frac{7^{13}}{3^{23}} = \frac{96\,889\,010\,407}{94\,143\,178\,827}$$

or about 49.772 cents.

Using perfect BP-tenths to form a diatonic BP-Pythagorean scale, we obtain the third column of the table above. Following Bohlen, we name the notes of the scale using the letters A–H and J. Note that our choice of the second degree of the diatonic scale differs from the choice made by Mathews and Pierce, and gives what Bohlen calls the Lambda scale.

To obtain a major 3:5:7 triad, we introduce a just major BP-sixth with a ratio of 5:3. This is very close to the BP 7/3-Pythagorean G, which gives rise to an interval called the BP-minor diesis, expressing the difference between these two versions of G. This interval, namely the difference between 5:3 and 81:49, is a ratio of 245:243 or about 14.191 cents.

The BP version of Eitz's notation works in a similar way to the octave version. We start with the BP 7/3-Pythagorean values for the notes and then adjust by a number of BP-minor dieses indicated by a superscript. So G^0 denotes the 81:49 version of G, while G^{+1} denotes the 5:3 version. The just scale given in the table above is then described by the following array:

$$\begin{array}{ccccc} & D^{+2} & & B^{+2} & \\ & J^{+1} & & G^{+1} & \\ E^0 & & C^0 & & A^0 \\ & H^{-1} & & F^{-1} & \end{array}$$

A reasonable way to fill this in to a thirteen tone just scale is as follows:

in our scale. Since $B\sharp^0$ is the twelfth note, we say that (12) is a unison vector. A periodicity block would then consist of a choice of 12 consecutive points on this lattice, to constitute a scale. Other choices of unison vector would include (53) and (665) (cf. Section 6.3).

Just intonation, at least the 5-limit version as we introduced it in Section 5.5, is really a 2-dimensional lattice, which we write as $\mathbb{Z}^2$. In Eitz's notation (see Section 5.9), here is a small part of the lattice with the origin circled.

$$F\sharp^{-2} \quad C\sharp^{-2} \quad G\sharp^{-2} \quad D\sharp^{-2}$$
$$D^{-1} \quad A^{-1} \quad E^{-1} \quad B^{-1} \quad F\sharp^{-1}$$
$$B\flat^0 \quad F^0 \quad \textcircled{C^0} \quad G^0 \quad D^0 \quad A^0$$
$$D\flat^{+1} \quad A\flat^{+1} \quad E\flat^{+1} \quad B\flat^{+1} \quad F^{+1} \quad C^{+1}$$
$$F\flat^{+2} \quad C\flat^{+2} \quad G\flat^{+2} \quad D\flat^{+2} \quad A\flat^{+2}$$

The same in ratio notation is as follows.

$$\frac{25}{18} \quad \frac{25}{24} \quad \frac{25}{16} \quad \frac{75}{64}$$
$$\frac{10}{9} \quad \frac{5}{3} \quad \frac{5}{4} \quad \frac{15}{8} \quad \frac{45}{32}$$
$$\frac{16}{9} \quad \frac{4}{3} \quad \textcircled{$\frac{1}{1}$} \quad \frac{3}{2} \quad \frac{9}{8} \quad \frac{27}{16}$$
$$\frac{16}{15} \quad \frac{8}{5} \quad \frac{6}{5} \quad \frac{9}{5} \quad \frac{27}{20} \quad \frac{81}{80}$$
$$\frac{32}{25} \quad \frac{48}{25} \quad \frac{36}{25} \quad \frac{27}{25} \quad \frac{81}{50}$$

We can choose a basis for this lattice, and write everything in terms of vectors with respect to this basis. This is the two dimensional version of our choice from two different ways of indexing the Pythagorean scale by the integers, but this time there are an infinite number of choices of basis.

For example, if our basis consists of G^0 and E^{-1} (i.e., $\frac{3}{2}$ and $\frac{5}{4}$) then here is the same part of the lattice in vector notation.

$$(-2, 2) \quad (-1, 2) \quad (0, 2) \quad (1, 2)$$
$$(-2, 1) \quad (-1, 1) \quad (0, 1) \quad (1, 1) \quad (2, 1)$$
$$(-2, 0) \quad (-1, 0) \quad (0, 0) \quad (1, 0) \quad (2, 0) \quad (3, 0)$$
$$(-1, -1) \quad (0, -1) \quad (1, -1) \quad (2, -1) \quad (3, -1) \quad (4, -1)$$
$$(0, -2) \quad (1, -2) \quad (2, -2) \quad (3, -2) \quad (4, -2)$$

The defining property of a basis is that every vector in the lattice has a unique expression as an integer combination of the basis vectors. The number of vectors in a basis is the dimension of the lattice.

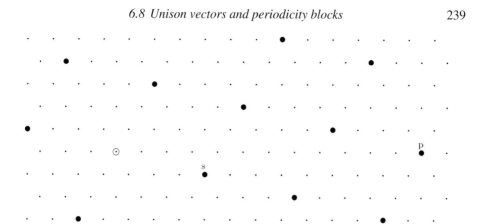

Figure 6.8 The unison sublattice.

Now we need to choose our unison vectors. The classical choice here is $(4, -1)$ and $(12, 0)$, corresponding to the syntonic comma and the Pythagorean comma. The sublattice *generated* by these unison vectors consists of all linear combinations

$$m(4, -1) + n(12, 0) = (4m + 12n, -4m)$$

with $m, n \in \mathbb{Z}$. This is called the *unison sublattice*. See Figure 6.8.

In this diagram, the syntonic comma and Pythagorean comma are marked with s and p respectively. Each vector (a, b) in the lattice may then be thought of as *equivalent* to the vectors

$$(a, b) + m(4, -1) + n(12, 0) = (a + 4m + 12n, b - 4m),$$

with $m, n \in \mathbb{Z}$, differing from it by vectors in the unison sublattice. So, for example, taking $m = -3$ and $n = 1$, we see that the vector $(0, 3)$ is in the unison sublattice. This corresponds to the fact that three just major thirds approximately make one octave.

There are many ways of choosing unison vectors generating a given sublattice. In the above example, $(4, -1)$ and $(0, 3)$ generate the same sublattice.

The set of vectors (or pitches) equivalent to a given vector is called a *coset*. The number of cosets is called the *index* of the unison sublattice in the lattice. It can be calculated by taking the *determinant* of the matrix formed from the unison vectors. So, in our example, the index of the unison sublattice is

$$\begin{vmatrix} 4 & -1 \\ 12 & 0 \end{vmatrix} = 12.$$

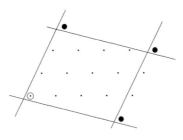

Figure 6.9.

The formula for the determinant of a 2×2 matrix is

$$\begin{vmatrix} a & b \\ c & d \end{vmatrix} = ad - bc.$$

If the determinant comes out negative, the index is the corresponding positive quantity. If two rows of a matrix are swapped, then the determinant changes sign, so the sign of the determinant is irrelevant to the index. It has to do with *orientation*, and will not be discussed here.

A *periodicity block* consists of a choice of one vector from each coset. In other words, we find a finite set of vectors with the property that each vector in the whole lattice is equivalent to a unique vector from the periodicity block. One way to do this is to draw a parallelogram using the unison vectors. We can then tile the plane using copies of this parallelogram, translated along unison vectors. In the above example, if we use the unison vectors $(4, -1)$ and $(0, 3)$ to generate the unison sublattice, then the parallelogram looks as in Figure 6.9.

This choice of periodicity block leads to the following just scale with twelve tones.

$$G\sharp^{-2} \quad D\sharp^{-2} \quad A\sharp^{-2} \quad E\sharp^{-2}$$
$$E^{-1} \quad B^{-1} \quad F\sharp^{-1} \quad C\sharp^{-1}$$
$$C^{0} \quad G^{0} \quad D^{0} \quad A^{0}$$

Of course, there are many other choices of periodicity block. For example, shifting this parallelogram one place to the left gives rise to Euler's monochord, described on page 178 (See Figure 6.10).

Periodicity blocks do not have to be parallelograms. For example, we can chop off a corner of the parallelogram, translate it through a unison vector, and stick it back on somewhere else to get a hexagon. Each of the just intonation scales in Section 5.10 may be interpreted as a periodicity block for the above choice of unison sublattice.

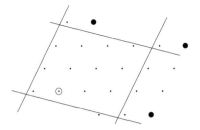

Figure 6.10 Euler's monochord.

Of course, there are other choices of unison sublattices. If we choose the unison vectors $(4, -1)$ and $(-1, 5)$, for example, then we get a scale of

$$\begin{vmatrix} 4 & -1 \\ -1 & 5 \end{vmatrix} = 19$$

tones. This gives rise to just scales approximating the equal tempered scale described at the beginning of Section 6.4. The choice of $(4, 2)$ and $(-1, 5)$ gives the Indian scale of 22 Srutis described in Section 6.1, corresponding to the calculation

$$\begin{vmatrix} 4 & 2 \\ -1 & 5 \end{vmatrix} = 22.$$

Taking the unison vectors $(4, -1)$ and $(3, 7)$ gives rise to 31 tone scales approximating 31 tone equal temperament, whose relationship with meantone is described in Section 6.4. This corresponds to the calculation

$$\begin{vmatrix} 4 & -1 \\ 3 & 7 \end{vmatrix} = 31.$$

The vectors $(8, 1)$ and $(-5, 6)$ correspond in the same way to just scales approximating the 53 tone equal tempered scale described in Section 6.3, corresponding to the calculation

$$\begin{vmatrix} 8 & 1 \\ -5 & 6 \end{vmatrix} = 53.$$

An example of a periodicity block for this choice of unison vectors can be found in Figure 6.4.

When we come to study groups and normal subgroups in Section 9.12, we shall make some more comments on how to interpret unison vectors and periodicity blocks in group theoretical language.

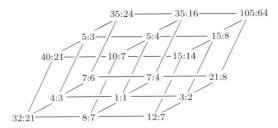

Figure 6.11 Septimal version of just intonation.

Further reading

Some of the material in this and the next section expresses ideas from an online article by
Paul Erlich (do a web search), together with the work of Fokker.

A. D. Fokker, Selections from the harmonic lattice of perfect fifths and major thirds
containing 12, 19, 22, 31, 41 or 53 notes, *Proc. Koninkl. Nederl. Akad.
Wetenschappen*, Series B, **71** (1968), 251–266.

6.9 Septimal harmony

Septimal harmony refers to 7-limit just intonation; in other words, just intonation
involving the primes 2, 3, 5 and 7. Taking octave equivalence into account, this
means that we need three dimensions, or $\mathbb{Z}^3$, to represent the septimal version of
just intonation, to take account of the primes 3, 5 and 7. It is harder to draw a three
dimensional lattice, but it can be done. In ratio notation, it will then look as in
Figure 6.11.

We take as our basis vectors the ratios $\frac{3}{2}, \frac{5}{4}$ and $\frac{7}{4}$. So the vector (a, b, c) represents
the ratio $3^a.5^b.7^c$, multiplied if necessary by a power of 2 so that it is between
1 and 2 (octave equivalence). The septimal comma introduced in Section 5.8 is
a ratio of 64 : 63, which corresponds to the vector $(-2, 0, -1)$. So it would be
reasonable to use the three commas $(4, -1, 0)$ (syntonic), $(12, 0, 0)$ (Pythagorean),
and $(-2, 0, -1)$ (septimal) as unison vectors.

The determinant of a 3×3 matrix

$$\begin{pmatrix} a & b & c \\ d & e & f \\ g & h & i \end{pmatrix}$$

is given by the formula

$$\begin{vmatrix} a & b & c \\ d & e & f \\ g & h & i \end{vmatrix} = aei + bfg + cdh - ceg - bdi - afh.$$

vector			ratio	cents	vector			ratio	cents
(−7	4	1)	$\frac{4375}{4374}$	0.40	(−2	0	−1)	$\frac{64}{63}$	27.26
(−1	−2	4)	$\frac{2401}{2400}$	0.72	(−3	−2	3)	$\frac{686}{675}$	27.99
(−8	2	5)	$\frac{420175}{419904}$	1.12	(−1	5	0)	$\frac{3125}{3072}$	29.61
(9	3	−4)	$\frac{2460375}{2458624}$	1.23	(−2	3	4)	$\frac{303125}{294912}$	30.33
(8	1	0)	$\frac{32805}{32764}$	1.95	(−1	−3	−3)	$\frac{131072}{128625}$	32.63
(1	5	1)	$\frac{65625}{65536}$	2.35	(−8	1	−2)	$\frac{327680}{321989}$	33.02
(0	3	5)	$\frac{2100875}{2097152}$	3.07	(−9	−1	2)	$\frac{100352}{98415}$	33.74
(−8	−6	2)	$\frac{102760448}{102515625}$	4.13	(0	2	−2)	$\frac{50}{49}$	34.98
(1	−3	−2)	$\frac{6144}{6125}$	5.36	(−1	0	2)	$\frac{49}{48}$	35.70
(0	−5	2)	$\frac{3136}{3125}$	6.08	(1	7	−1)	$\frac{234375}{229376}$	37.33
(−7	−1	3)	$\frac{10976}{10935}$	6.48	(7	1	2)	$\frac{535815}{524288}$	37.65
(2	2	−1)	$\frac{225}{224}$	7.71	(0	5	3)	$\frac{1071875}{1048576}$	38.05
(8	−4	2)	$\frac{321489}{320000}$	8.04	(1	−1	−4)	$\frac{12278}{12005}$	40.33
(−5	6	0)	$\frac{15625}{15552}$	8.11	(0	−3	0)	$\frac{128}{125}$	41.06
(1	0	3)	$\frac{1029}{1024}$	8.43	(−7	1	1)	$\frac{2240}{2187}$	41.45
(3	7	0)	$\frac{2109375}{2083725}$	10.06	(2	4	−1)	$\frac{5625}{5488}$	42.69
(−5	−2	−3)	$\frac{2097152}{2083725}$	11.12	(1	2	1)	$\frac{525}{512}$	43.41
(3	−1	−3)	$\frac{1728}{1715}$	13.07	(0	0	5)	$\frac{16807}{16384}$	44.13
(−4	3	−2)	$\frac{4000}{3969}$	13.47	(1	−6	−2)	$\frac{786432}{765625}$	46.42
(2	−3	1)	$\frac{126}{125}$	13.79	(−6	−2	−1)	$\frac{131072}{127575}$	46.81
(−5	1	2)	$\frac{245}{243}$	14.19	(2	−1	−1)	$\frac{36}{35}$	48.77
(10	−2	1)	$\frac{413343}{409600}$	15.75	(−6	1	4)	$\frac{12005}{11664}$	49.89
(3	2	2)	$\frac{33075}{32768}$	16.14	(2	2	4)	$\frac{540225}{524288}$	51.84
(−3	0	−4)	$\frac{65536}{64827}$	18.81	(2	−6	1)	$\frac{16128}{15625}$	54.85
(3	−6	−1)	$\frac{110592}{109375}$	19.16	(−5	−2	2)	$\frac{6272}{6075}$	55.25
(−4	−2	0)	$\frac{2048}{2025}$	19.55	(4	1	−2)	$\frac{405}{392}$	56.48
(5	1	−4)	$\frac{2430}{2401}$	20.79	(3	−1	2)	$\frac{1323}{1280}$	57.20
(4	−1	0)	$\frac{81}{80}$	21.51	(−4	3	3)	$\frac{42875}{42472}$	57.60
(−3	3	1)	$\frac{875}{864}$	21.90	(4	−4	0)	$\frac{648}{625}$	62.57
(12	0	0)	$\frac{531441}{524288}$	23.46	(−3	0	1)	$\frac{28}{27}$	62.96
(5	4	1)	$\frac{1063125}{1048576}$	23.86	(−1	2	0)	$\frac{25}{24}$	70.72
(4	2	5)	$\frac{34034175}{33554432}$	24.58	(1	−1	1)	$\frac{21}{20}$	84.42
(−3	−5	−2)	$\frac{4194302}{4134375}$	24.91	(3	1	0)	$\frac{135}{128}$	92.23
(−10	−1	−1)	$\frac{2097152}{2066715}$	25.31	(−3	3	1)	$\frac{3584}{3575}$	104.02
(5	−4	−2)	$\frac{31104}{30625}$	26.87	(−1	4	−2)	$\frac{625}{588}$	105.65

Figure 6.12 Unison vectors in 7-limit just intonation.

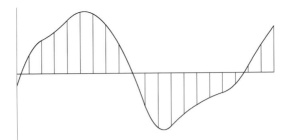

Figure 7.1 Analogue signal.

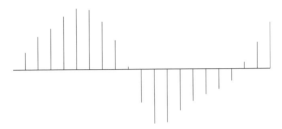

Figure 7.2 Sampled signal.

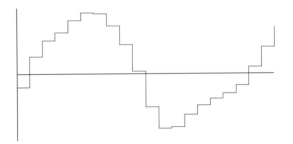

Figure 7.3 Sampled and held.

signal. We shall do this by systematically making use of Dirac delta functions, starting in Section 7.5.

Finally, if we digitize the samples, each sample value gets adjusted to the nearest allowed value. See Figure 7.4.

This part of the process of turning an analogue signal into a digital signal does entail some loss of information, even if the original signal contains no frequency component above half the sample rate. To see this, think what happens to a very low level signal. It will simply get reported as zero. There is a method for overcoming this limitation to a certain extent, called *dithering*, and it is described in Section 7.2.

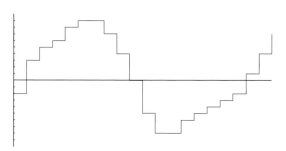

Figure 7.4 Digitized signal.

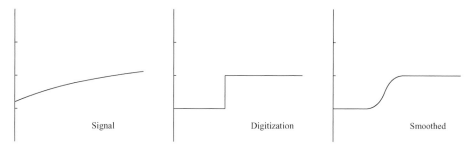

Signal   Digitization   Smoothed

Figure 7.5 Signal plus digitization.

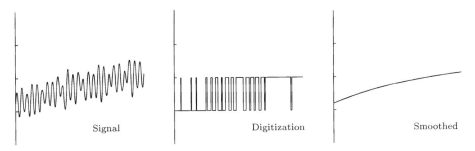

Signal   Digitization   Smoothed

Figure 7.6 Noise added.

Further reading

Ken C. Pohlmann, *The Compact Disc Handbook*, 2nd edition, A-R Editions, Inc., 1992.

7.2 Dithering

Dithering is a method of decreasing the distortion of a low level signal due to digitization of signal level. This is based on the audacious proposition that adding a low level source of random noise to a signal can increase the signal resolution. This works best when the sample rate is high in comparison with the rate at which the signal is changing.

To see how this works, consider a slowly varying signal and its digitization, as shown in Figure 7.5.

Now if we add noise to the original signal at amplitude roughly one half the step size in the digitization process, Figure 7.6 shows what the outcome looks like.

If the digitized signal is put through a resistor–capacitor circuit to smooth it out, some reasonable approximation to the original signal can be recovered. There is no theoretical limit to the accuracy possible with this method, as long as the sampling rate is high enough.

Further reading

J. Vanderkooy and S. Lipshitz, Resolution below the least significant bit in digital audio systems with dither, *J. Audio Eng. Soc.* **32** (3) (1984), 106–113; Correction, *J. Audio Eng. Soc.* **32** (11) (1984), 889.

7.3 WAV and MP3 files

A common format for digital sound files on a computer is the WAV format. This is an example of Resource Interchange File Format (RIFF) for multimedia files; another example of RIFF is the AVI movie format. Here is what a WAV file looks like. The file begins with some header information, which comes in a 12 byte RIFF chunk and a 24 byte FORMAT chunk, and then the actual wave data, which comes in a DATA chunk occupying the rest of the file.

The binary numbers in a WAV file are always *little endian*, which means that the least significant byte comes first, so that the bytes are in the reverse of what might be thought of as the normal order. We shall represent binary numbers using *hexadecimal* format, or base 16. Each hexadecimal digit encodes four binary digits, so that there are two hexadecimal digits in a byte. The sixteen symbols used are 0–9 and A–F. So, for example, in little endian format, 4E 02 00 00 would represent the hexadecimal number 24E, which is the binary number 0010 0100 1110, or in decimal, 590.

The first 12 bytes are called the RIFF chunk. Bytes 0–3 are 52 49 46 46, the ascii characters 'RIFF'. Bytes 4–7 give the total number of bytes in the remaining part of the entire WAV file (byte 8 onward), in little endian format as described above. Bytes 8–11 are 57 41 56 45, the ascii characters 'WAVE' to indicate the RIFF file type.

The next 24 bytes are called the FORMAT chunk. Bytes 0–3 are 66 6D 74 20, the ascii characters 'fmt'. Bytes 4–7 give the length of the remainder of the FORMAT chunk, which for a WAV file will always be 10 00 00 00 to indicate 16 bytes. Bytes 8–9 are always 01 00, don't ask me why. Bytes 10–11 indicate the number of channels, 01 00 for mono and 02 00 for stereo. Bytes 12–15 give the sample rate, which is measured in Hz. So, for example 44 100 Hz comes out as 44 AC 00 00. Bytes 16–19 give the number of bytes per second. This can be found by multiplying the sample rate with the number of bytes representing each sample. Bytes 20–21 give the number of bytes per sample, so 01 00 for 8-bit mono, 02 00 for 8-bit stereo or 16-bit mono, and 04 00 for 16-bit stereo. Finally, bytes 22–23 give the number of bits per sample, which is eight times as big as bytes 20–21.

For 16-bit CD quality stereo audio, the number of bytes per second is $44\,100 \times 2 \times 2 = 176\,400$, which in little endian hexadecimal is 10 B1 02 00. So the RIFF and FORMAT chunk would be as follows.

```
52 49 46 46 xx xx xx xx-57 41 56 45 66 6D 74 20
10 00 00 00 01 00 02 00-44 AC 00 00 10 B1 02 00
04 00 20 00
```

Here, xx xx xx xx represents the total length of the WAV file after the first eight bytes.

Finally, for the DATA chunk, bytes 0–3 are 64 61 74 61 for ascii 'data'. Bytes 4–7 give the length of the remainder of the DATA chunk, in bytes. Bytes 8 onwards are the actual data samples, in little endian binary as always. The data come in pieces called *sample frames*, each representing the data to be played at a particular point in time. So, for example, for a 16-bit stereo signal, each sample frame would consist of two bytes for the left channel followed by two bytes for the right channel. Since both positive and negative numbers are to be encoded in the binary data, the format used is *two's complement*. So positive numbers from 0 to 32 767 begin with a binary digit zero, and negative numbers from $-32\,768$ to -1 are represented by adding 65 536, so that they begin with a binary digit one. For example, the number -1 is represented by the little endian hexadecimal bytes FF FF, $-32\,768$ is represented by 00 80, and 32 767 is represented by FF 7F.

Unfortunately, two's complement is only used when the samples are more than 8 bits long; 8-bit samples are represented using the numbers from 0 to 255, with no negative numbers. So 128 is the neutral position for the signal.

Other digital audio formats similar in nature to the WAV file include the AIFF format (Audio Interchange File Format), commonly used on Macintosh computers, and the AU format, developed by Sun Microsystems and commonly used on UNIX computers.

The MP3 format[2] is different from the WAV format in that it uses *data compression*. The file needs to be uncompressed as it is played.

---

[2] MP3 stands for 'MPEG I/II Layer 3.' MPEG is itself an acronym for 'Motion Picture Experts Group', which is a family of standards for encoding audio-visual information such as movies and music.

There are two kinds of compression: lossless and lossy. Lossless compression gives rise to a shorter file from which the original may be reconstructed exactly. For example, the ZIP file format is a lossless compression format. This can only work with non-random data. The more random the data, the less it can be compressed. For example, if the data to be compressed consists of 10 000 consecutive copies of the binary string 01001000, then that information can be imparted in a lot less than 10 000 bytes. In information theory, this is captured by the concept of *entropy*. The entropy of a signal is defined to be the logarithm to base two of the number of different possibilities for the signal. The less random the signal, the fewer possibilities are allowed for the data in the signal, and hence the smaller the entropy. The entropy measures the smallest number of binary bits the signal could be compressed into.

Lossy compression retains the most essential features of the file, and allows some degeneration of the data. The kind of degeneration allowed must always depend on the context. For an audio file, for example, we can try to decide which aspects of the signal make little difference to the perception of the sound, and allow these aspects to change. This is precisely what the MP3 format does.

The actual algorithm is very complicated, and makes use of some subtle psychoacoustics. Here are some of the techniques used for encoding MP3 formats.

(i) The threshold of hearing depends on frequency, and the ear is most sensitive in the middle of the audio frequency range. This is described using the Fletcher–Munson curves, as explained on page 15. So low amplitude sounds at the extremes of the frequency range can be ignored unless there is no other sound present.

(ii) The phenomenon of masking means that some sounds will be present but will not be perceived because of the existence of some other component of the sound. These masked sounds are omitted from the compressed signal.

(iii) A system of borrowing is used, so that a passage which needs more bytes to represent the sound in a perceptually accurate way can use them at the expense of using fewer bytes to represent perceptually simpler passages.

(iv) A stereo signal often does not contain much more information than each channel alone, and joint stereo encoding makes use of this.

(v) The MP3 format also makes use of Huffman coding, in which strings of information which occur with higher probability are coded using a shorter string of bits.

Newer formats take the same idea further. For example, Apple's iTunes now uses MPEG 4 Audio as its standard, also known as AAC (Advanced Audio Coding). This produces better audio quality than MP3 with smaller file sizes. Files produced using this standard have names of the form <filename>.m4a.

7.4 MIDI

Most synthesizers these days talk to each other and to computers via MIDI cables. MIDI stands for 'Musical Instrument Digital Interface'. It is an internationally agreed data transmission protocol, introduced in 1982, for the transmission of musical information between different digital devices. It is important to realise that in general there is no waveform information present in MIDI, unless the message is a 'sample dump'. Instead, most MIDI messages give a short list of abstract parameters for an event.

There are three basic types of MIDI message: note messages, controller messages, and system exclusive messages. Note messages carry information about the starting time and stopping time of notes, which patch (or voice) should be used, the volume level, and so on. Controller messages change parameters like chorus, reverb, panning, master volume, etc. System exclusive messages are for transmitting information specific to a given instrument or device. They start with an identifier for the device, and the body can contain any kind of information in a format proprietary to that device. The commonest kind of system exclusive messages are for transmitting the data for setting up a patch on a synthesizer.

The MIDI standard also includes some hardware specifications. It specifies a baud rate of 31.25 KBaud. For modern machines this is very slow, but for the moment we are stuck with this standard. One of the results of this is that systems often suffer from MIDI 'bottlenecks', which can cause audibly bad timing. The problem is especially bad with MIDI data involving continually changing values of a control variable such as volume or pitch.

Further reading

S. de Furia and J. Scacciaferro (1989), *MIDI Programmer's Handbook*.

F. Richard Moore, The dysfunctions of MIDI, *Computer Music Journal* **12** (1) (1988), 19–28.

Joseph Rothstein (1992), *MIDI, a Comprehensive Introduction*.

Eleanor Selfridge-Field (editor), Donald Byrd (contributor), David Bainbridge (contributor), *Beyond MIDI: The Handbook of Musical Codes*, M. I. T. Press (1997).

7.5 Delta functions and sampling

One way to represent the process of sampling a signal is as multiplication by a stream of Dirac delta functions (see Section 2.17). Let N denote the sample rate, measured in samples per second, and let $\Delta t = 1/N$ denote the interval between sample times. So, for example, for compact disc recording we want $N = 44\,100$ samples per second, and $\Delta t = 1/44\,100$ seconds. We define the *sampling function*

$\delta_s(t)$

$\leftarrow\!\Delta t\!\rightarrow$

Figure 7.7 Sampling function.

(See Figure 7.7) with spacing Δt to be

$$\delta_s(t) = \sum_{n=-\infty}^{\infty} \delta(t - n\Delta t). \tag{7.5.1}$$

If $f(t)$ represents an analogue signal, then

$$f(t)\delta_s(t) = \sum_{n=-\infty}^{\infty} f(t)\delta(t - n\Delta t) = \sum_{n=-\infty}^{\infty} f(n\Delta t)\delta(t - n\Delta t)$$

represents the sampled signal. This has been digitized with respect to time, but not with respect to signal amplitude. The integral of the digitized signal $f(t)\delta_s(t)$ over any period of time approximates the integral of the analogue signal $f(t)$ over the same time interval, multiplied by the sample rate $N = 1/\Delta t$.

We give two different expressions for the Fourier transform of a sampled signal in Theorem 7.5.1 and Corollary 7.5.4. Both of these expressions show that the Fourier transform is periodic, with period equal to the sample rate $N = 1/\Delta t$.

Theorem 7.5.1 *The Fourier transform of a sampled signal is given by*

$$\widehat{f.\delta_s}(\nu) = \sum_{n=-\infty}^{\infty} f(n\Delta t)e^{-2\pi i\nu n\Delta t}.$$

Proof Using the definition (2.13.1) of the Fourier transform, we have

$$\widehat{f.\delta_s}(\nu) = \int_{-\infty}^{\infty} f(t)\delta_s(t)e^{-2\pi i\nu t}\,dt$$

$$= \int_{-\infty}^{\infty} \left(\sum_{n=-\infty}^{\infty} f(n\Delta t)\delta(t - n\Delta t) \right) e^{-2\pi i\nu t}\,dt$$

$$= \sum_{n=-\infty}^{\infty} f(n\Delta t) \int_{-\infty}^{\infty} \delta(t - n\Delta t)e^{-2\pi i\nu t}\,dt$$

$$= \sum_{n=-\infty}^{\infty} f(n\Delta t)e^{-2\pi i\nu n\Delta t}.$$

□

The key to understanding the other expression for the Fourier transform of a digitized signal is Poisson's summation formula from Fourier analysis.

Theorem 7.5.2

$$\sum_{n=-\infty}^{\infty} f(n\Delta t) = \frac{1}{\Delta t} \sum_{n=-\infty}^{\infty} \hat{f}\left(\frac{n}{\Delta t}\right). \tag{7.5.2}$$

Proof This follows from the Poisson summation formula (2.16.1), using Exercise 3 of Section 2.13. □

Corollary 7.5.3 *The Fourier transform of the sampling function $\delta_s(t)$ is another sampling function in the frequency domain,*

$$\widehat{\delta_s}(v) = \frac{1}{\Delta t} \sum_{n=-\infty}^{\infty} \delta\left(v - \frac{n}{\Delta t}\right).$$

Proof If $f(t)$ is a test function, then the definition of $\delta_s(t)$ gives

$$\int_{-\infty}^{\infty} f(t)\delta_s(t)\,dt = \sum_{n=-\infty}^{\infty} f(n\Delta t).$$

Applying Parseval's identity (2.15.1) to the left hand side (and noting that the sampling function is real, so that $\delta_s(t) = \overline{\delta_s(t)}$), and applying Formula (7.5.2) to the right hand side, we obtain

$$\int_{-\infty}^{\infty} \hat{f}(v)\overline{\widehat{\delta_s}(v)}\,dv = \frac{1}{\Delta t} \sum_{n=-\infty}^{\infty} \hat{f}\left(\frac{n}{\Delta t}\right).$$

The required formula for $\widehat{\delta_s}(v)$ follows. □

Corollary 7.5.4 *The Fourier transform of a digital signal $f(t)\delta_s(t)$ is*

$$\widehat{f.\delta_s}(v) = \frac{1}{\Delta t} \sum_{n=-\infty}^{\infty} \hat{f}\left(v - \frac{n}{\Delta t}\right),$$

which is periodic in the frequency domain, with period equal to the sampling frequency $1/\Delta t$.

Proof By Theorem 2.18.1(ii), we have

$$\widehat{f.\delta_s}(v) = (\hat{f} * \widehat{\delta_s})(v),$$

and by Corollary 7.5.3, this is equal to

$$\int_{-\infty}^{\infty} \hat{f}(u) \frac{1}{\Delta t} \sum_{n=-\infty}^{\infty} \delta\left(v - \frac{n}{\Delta t} - u\right) du = \frac{1}{\Delta t} \sum_{n=-\infty}^{\infty} \hat{f}\left(v - \frac{n}{\Delta t}\right).$$

□

7.6 Nyquist's theorem

Nyquist's theorem[3] states that the maximum frequency that can be represented when digitizing an analogue signal is exactly half the sampling rate. Frequencies above this limit will give rise to unwanted frequencies below the *Nyquist frequency* of half the sampling rate. What happens to signals at exactly the Nyquist frequency depends on the phase. If the entire frequency spectrum of the signal lies below the Nyquist frequency, then the *sampling theorem* states that the signal can be reconstructed exactly from its digitization.

To explain the reason for Nyquist's theorem, consider a pure sinusoidal wave with frequency v, for example

$$f(t) = A \cos(2\pi v t).$$

Given a sample rate of $N = 1/\Delta t$ samples per second, the height of the function at the Mth sample is given by

$$f(M/N) = A \cos(2\pi v M/N).$$

If v is greater than $N/2$, say $v = N/2 + \alpha$, then

$$\begin{aligned} f(M/N) &= A \cos(2(N/2 + \alpha)M\pi/N) \\ &= A \cos(M\pi + 2\alpha M\pi/N) \\ &= (-1)^M A \cos(2\alpha M\pi/N). \end{aligned}$$

Changing the sign of α makes no difference to the outcome of this calculation, so this gives exactly the same answer as the waveform with $v = N/2 - \alpha$ instead of $v = N/2 + \alpha$. To put it another way, the sample points in this calculation are exactly the points where the graphs of the functions $A \cos(2(N/2 + \alpha)\pi t)$ and $A \cos(2(N/2 - \alpha)\pi t)$ cross, as in Figure 7.8.

The result of this is that a frequency which is greater than half the sample frequency gets reflected through half the sample frequency, so that it sounds like a frequency of the corresponding amount less than half. This phenomenon is called *aliasing*. In the above diagram, the sample points are represented by black dots. The two waves have frequency slightly more and slightly less than half the sample frequency. It is easy to see from the diagram why the sample values are equal. That is, the sample points are simply the points where the two graphs cross.

For waves at exactly half the sampling frequency, something interesting occurs. Cosine waves survive intact, but sine waves disappear altogether. This means that phase information is lost, and amplitude information is skewed.

---

[3] Harold Nyquist, Certain topics in telegraph transmission theory, *Trans. Amer. Inst. Elec. Eng.*, April 1928. Nyquist retired from Bell Labs in 1954 with about 150 patents to his name. He was renowned for his ability to take a complex problem and produce a simple minded solution that was far superior to other, more complicated approaches.

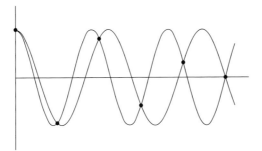

Figure 7.8 Aliasing.

The upshot of Nyquist's theorem is that before digitizing an analogue signal, it is essential to pass it through a low pass filter to cut off frequencies above half the sample frequency. Otherwise, each frequency will come paired with its reflection.

In the case of digital compact discs, the cutoff frequency is half of 44.1 KHz, or 22.05 KHz. Since the limit of human perception is usually below 20 KHz, this may be considered satisfactory, but only by a small margin.

We can also explain Nyquist's theorem in terms of Corollary 7.5.4. Namely, the Fourier transform

$$\widehat{f\delta_s}(v) = \frac{1}{\Delta t} \sum_{n=-\infty}^{\infty} \hat{f}\left(v - \frac{n}{\Delta t}\right)$$

is periodic with period equal to the sampling frequency $N = 1/\Delta t$. The term with $n = 0$ in this sum is the Fourier transform of $f(t)$. The remaining terms with $n \neq 0$ appear as aliased artifacts, consisting of frequency components shifted in frequency by multiples of the sampling frequency $N = 1/\Delta t$. If $f(t)$ has a nonzero part of its spectrum at frequency greater than $N/2$, then its Fourier transform will be nonzero at plus and minus this quantity. Then adding or subtracting N will result in an artifact at the corresponding amount less than $N/2$, the other side of the origin.

Another remarkable fact comes out of Corollary 7.5.4, namely the *sampling theorem*. Provided the original signal $f(t)$ satisfies $\hat{f}(v) = 0$ for $v \geq N/2$, in other words, provided that the entire spectrum lies below the Nyquist frequency, the original signal can be reconstructed exactly from the sampled signal, without any loss of information. To reconstruct $\hat{f}(v)$, we begin by by truncating $\widehat{f\delta_s}(v)$, and then $f(t)$ is reconstructed using the inverse Fourier transform. Carrying this out in practice is a different matter, and requires very accurate analogue filters.

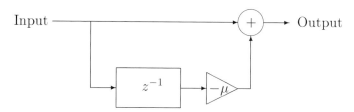

Input ——————————————→ (+) ——→ Output

Figure 7.11 Digital filter.

Figure 7.12 Impulse response.

If $f(n\Delta t)$ is the input and $g(n\Delta t)$ is the output, then the relation represented by the above diagram is

$$g(n\Delta t) = f(n\Delta t) - \mu f((n-1)\Delta t). \tag{7.8.1}$$

So the relation between the z-transforms is

$$G(z) = F(z) - \mu z^{-1} F(z) = (1 - \mu z^{-1})F(z).$$

This tells us about the frequency response of the filter. A given frequency ν corresponds to the points $z = e^{\pm 2\pi i \nu \Delta t}$ on the unit circle in the complex plane, with half the sampling frequency corresponding to $e^{\pi i} = -1$.

At a particular point on the unit circle, the value of $1 - \mu z^{-1}$ gives the frequency response. Namely, the amplification is $|1 - \mu z^{-1}|$, and the phase shift is given by the argument of $1 - \mu z^{-1}$.

More generally, if the relationship between the z-transforms of the input and output signal, $F(z)$ and $G(z)$, is given by

$$G(z) = H(z)F(z),$$

then the function $H(z)$ is called the *transfer function* of the filter. The interpretation of the transfer function, for example $1 - \mu z^{-1}$ in the above filter, is that it is the z-transform of the *impulse response* of the filter. See Figure 7.12.

The impulse response is defined to be the output resulting from an input which is zero except at the one sample point $t = 0$, where its value is one, namely

$$f(n\Delta t) = \begin{cases} 1 & n = 0, \\ 0 & n \neq 0. \end{cases}$$

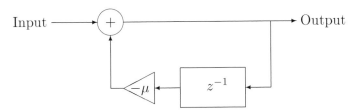

Figure 7.13 Filter with feedback.

Figure 7.14 Impulse response of filter with feedback.

The sampled function $f\delta_s$ is then a Dirac delta function.

For digital signals, the convolution of f_1 and f_2 is defined to be

$$(f_1 * f_2)(n\Delta t) = \sum_{m=-\infty}^{\infty} f_1((n-m)\Delta t) f_2(m\Delta t).$$

Multiplication of z-transforms corresponds to convolution of the original signals. This is easy to see in terms of how power series in z^{-1} multiply. So, in the above example, the impulse response is 1 at $n = 0$, $-\mu$ at $n = 1$, and zero for $n \neq 0, 1$. Convolution of the input signal $f(n\Delta t)$ with the impulse response gives the output signal $g(n\Delta t)$ according to Equation (7.8.1).

As a second example, consider a filter with feedback, as in Figure 7.13. The relation between the input $f(n\Delta t)$ and the output $g(n\Delta t)$ is now given by

$$g(n\Delta t) = f(n\Delta t) - \mu g((n-1)\Delta t).$$

This time, the relation between the z-transforms is

$$G(z) = F(z) - \mu z^{-1} G(z),$$

or

$$G(z) = \frac{1}{1 + \mu z^{-1}} F(z).$$

Notice that this is unstable when $|\mu| > 1$, in the sense that the signal grows without bound. Even when $|\mu| = 1$, the signal never dies away, so we say that this filter is *stable* provided $|\mu| < 1$. This is easiest to see in terms of the impulse response of this filter, which is

$$\frac{1}{1+\mu z^{-1}} = 1 - \mu z^{-1} + \mu^2 z^{-2} - \mu^3 z^{-3} + \cdots$$

Filters are usually designed in such a way that the output $g(n\Delta t)$ depends linearly on $f((n-m)\Delta t)$ for a finite set of values of $m \geq 0$ and on $g((n-m)\Delta t)$ for a finite set of values of $m > 0$. For such a filter, the z-transform of the impulse response is a rational function of z, which means that it is a ratio of two polynomials

$$\frac{p(z)}{q(z)} = a_0 + a_1 z^{-1} + a_2 z^{-2} + \cdots$$

The coefficients $a_0, a_1, a_2, \ldots$ are the values of the impulse response at $t = 0$, $t = \Delta t, t = 2\Delta t, \ldots$

The coefficients a_n tend to zero as n tends to infinity, if and only if the poles μ of $p(z)/q(z)$ satisfy $|\mu| < 1$. This can be seen in terms of the complex partial fraction expansion of the function $p(z)/q(z)$.

The location of the poles inside the unit circle has a great deal of effect on the frequency response of the filter. If there is a pole near the boundary, it will cause a local maximum in the frequency response, which is called a *resonance*. The frequency is given in terms of the argument of the position of the pole by

$$\nu = \text{(sample rate)} \times \text{(argument)}/2\pi.$$

Decay time

The decay time of a filter for a particular frequency is defined to be the time it takes for the amplitude of that frequency component to reach $1/e$ of its initial value. To understand the effect of the location of a pole on the decay time, we examine the transfer function

$$H(z) = \frac{1}{z-a} = \frac{z^{-1}}{1-az^{-1}} = z^{-1} + az^{-2} + a^2 z^{-3} + \cdots$$

So, in a period of n sample times, the amplitude is multiplied by a factor of a^n. So we want $|a|^n = 1/e$, or $n = -1/\ln|a|$. So the formula for decay time is

$$\boxed{\text{Decay time} = \frac{-\Delta t}{\ln|a|} = \frac{-1}{N \ln|a|},} \qquad (7.8.2)$$

where $N = 1/\Delta t$ is the sample rate. So the decay time is inversely proportional to the logarithm of the absolute value of the location of the pole. The further the pole is inside the unit circle, the smaller the decay time, and the faster the decay. A pole near the unit circle gives rise to a slow decay.

(a) Design a digital filter whose transfer function is $z^2/(z^2 + z + \frac{1}{2})$, using the symbol z^{-1} in a box to denote a delay of one sample time, as above.
(b) Compute the frequency response of this filter. Let N denote the number of sample points per second, so that the answer should be a function of v for $-N/2 < v < N/2$.
(c) Is this filter stable?

Further reading

R. W. Hamming (1989), *Digital Filters*.
Bernard Mulgrew, Peter Grant and John Thompson (1999), *Digital Signal Processing*.

7.9 The discrete Fourier transform

How do we describe the frequency components of a sampled signal of finite length, such as a small window from a digital recording? We have already seen in Section 7.7 that for a potentially infinite sampled signal, the frequency spectrum forms a circle in the z-plane, where $z = e^{2\pi i v \Delta t}$. The effect of restricting the length of the sampled signal is to restrict the frequency spectrum to a discrete set of points on the circle.

Suppose that the length of the signal is M, and let's index so that $f(n\Delta t) = 0$ except when $0 \le n < M$. Then the Fourier transform given in Theorem 7.5.1 becomes

$$\widehat{f.\delta_s}(v) = \sum_{n=0}^{M-1} f(n\Delta t)e^{-2\pi i v n \Delta t}.$$

This is a periodic function of v with period $N = 1/\Delta t$, as we observed before.

Since there are only M pieces of information in the signal, we might expect to be able to reconstruct it from the value of the Fourier transform at M different values of v. Let's try spacing them equally around the circle in the z plane, or in other words, just look at the values when $v = k/(M\Delta t)$ for $0 \le k < M$. So we set

$$F(k) = \widehat{f.\delta_s}\left(\frac{k}{M\Delta t}\right) = \sum_{n=0}^{M-1} f(n\Delta t)e^{-2\pi i n k/M}.$$

See Exercise 4 in Section 9.7 for an interpretation of this formula in terms of characters of cyclic groups.

We shall see that $f(n\Delta t)$ ($0 \le n < M$) can indeed be reconstructed from $F(k)$ ($0 \le k < M$). We can see this using the following *orthogonality relation*, without knowing any of the Fourier theory we have developed.

Proposition 7.9.1 *Let M be a positive integer and let k be an integer in the range* $0 \le k < M$. *Then*

$$\sum_{n=0}^{M-1} e^{2\pi ink/M} = \begin{cases} M & \text{if } k \text{ is divisible by } M, \\ 0 & \text{otherwise.} \end{cases}$$

Proof We have

$$e^{2\pi ik/M} \cdot \sum_{n=0}^{M-1} e^{2\pi ink/M} = \sum_{n=0}^{M-1} e^{2\pi i(n+1)k/M} = \sum_{n=1}^{M} e^{2\pi ink/M}.$$

But the term with $n = M$ is equal to the term with $n = 0$ because $e^{2\pi ik} = 1 = e^0$. So the sum remains unchanged when multiplied by $e^{2\pi ik/M}$. It follows that

$$\left(e^{2\pi ik/M} - 1\right) . \sum_{n=0}^{M-1} e^{2\pi ink/M} = 0.$$

If k is not divisible by M then $e^{2\pi ik/M} \neq 1$, and so we can divide by $e^{2\pi ik/M} - 1$ to see that the sum is zero. On the other hand, if k is divisible by M then all the terms in the sum are equal to one. So the sum is equal to the number M of terms. □

Theorem 7.9.2 (discrete Fourier transform) *If* $f(n\Delta t) = 0$ *except when* $0 \le n < M$, *then the digital signal* $f(n\Delta t)$ *can be recovered from the values of*

$$F(k) = \sum_{n=0}^{M-1} f(n\Delta t)e^{-2\pi ink/M} \tag{7.9.1}$$

for $0 \le k < M$ *by the formula*

$$f(n\Delta t) = \frac{1}{M} \sum_{k=0}^{M-1} F(k)e^{2\pi ink/M}. \tag{7.9.2}$$

Proof Substituting in the definition of $F(k)$, we get

$$\frac{1}{M} \sum_{k=0}^{M-1} F(k)e^{2\pi ink/M} = \frac{1}{M} \sum_{k=0}^{M-1} \left(\sum_{m=0}^{M-1} f(m\Delta t)e^{-2\pi imk/M} \right) e^{2\pi ink/M}$$

$$= \sum_{m=0}^{M-1} f(m\Delta t) \left(\frac{1}{M} \sum_{k=0}^{M-1} e^{2\pi i(n-m)k/M} \right).$$

We can now apply Proposition 7.9.1 to the inside sum, to see that it is equal to one when $m = n$ and to zero when $m \neq n$ (notice that both m and n lie between zero and $M - 1$, so their difference is less than M in magnitude). So the outside sum has only one nonzero term, namely the one where $m = n$, and this gives $f(n\Delta t)$ as desired. □

Example 7.9.3 Consider the case $M = 4$. In this case, the numbers $e^{2\pi ik/M}$ are equally spaced around the unit circle in the complex plane, so that they are:

$$e^0 = 1 \qquad (k = 0),$$
$$e^{2\pi i/4} = i \qquad (k = 1),$$
$$e^{4\pi i/4} = -1 \qquad (k = 2),$$
$$e^{6\pi i/4} = -i \qquad (k = 3).$$

The formulas in the theorem reduce to

$$F(0) = f(0) + f(\Delta t) + f(2\Delta t) + f(3\Delta t),$$
$$F(1) = f(0) - if(\Delta t) - f(2\Delta t) + if(3\Delta t),$$
$$F(2) = f(0) - f(\Delta t) + f(2\Delta t) - f(3\Delta t),$$
$$F(3) = f(0) + if(\Delta t) - f(2\Delta t) - if(3\Delta t)$$

and

$$f(0) = \tfrac{1}{4}(F(0) + F(1) + F(2) + F(3)),$$
$$f(\Delta t) = \tfrac{1}{4}(F(0) + iF(1) - F(2) - iF(3)),$$
$$f(2\Delta t) = \tfrac{1}{4}(F(0) - F(1) + F(2) - F(3)),$$
$$f(3\Delta t) = \tfrac{1}{4}(F(0) - iF(1) - F(2) + iF(3)).$$

For a long signal, the usual process is to choose for M a number that is used as a window size for a moving window in the signal. So the discrete Fourier transform is really a digitized version of the windowed Fourier transform.

7.10 The fast Fourier transform

The *fast Fourier transform* (often abbreviated to FFT) or *Cooley–Tukey algorithm* is a way to organise the work of computing the discrete Fourier transform in such a way that fewer arithmetic operations are necessary than just using Equation (7.9.1) in the obvious straightforward way.

To explain how it works, let's suppose that M is even. Then we can split up the sum (7.9.1) into the even numbered and the odd numbered terms:

$$F(k) = \sum_{n=0}^{\frac{M}{2}-1} f(2n\Delta t)e^{-2\pi i(2n)k/M} + \sum_{n=0}^{\frac{M}{2}-1} f((2n+1)\Delta t)e^{-2\pi i(2n+1)k/M}.$$

The crucial observation is that the value of $F(k + \frac{M}{2})$ is very similar to the value of $F(k)$. Observing that $e^{-\pi i(2n)k} = 1$ and $e^{-\pi i(2n+1)k} = (-1)^k$, we get

$$F\left(k + \tfrac{M}{2}\right) = \sum_{n=0}^{\frac{M}{2}-1} f(2n\Delta t)e^{-2\pi i(2n)k/M} + (-1)^k \sum_{n=0}^{\frac{M}{2}-1} f((2n+1)\Delta t)e^{-2\pi i(2n+1)k/M}.$$

So we can compute the values of $F(k)$ and $F(k + \frac{M}{2})$ at the same time for half the work it would otherwise have taken, plus a slight overhead for the additions and subtractions of the answers. The two sums we're calculating are themselves discrete Fourier transforms (with the right hand one multiplied by $e^{-2\pi ik/M}$) for $M/2$ points instead of M points, so if $M/2$ is even, we can repeat the division of labour.

In the example of the previous section with $M = 4$, this rearranges the computation as follows:

$$F(0) = (f(0) + f(2\Delta t)) + (f(\Delta t) + f(3\Delta t)),$$
$$F(2) = (f(0) + f(2\Delta t)) - (f(\Delta t) + f(3\Delta t)),$$
$$F(1) = (f(0) - f(2\Delta t)) - i(f(\Delta t) - f(3\Delta t)),$$
$$F(3) = (f(0) - f(2\Delta t)) + i(f(\Delta t) - f(3\Delta t)).$$

If M is a power of 2, then this method can be used to compute the discrete Fourier transform using $2M \log_2 M$ operations rather than M^2. With slight adjustment, the method can be made to work for any highly composite value of M, but it is most efficient for a power of 2.

Notice also that Formula (7.9.2) for reconstructing the digital signal from the discrete Fourier transform is just a lightly disguised version of the same process, and so the same method can be used.

Further reading

G. D. Bergland, A guided tour of the fast Fourier transform, *IEEE Spectrum* **6** (1969), 41–52.

James W. Cooley and John W. Tukey, An algorithm for the machine calculation of complex Fourier series, *Math. Comp.* **19** (1965), 297–301. This is usually regarded as the original article announcing the fast Fourier transform as a practical algorithm, although the method appears in the work of Gauss in the nineteenth century (see the next reference).

M. T. Heideman, D. H. Johnson and C. S. Burrus, Gauss and the history of the fast Fourier transform, *Archive for History of Exact Sciences* **34** (3) (1985), 265–277.

David K. Maslen and Daniel N. Rockmore, The Cooley–Tukey FFT and group theory, *Not. AMS* **48** (10) (2001), 1151–1160. Reprinted in *Modern Signal Processing*, MSRI Publications, Vol. 46, Cambridge University Press (2004), 281–300.

Bernard Mulgrew, Peter Grant and John Thompson (1999), *Digital Signal Processing*. Chapters 9 and 10 of this book explain in detail how to set up a fast Fourier transform, and give an analysis of the effect of various window shapes.

8

Synthesis

8.1 Introduction

In this chapter, we investigate synthesis of musical sounds. We pay special attention to Frequency Modulation (or FM) synthesis, not because it is a particularly
important method of synthesis, but rather because it is easy to use FM synthesis as
a vehicle for conveying general principles.

Interesting musical sounds do not in general have a static frequency spectrum.
The development with time of the spectrum of a note can be understood to some
extent by trying to mimic the sound of a conventional musical instrument synthet-
ically. This exercise focuses our attention on what are usually referred to as the
attack, decay, sustain and release parts of a note (ADSR). Not only does the amplitude change during these intervals, but also the frequency spectrum. Synthesizing
sounds which do not sound mechanical and boring turns out to be harder than one
might guess. The ear is very good at picking out the regular features produced by
simple minded algorithms and identifying them as synthetic. This way, we are led
to an appreciation of the complexity of even the simplest of sounds produced by
conventional instruments.

Of course, the real strength of synthesis is the ability to produce sounds not
previously attainable, and to manipulate sounds in ways not previously possible.
Most music, even in today's era of the availability of cheap and powerful digital
synthesizers, seems to occupy only a very small corner of the available sonic pallette.
The majority of musicians who use synthesizers just punch the presets until they
find the ones they like, and then use them without modification. Exceptions to this
rule stand out from the crowd; listening to a recording by the Japanese synthesist
Tomita, for example, one is struck immediately by the skill expressed in the shaping
of the sound.

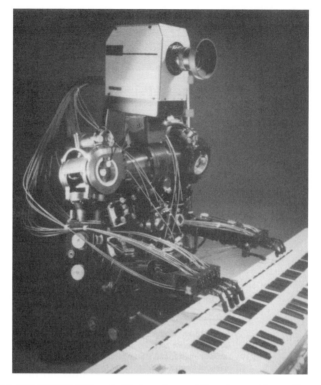

Figure 8.1 WABOT-2 (Waseda University and Sumitomo Corp., Japan 1985).

Further listening (See Appendix G)

Isao Tomita, *Pictures at an Exhibition* (Mussorgsky).

8.2 Envelopes and LFOs

Whatever method is used to synthesize sounds, attention has to be paid to envelopes, so we discuss these first. Very few sounds just consist of a spectrum, static in time. If we hear a note on almost any instrument, there is a clearly defined attack at the beginning of the sound, followed by a decay, then a sustained part in the middle, and finally a release. In any particular instrument, some of these may be missing, but the basic structure is there. Synthesis follows the same pattern. The commonly used abbreviation is ADSR envelope, for attack/decay/sustain/release envelope. See Figure 8.2.

It was not really understood properly until the middle of the twentieth century, when electronic synthesis was taking its first tentative steps, that the attack portion of a note is the most vital to the human ear in identifying the instrument. The

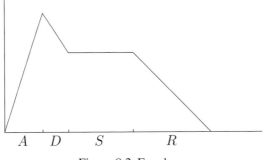

Figure 8.2 Envelope.

transients at the beginning are much more different from one instrument to another than the steady part of the note.

On a typical synthesizer, there are a number of envelope generators. Each one determines how the amplitude of the output of some component of the system varies with time. It is important to understand that amplitude of the final signal is not the only attribute which is assigned an envelope. For example, when a bell sounds, initially the frequency spectrum is very rich, but many of the partials die away very quickly leaving a purer sound. Mimicking this sort of behaviour using FM synthesis turns out to be relatively easy, by assigning an envelope to a modulating signal, which controls timbre. We shall discuss this further when we discuss FM synthesis, but for the moment we note that aspects of timbre are often controlled with an envelope generator. When the synthesizer is controlled by a keyboard, as is often the case, it is usual to arrange that depressing a key initiates the attack, and releasing the key initiates the release portion of the envelope.

An envelope generator produces an envelope whose shape is determined by a number of programmable parameters. These parameters are usually given in terms of levels and rates. Here is an example of how an envelope might work in a typical keyboard synthesizer or other MIDI controlled environment. Level 0 is the level of the envelope at the 'key on' event. Rate 1 then determines how fast the level changes, until it reaches level 1. Then it switches to rate 2 until level 2 is reached, and then rate 3 until level 3 is reached. Level 3 is then in effect until the 'key off' event, when rate 4 takes effect until level 4 is reached. Finally, level 4 is the same as level 0, so that we are ready for the next 'key on' event. In this example, there are two separate components to the decay phase of the envelope. Some synthesizers make do with only one, and some have even more.

Similar in concept to the envelope is the *low frequency oscillator* or LFO. This produces an output which is usually in the range 0.1–20 Hz, and whose waveform is usually something like triangle, sawtooth (up or down), sine, square or random. The LFO is used to produce repeating changes in some controllable parameter.

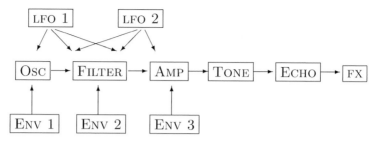

Figure 8.3 Block diagram for an analogue synthesizer.

Examples include pitch control for vibrato, and amplitude or timbre control for tremolo. The LFO can also be used to control less obvious parameters such as the cutoff and resonance of a filter, or the pulse width of a square wave (pulse width modulation, or PWM), see Exercise 6 in Section 2.4.

The parameters associated with an LFO are rate (or frequency), depth (or amplitude), waveform, and attack time. Attack time is used when the effect is to be introduced gradually at the beginning of the note. Figure 8.3 shows a block diagram for a typical analogue synthesizer.

The oscillator (Osc) generates the basic waveform, which can be chosen from sine wave, square wave, triangular wave, sawtooth, noise, etc. The envelope (Env 1) specifies how the pitch changes with time. The filter specifies the 'brightness' of the sound. It can be chosen from high pass, low pass and band pass. The envelope (Env 2) specifies how the brightness varies with time. Also, a *resonance* is specified, which determines the emphasis applied to the region at the cutoff frequency. The amplifier (Amp) specifies the volume, and the envelope (Env 3) specifies how the volume changes with time. The tone control (Tone) adjusts the overall tone, the delay unit (Echo) adds an echo effect, and the effects unit (FX) can be used to add reverberation, chorus and so on. Low frequency oscillators (LFO 1 and LFO 2) are provided, which can be used to modulate the oscillator, filter or amplifier.

8.3 Additive synthesis

The easiest form of synthesis to understand is additive synthesis, which is in effect the opposite of Fourier analysis of a signal. To synthesize a periodic wave, we generate its Fourier components at the correct amplitudes and mix them. This is a comparatively inefficient method of synthesis, because, in order to produce a note with a large number of harmonics, a large number of sine waves will need to be mixed together. Each will be assigned a separate envelope in order to create the development of the note with time. This way, it is possible to control the development of timbre with time, as well as the amplitude. So, for example, if it is desired to create a waveform whose attack phase is rich in harmonics and which then decays

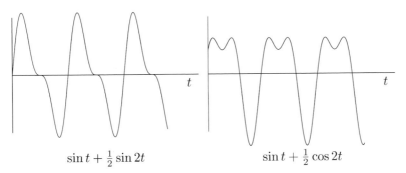

$$\sin t + \tfrac{1}{2} \sin 2t \qquad\qquad \sin t + \tfrac{1}{2} \cos 2t$$

Figure 8.4.

to a purer tone, then the components of higher frequency will have a more rapidly decaying envelope than the lower frequency components.

Phase is unimportant to the perception of steady sounds, but more important in the perception of transients. So, for steady sounds, the graph representing the waveform is not very informative. For example, Figure 8.4 shows the graphs of the functions $\sin t + \tfrac{1}{2} \sin 2t$ and $\sin t + \tfrac{1}{2} \cos 2t$.

The only difference between these functions is that the second partial has had its phase changed by an angle of $\pi/2$, so as steady sounds, these will sound identical. With more partials, it becomes extremely hard to tell whether two waveforms represent the same steady sound. It is for this reason that the waveform is not a very useful way to represent the sound, whereas the spectrum, and its development with time, are much more useful.

In some ways, additive synthesis is a very old idea. A typical cathedral or church organ has a number of register stops, determining which sets of pipes are used for the production of the note. The effect of this is that depressing a single key can be made to activate a number of harmonically related pipes, typically a mixture of octaves and fifths. Early electronic instruments such as the Hammond organ (Figure 8.5) operated on exactly the same principle.

More generally, additive synthesis may be used to construct sounds whose partials are not multiples of a given fundamental. This will give non-periodic waveforms which nevertheless sound like steady tones.

Exercises

1. Explain how to use additive synthesis to construct a square wave out of pure sine waves. (Hint: look at Section 2.2.)
2. Explain in terms of the human ear (Section 1.2) why the phases of the harmonic components of a steady waveform should not have a great effect on the way the sound is perceived.

Figure 8.5 Hammond B3 organ.

Further reading

F. de Bernardinis, R. Roncella, R. Saletti, P. Terreni and G. Bertini, A new VLSI
   implementation of additive synthesis, *Computer Music Journal* **22** (3) (1998),
   49–61.

8.4 Physical modelling

The idea of physical modelling is to take a physical system, such as a musical
instrument, and to mimic it digitally. We give one simple example to illustrate the
point. We examined the wave equation for the vibrating string in Section 3.2, and
found d'Alembert's general solution,

$$y = f(x + ct) + g(x - ct).$$

Given that time is quantized with sample points at spacing Δt, it makes sense to
quantize the position along the string at intervals of $\Delta x = c\Delta t$. Then at time $n\Delta t$

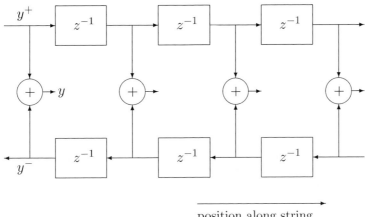

position along string

Figure 8.6 Delay lines.

and position $m\Delta x$, the value of y is

$$y = f(m\Delta x + nc\Delta t) + g(m\Delta x - nc\Delta t)$$
$$= f((m+n)c\Delta t) + g((m-n)c\Delta t).$$

To simplify the notation, we write

$$y^-(n) = f(nc\Delta t), \qquad y^+(n) = g(nc\Delta t),$$

so that y^- and y^+ represent the parts of the wave travelling left and right, respectively, along the string. Then at time $n\Delta t$ and position $m\Delta x$ we have

$$y = y^-(m+n) + y^+(m-n).$$

This can be represented by two delay lines moving left and right, as in Figure 8.6.

It is a good idea to make the string an integer number of sample points long, let us say $l = L\Delta x$. Then the boundary conditions at $x = 0$ and $x = l$ (see Equations (3.2.3) and (3.2.4)) say that

$$y^-(n) = -y^+(-n)$$

and that

$$y^+(n + 2L) = y^+(n).$$

This means that at the ends of the string, the signal gets negated and passed round to the other set of delays. Then the initial pluck or strike is represented by setting the values of $y^-(n)$ and $y^+(n)$ suitably at $t = 0$, for $0 \le n < 2L$.

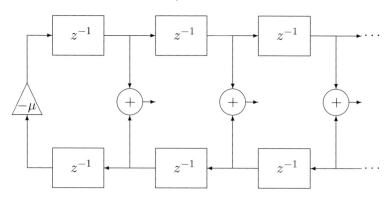

Figure 8.7 Delay lines with energy loss.

Thinking in terms of digital filters, the z-transform of the y^+ signal

$$Y^+(z) = y^+(0) + y^+(1)z^{-1} + y^+(2)z^{-2} + \cdots$$

satisfies

$$Y^+(z) = z^{-2L}Y^+(z) + (y^+(0) + y^+(1)z^{-1} + \cdots + y^+(2L-1))$$

or

$$Y^+(z) = \frac{y^+(0)z^{2L} + y^+(1)z^{2L-1} + \cdots + y^+(2L-1)z}{z^{2L} - 1}.$$

The poles are equally spaced on the unit circle, so the resonant frequencies are multiples of $N/2L$, where N is the sample frequency. Since the poles are actually *on* the unit circle, the resonant frequencies never decay.

To make the string more realistic, we can put in energy loss at one end, represented by multiplication by a fixed constant factor $-\mu$ with $0 < \mu \le 1$, instead of just negating. See Figure 8.7.

The effect of this on the filter analysis is to move the poles slightly inside the unit circle:

$$Y^+(z) = \frac{y^+(0)z^{2L} + y^+(1)z^{2L-1} + \cdots + y^+(2L-1)z}{z^{2L} - \mu}.$$

The absolute values of the location of the poles are all equal to $|\mu|^{\frac{1}{2L}}$. The decay time is given by Equation (7.8.2) as

$$\text{decay time} = \frac{-2L}{N \ln |\mu|}.$$

The above model is still not very sophisticated, because decay time is independent of frequency. But it is easy to modify by replacing the multiplication by μ by a more complicated digital filter. We shall see a particular example of this idea in the next section. Another easy modification is to have two or more strings cross-coupled, by adding a small multiple of the signal at the end of each into the end of the others. Adding a model of a sounding board is not so easy, but it can be done.

Further reading

Eric Ducasse, A physical model of a single-reed instrument, including actions of the player, *Computer Music Journal* **27** (1) (2003), 59–70.

G. Essl, S. Serafin, P. R. Cook and J. O. Smith, Theory of banded waveguides, *Computer Music Journal* **28** (1) (2004), 37–50.

G. Essl, S. Serafin, P. R. Cook and J. O. Smith, Musical applications of banded waveguides, *Computer Music Journal* **28** (1) (2004), 51–62.

M. Laurson, C. Erkut, V. Välimäki and M. Kuuskankare, Methods for modeling realistic playing in acoustic guitar synthesis, *Computer Music Journal* **25** (3) (2001), 38–49.

Julius O. Smith III, Physical modeling using digital waveguides, *Computer Music Journal* **16** (4) (1992), 74–87.

Julius O. Smith III (1997), Acoustic modeling using digital waveguides, article 7 in Roads *et al.*, pages 221–263.

Vesa Välimämi, Mikael Laurson and Cumhur Erkut, Commuted waveguide synthesis of the clavichord, *Computer Music Journal* **27** (1) (2003), 71–82.

8.5 The Karplus–Strong algorithm

The Karplus–Strong algorithm gives very good plucked strings and percussion instruments. The basic technique is a modification of the technique described in the last section, and consists of a digital delay followed by an averaging process. Denote by $g(n\Delta t)$ the value of the nth sample point in the digital output signal for the algorithm. A positive integer p is chosen to represent the delay, and the recurrence relation

$$g(n\Delta t) = \tfrac{1}{2}(g((n-p)\Delta t) + g((n-p-1)\Delta t))$$

is used to define the signal after the first $p+1$ sample points. The first $p+1$ values to feed into the recurrence relation are usually chosen by some random algorithm, and then the feedback loop is switched in. This is represented by an input signal $f(n\Delta t)$ which is zero outside the range $0 \leq n \leq p$. See Figure 8.8.

Computationally, this algorithm is very efficient. Each sample point requires one addition operation. Halving does not need a multiplication, only a shift of the binary digits.

Let us analyze the algorithm by regarding it as a digital filter, and using the z-transform, as described in Section 7.8. Let $G(z)$ be the z-transform of the signal $g(n\Delta t)$, and $F(z)$ be the z-transform of the signal given for the first $p+1$ sample

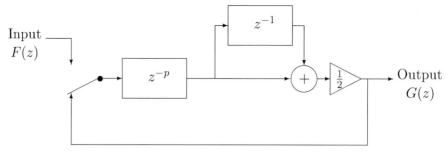

Figure 8.8 The Karplus–Strong algorithm.

points, $f(n\Delta t)$. We have

$$G(z) = \tfrac{1}{2}(1 + z^{-1})z^{-p}(F(z) + G(z)).$$

This gives

$$G(z) = \frac{z+1}{2z^{p+1} - z - 1} F(z),$$

and so the z-transform of the impulse response is $(z + 1)/(2z^{p+1} - z - 1)$. The poles are the solutions of the equation

$$2z^{p+1} - z - 1 = 0.$$

These are roughly equally spaced around the unit circle, at amplitude just less than one. The solution with smallest argument corresponds to the fundamental of the vibration, with argument roughly $2\pi/(p + \tfrac{1}{2})$. A more precise analysis is given in Section 8.6.

The effect of this is a plucked string sound with pitch determined by the formula

$$\text{pitch} = (\text{sample rate})/\left(p + \tfrac{1}{2}\right).$$

Since p is constrained to be an integer, this restricts the possible frequencies of the resulting sound in terms of the sample rate. Changing the value of p without introducing a new inital values results in a slur, or tie between notes.

A simple modification of the algorithm gives drumlike sounds. Namely, a number b is chosen with $0 \le b \le 1$, and

$$g(n\Delta t) = \begin{cases} +\tfrac{1}{2}(g((n - p)\Delta t) + g((n - p - 1)\Delta t) & \text{with probability } b, \\ -\tfrac{1}{2}(g((n - p)\Delta t) + g((n - p - 1)\Delta t)) & \text{with probability } 1 - b. \end{cases}$$

The parameter b is called the *blend factor*. Taking $b = 1$ gives the original plucked string sound. The value $b = \tfrac{1}{2}$ gives a drumlike sound. With $b = 0$, the period is doubled and only odd harmonics result. This gives some interesting sounds, and at high pitches this gives what Karplus and Strong describe as a *plucked bottle* sound.

Another variation described by Karplus and Strong is what they call *decay stretching*. In this version, the recurrence relation

$$g(n\Delta t) = \begin{cases} g((n-p)\Delta t) & \text{with probability } 1 - \alpha, \\ \frac{1}{2}(g((n-p)\Delta t) + g((n-p-1)\Delta t)) & \text{with probability } \alpha. \end{cases}$$

The *stretch factor* for this version is $1/\alpha$, and the pitch is given by

$$\text{pitch} = (\text{sample rate})/\left(p + \tfrac{\alpha}{2}\right).$$

Setting $\alpha = 0$ gives a non-decaying periodic signal, while setting $\alpha = 1$ gives the original algorithm described above.

There are obviously a lot of variations on these algorithms, and many of them give interesting sounds.

8.6 Filter analysis for the Karplus–Strong algorithm

We saw in the last section that in order to understand the Karplus–Strong algorithm in its simplest form, we need to locate the zeros of the polynomial $2z^{p+1} - z - 1$, where p is a positive integer. In order to do this, we begin by rewriting the equation as

$$2z^{p+\frac{1}{2}} = z^{\frac{1}{2}} + z^{-\frac{1}{2}}.$$

Since we expect z to have absolute value close to one, the imaginary part of $z^{\frac{1}{2}} + z^{-\frac{1}{2}}$ will be very small. If we ignore this imaginary part, then the nth zero of the polynomial around the unit circle will have argument equal to $2n\pi/(p + \frac{1}{2})$. So we write

$$z = (1 - \varepsilon)e^{2n\pi i/(p+\frac{1}{2})}$$

and calculate ε, ignoring terms in ε^2 and higher powers. Already, from the form of this approximation, we see that the resonant frequency corresponding to the nth pole is equal to $nN/(p + \frac{1}{2})$, where N is the sample frequency. This means that the different resonant frequencies are at multiples of a fundamental frequency of $N/(p + \frac{1}{2})$.

We have

$$2z^{p+\frac{1}{2}} = 2(1 - \varepsilon)^{p+\frac{1}{2}} \approx 2 - 2\left(p + \tfrac{1}{2}\right)\varepsilon,$$

and

$$z^{\frac{1}{2}} + z^{-\frac{1}{2}} = (1 - \varepsilon)^{\frac{1}{2}} e^{n\pi i/(p+\frac{1}{2})} + (1 - \varepsilon)^{-\frac{1}{2}} e^{-n\pi i/(p+\frac{1}{2})}$$

$$\approx \left(1 - \tfrac{1}{2}\varepsilon\right)\left(1 + \tfrac{1}{2}i\left(\tfrac{2n\pi}{p+\frac{1}{2}}\right) - \tfrac{1}{8}\left(\tfrac{2n\pi}{p+\frac{1}{2}}\right)^2\right)$$

$$+ \left(1 + \tfrac{1}{2}\varepsilon\right)\left(1 - \tfrac{1}{2}i\left(\tfrac{2n\pi}{p+\frac{1}{2}}\right) - \tfrac{1}{8}\left(\tfrac{2n\pi}{p+\frac{1}{2}}\right)^2\right)$$

$$\approx 2 - \left(\tfrac{n\pi}{p+\frac{1}{2}}\right)^2 + \tfrac{1}{2}i\varepsilon\left(\tfrac{2n\pi}{p+\frac{1}{2}}\right).$$

So, equating the real parts, we find that the approximate value of ε is

$$\varepsilon \approx \frac{n^2\pi^2}{2(p+\frac{1}{2})^3}.$$

Using the approximation $\ln(1 - \varepsilon) \approx -\varepsilon$, Equation (7.8.2) gives

$$\text{decay time} \approx \frac{2(p+\frac{1}{2})^3}{Nn^2\pi^2},$$

where N is the sample rate. This means that the lower harmonics are decaying more slowly than the higher harmonics, in accordance with the behaviour of a plucked string.

Further reading

D. A. Jaffe and J. O. Smith III, Extensions of the Karplus–Strong plucked string algorithm, *Computer Music Journal* **7** (2) (1983), 56–69. Reprinted in Roads (1989), 481–494.

M. Karjalainen, V. Välimäki and T. Tolonen, Plucked-string models: from the Karplus–Strong algorithm to digital waveguides and beyond, *Computer Music Journal* **22** (3) (1998), 17–32.

K. Karplus and A. Strong, Digital synthesis of plucked string and drum timbres, *Computer Music Journal* **7** (2) (1983), 43–55. Reprinted in Roads (1989), 467–479.

F. Richard Moore (1990), *Elements of Computer Music*, page 279.

Curtis Roads (1996), *The Computer Music Tutorial*, page 293.

C. Sullivan, Extending the Karplus–Strong plucked-string algorithm to synthesize electric guitar timbres with distortion and feedback, *Computer Music Journal* **14** (3), 26–37.

8.7 Amplitude and frequency modulation

The familiar context for amplitude and frequency modulation is as a way of carrying audio signals on a radio frequency carrier (AM and FM radio). In the case of AM radio, the carrier frequency is usually in the range 500–2000 KHz, which is much greater than the frequency of the carried signal. The latter is encoded in the amplitude of the carrier. So, for example, a 700 KHz carrier signal modulated by a 440 Hz

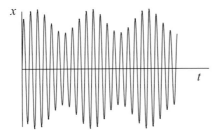

Figure 8.9 Amplitude modulation.

sine wave would be represented by the function

$$x = (A + B\sin(880\pi t))\sin(1400\,000\pi t),$$

where A is an offset to allow both positive and negative values of the waveform to be decoded. See Figure 8.9.

Decoding the received signal is easy. A diode is used to allow only the positive part of the wave through, and then a capacitor is used to smooth it out and remove the high frequency carrier wave. The resulting audio signal may then be amplified and put through a loudspeaker.

In the case of frequency modulation, the carrier frequency is normally around 90–120 MHz, which is even greater in comparison to the frequency of the carried signal. The latter is encoded in variations in the frequency of the carrier. So, for example, a 100 MHz carrier signal modulated by a 440 Hz sine wave would be represented by the function

$$x = A\sin(10^8.2\pi t + B\sin(880\pi t)).$$

The amplitude A is associated with the carrier wave, while the amplitude B is associated with the audio wave. More generally, an audio wave represented by $x = f(t)$, carried on a carrier of frequency ν and amplitude A, is represented by

$$x = A\sin(2\pi \nu t + Bf(t)).$$

See Figure 8.10.

Decoding frequency modulated signals is harder than amplitude modulated signals, and will not be discussed here. But the big advantage is that it is less susceptible to noise, and so it gives cleaner radio reception.

An example of the use of amplitude modulation in the theory of synthesis is ring modulation. A ring modulator takes two inputs, and the output contains only the sum and difference frequencies of the partials of the inputs. This is generally used to construct waveforms with inharmonic partials, so as to impart a metallic or bell-like timbre. The method for constructing the sum and difference frequencies

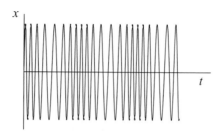

Figure 8.10 Frequency modulation.

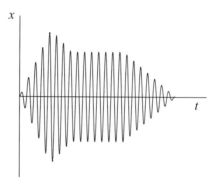

Figure 8.11 Amplitude modulation using an envelope.

is to multiply the incoming amplitudes. Equations (1.8.4), (1.8.7) and (1.8.8) explain how this has the desired result. The origin of the term 'ring modulation' is that in order to deal with both positive and negative amplitudes on the inputs and get the right sign for the outputs, four diodes were connected head to tail in a ring.

   Another example of amplitude modulation is the application of envelopes, as discussed in Section 8.2. The waveform is multiplied by the function used to describe the envelope. See Figure 8.11.

   A great breakthrough in synthesis was achieved in the late nineteen sixties when John Chowning (Figure 8.12) developed the idea of using frequency modulation instead of additive synthesis.

   The idea behind FM synthesis or *frequency modulation synthesis* is similar to FM radio, but the carrier and the signal are both in the audio range, and usually related by a small rational frequency ratio. So, for example, a 440 Hz carrier and 440 Hz modulator would be represented by the function

$$x = A \sin(880\pi t + B \sin(880\pi t)).$$

Figure 8.12 John Chowning.

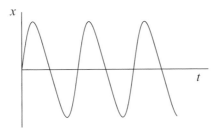

Figure 8.13.

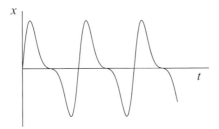

Figure 8.14.

The resulting wave is still periodic with frequency 440 Hz, but has a richer harmonic spectrum than a pure sine wave. For small values of B, the wave is nearly a sine wave, as in Figure 8.13, whereas for larger values of B the harmonic content grows richer (Figure 8.14) and richer (Figure 8.15).

This gives a way of making an audio signal with a rich harmonic content relatively simply. If we wanted to synthesize these waves using additive synthesis, it would be much harder.

Figure 8.15.

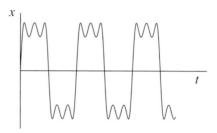

Figure 8.16.

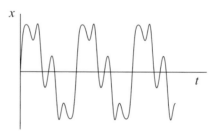

Figure 8.17.

Figures 8.16 and 8.17 show examples of frequency modulated waves in which the modulating frequency is twice the carrier frequency (8.16) and three times the carrier frequency (8.17).

In the next section, we discuss the Fourier series for a frequency modulated signal. The Fourier coefficients are called Bessel functions, for which the groundwork was laid in Section 2.8. We shall see that the Bessel functions may be interpreted as giving the amplitudes of side bands in a frequency modulated signal.

8.8 The Yamaha DX7 and FM synthesis

The Yamaha DX7 (Figure 8.18), which came out in the autumn of 1983, was the first affordable commercially available digital synthesizer. This instrument was the

Figure 8.18 Yamaha DX7.

result of a long collaboration between John Chowning and Yamaha Corporation through the nineteen seventies. It works by FM synthesis, with six configurable 'operators'. An operator produces as output a frequency modulated sine wave, whose frequency is determined by the level of a modulating input, and whose envelope is determined by another input. The power of the method comes from hooking up the output of one such operator to the modulating input of another. In this section, we shall investigate FM synthesis in detail, using the Yamaha DX7 for the details of the examples. Most of the discussion translates easily to any other FM synthesizer. Later on, in Sections 8.11–8.12, we shall also investigate FM synthesis using the CSound computer music language.

The DX7 calculates the sine function in the simplest possible way. It has a digital lookup table of values of the function. This is much faster than any conceivable formula for calculating the function, but this is at the expense of having to commit a block of memory to this task.

Let us begin by examining a frequency modulated signal of the form

$$\sin(\omega_c t + I \sin \omega_m t). \tag{8.8.1}$$

Here, $\omega_c = 2\pi f_c$, where f_c denotes the carrier frequency, $\omega_m = 2\pi f_m$, where f_m denotes the modulating frequency, and I is the index of modulation.

We first discuss the relationship between the index of modulation I, the maximal frequency deviation d of the signal, and the frequency f_m of the modulating wave. For this purpose, we make a linear approximation to the modulating signal at any particular time, and use this to determine the instantaneous frequency, to the extent that this makes sense. When $\sin \omega_m t$ is at a peak or a trough, namely when its derivative with respect to t vanishes, the linear approximation is a constant function, which then acts as a phase shift in the modulated signal. So, at these points, the frequency is f_c. The maximal frequency deviation occurs when $\sin \omega_m t$ is varying most rapidly. This function *increases* most rapidly when $\omega_m t = 2n\pi$ for some integer n. Since the derivative of $\sin \omega_m t$ with respect to t is $\omega_m \cos \omega_m t$,

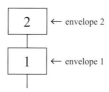

Figure 8.19 Block diagram.

which takes the value ω_m at these values of t, the linear approximation around these values of t is $\sin \omega_m t \simeq \omega_m t - 2n\pi$. So the function (8.8.1) approximates to

$$\sin(\omega_c t + I\omega_m(t - 2\pi)) = \sin((\omega_c + I\omega_m)t - 2\pi I\omega_m).$$

So the instantaneous frequency is $f_c + I f_m$. Similarly, $\sin \omega_m t$ *decreases* most rapidly when $\omega_m t = (2n + 1)\pi$ for some integer n, and a similar calculation shows that the instantaneous frequency is $f_c - I f_m$. It follows that the maximal deviation in the frequency is given by

$$d = I f_m. \tag{8.8.2}$$

The Fourier series for functions of the form (8.8.1) were analyzed in Section 2.8 in terms of the Bessel functions.

Putting $\phi = \omega_c t$, $z = I$ and $\theta = \omega_m t$ in Equation (2.8.9), we obtain the fundamental equation for frequency modulation:

$$\sin(\omega_c t + I \sin \omega_m t) = \sum_{n=-\infty}^{\infty} J_n(I) \sin(\omega_c + n\omega_m)t. \tag{8.8.3}$$

The interpretation of this equation is that for a frequency modulated signal with carrier frequency f_c and modulating frequency f_m, the frequencies present in the modulated signal are $f_c + n f_m$. Notice that positive and negative values of n are allowed here. The component with frequency $f_c + n f_m$ is called the *n*th *side band* of the signal. Thus the Bessel function $J_n(I)$ is giving the amplitude of the *n*th side band in terms of the index of modulation. The block diagram on the DX7 for frequency modulating a sine wave in this fashion is as shown in Figure 8.19.

The box marked '1' represents the operator producing the carrier signal and the box marked '2' represents the operator producing the modulating signal.

Each operator has its own envelope, which determines how its amplitude develops with time. So envelope 1 determines how the amplitude of the final signal varies with time, but it is less obvious what envelope 2 is determining. Since the output of operator 2 is frequency modulating operator 1, the amplitude of the output can be interpreted as the index of modulation I. For small values of I, $J_0(I)$ is much larger than any other $J_n(I)$ (see the graphs in Section 2.8), and so operator 1 is producing

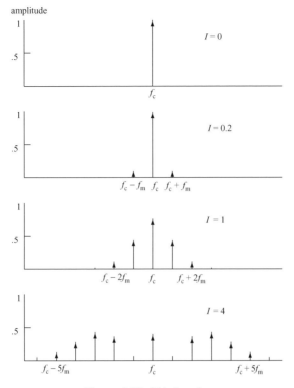

Figure 8.20 Side bands.

an output which is nearly a pure sine wave, but with other frequencies present with small amplitudes. However, for larger values of I, the spectrum of the output of operator 1 grows richer in harmonics. For any particular value of I, as n gets larger, the amplitudes $J_n(I)$ eventually tend to zero. But the point is that for small values of I, this happens more quickly than for larger values of I, so the harmonic spectrum gives a purer note for small values of I and a richer sound for larger values of I. So envelope 2 is controlling the *timbre* of the output of operator 1. See Figure 8.20.

Example

Suppose that we have a carrier frequency of 3ν and a modulating frequency of 2ν. Then the zeroth side band has frequency 3ν, the first 5ν, the second 7ν, and so on. But there are also side bands corresponding to negative values of n. The minus first side band has frequency ν. But there's no reason to stop there, just because the next side band has negative frequency $-\nu$. The point is that a sine wave with frequency $-\nu$ is just the same as a sine wave with frequency ν but with the amplitude negated. So really the way to think of it is that the side bands with negative frequency undergo reflection to make the corresponding positive frequency.

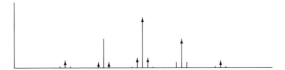

Figure 8.23 Multiplying side band amplitudes.

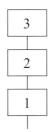

Figure 8.24 Block diagram for a cascade.

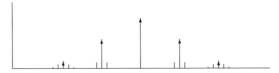

Figure 8.25 Side bands for a cascade.

The corresponding formula is obtained by feeding Formula (8.8.3) into itself, giving

$$\sin(\omega_1 t + I_2 \sin(\omega_2 t + I_3 \sin \omega_3 t))$$

$$= \sum_{n_2=-\infty}^{\infty} J_{n_2}(I_2) \sin(\omega_1 t + n_2 \omega_2 t + n_2 I_3 \sin \omega_3 t)$$

$$= \sum_{n_2=-\infty}^{\infty} \sum_{n_3=-\infty}^{\infty} J_{n_2}(I_2) J_{n_3}(n_2 I_3) \sin(\omega_1 + n_2 \omega_2 + n_3 \omega_3)t.$$

Here, the subscripts 2 and 3 correspond to the numbering on the oscillators in the diagram. Again, the frequencies of the side bands are given by adding positive and negative multiples of the two modulating frequencies to the carrier frequency in all possible ways. But, this time, the amplitudes of the side bands are given by the more complicated formula $J_{n_2}(I_2) J_{n_3}(n_2 I_3)$. The effect of this is that the number of the side band on the second operator is used to scale the size of the index of modulation of the third operator. In particular, the original frequency has no side bands corresponding to the third operator, while the more remote side bands of the second are more heavily modulated. See Figure 8.25.

Exercise

Find the amplitudes of the first few frequency components of the frequency modulated wave

$$y = \sin(440(2\pi t) + \tfrac{1}{10}\sin 660(2\pi t)).$$

Stop when the frequency components are attenuated by at least 100dB from the
strongest one.

Remember that power is proportional to square of amplitude, so that dividing
the amplitude by 10 attenuates the signal by 20dB.

8.9 Feedback, or self-modulation

One final twist in FM synthesis is feedback, or self-modulation. This involves the
output of an oscillator being wrapped back round and used to modulate the input
of the same oscillator. This corresponds to the block diagram below,

and the corresponding equation is given by

$$f(t) = \sin(\omega_c t + I f(t)). \tag{8.9.1}$$

We saw in Section 2.11 that this equation only has a unique solution provided
$|I| \le 1$, and that then it defines a periodic function of t. The Fourier series is given
in Equation (2.11.4) as

$$f(t) = \sum_{n=1}^{\infty} \frac{2 J_n(nI)}{nI} \sin(n\omega_c t).$$

For values of I satisfying $|I| > 1$, Equation (8.9.1) no longer has a single valued
continuous solution (see Section 2.11), but it still makes sense in the form of a
recursion defining the next value of $f(t)$ in terms of the previous one,

$$f(t_n) = \sin(\omega_c t_n + I f(t_{n-1})). \tag{8.9.2}$$

Here, t_n is the nth sample time, and the sample times are usually taken to be equally
spaced. The effect of this equation is not quite intuitively obvious. As might be
expected, the graph of this function stays close to the solution to Equation (8.9.1)
when this is unique. When it is no longer unique, it continues going along the

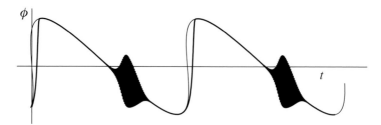

Figure 8.26 Solutions to Equations (8.9.1) and (8.9.2).

same branch of the function as long as it can, and then jumps suddenly to the one remaining branch when it no longer can. But the feature which it is easy to overlook is that there is a slightly delayed instability for small values of $f(t)$. Figure 8.26 is a graph of the solutions to Equations (8.9.1) and (8.9.2) superimposed.

The effect of the instability is to introduce a wave packet whose frequency is roughly half the sampling frequency. Usually the sampling frequency is high enough that the effect is inaudible, but this does make it desirable to pass the resulting signal through a low-pass filter at slightly below the Nyquist frequency.

Feedback for a stack of two or more oscillators is also used. It seems hard to analyze this mathematically, and often the result is perceived as 'noise'. According to Slater (reference given on page 291), as the index of modulation increases, the behaviour of a stack of two FM oscillators with different frequencies, each modulating the other, exhibits the kind of bifurcation that is characteristic of chaotic dynamical systems. This subject needs to be investigated further.

In the DX7, there are a total of six oscillators. The process of designing a patch[1] begins with a choice of one of 32 given configurations, or 'algorithms' for these oscillators. Each oscillator is given an envelope whose parameters are determined by the patch, so that the amplitude of the output of each oscillator varies with time in a chosen manner. Figure 8.27 is a table of the 32 available algorithms.

Not all the operators have to be used in a given patch. The operators which are not used can just be switched off. Output level is an integer in the range 0–99; index of modulation is not a linear function of output level, but rather there is a complicated recipe for causing an approximately exponential relationship. A table showing this relationship for various different FM synthesizers can be found in Appendix B.

We now start discussing how to use FM synthesis to produce various recognisable kinds of sounds. In order to sound like a brass instrument such as a trumpet, it is necessary for the very beginning of the note to be an almost pure sine wave. Then the harmonic spectrum grows rapidly richer, overshooting the steady spectrum by

---

[1] Yamaha uses the nonstandard terminology 'voice' instead of the more usual 'patch'.

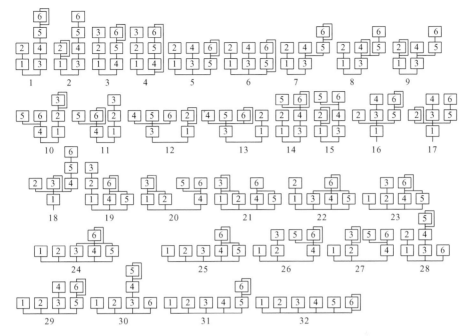

Figure 8.27 Algorithms in the DX7.

some way, and then returning to a reasonably rich spectrum. When the note stops, the spectrum decays rapidly to a pure note and then disappears altogether. This effect may be achieved with FM synthesis by using two operators, one modulating the other. The modulating operator is given an envelope looking like the one in Figure 8.2. The carrier operator uses a very similar envelope to control the amplitude.

Next, we discuss woodwind instruments such as the flute, as well as organ pipes. At the beginning of the note, in the attack phase, higher harmonics dominate. They then decrease in amplitude until, in the steady state, the fundamental dominates and the higher harmonics are not very strong. This can be achieved either by making the modulating operator have an envelope looking like the one in Figure 8.2 only upside down, or by making the carrier frequency a small integer multiple of the modulating frequency, so that for small values of the index of modulation, this higher frequency dominates. In any case, the decay phase for the modulating operator should be omitted for a more realistic sound. For some woodwind instruments such as the clarinet, it is necessary to make sure that predominantly odd harmonics are present. This can be achieved, as in the example on page 283, by setting $f_c = 3f$ and $f_m = 2f$, or some variation on this idea.

Percussive sounds have a very sharp attack and a roughly exponential decay. So an envelope looking like the graph of $x = e^{-t}$ is appropriate for the amplitude.

Usually a percussion instrument will have an inharmonic spectrum, so that it is appropriate to make sure that f_c and f_m are not in a ratio which can be expressed as a ratio of small integers. We saw in Exercise 1 of Section 6.2 that the golden ratio is in some sense the number furthest from being able to be approximated well by ratios of small integers, so this is a good choice for producing spectra which will be perceived as inharmonic. Alternatively, the analysis carried out in Section 3.6 can be used to try to emulate the frequency spectrum of an actual drum.

Section 8.10 and the ones following it consist of an introduction to the public domain computer music language CSound. One of our goals will be to describe explicit implementations of two operator FM synthesis realizing the above descriptions.

Further reading on FM synthesis

J. Bate, The effect of modulator phase on timbres in FM synthesis, *Computer Music Journal* **14** (3) (1990), 38–45.

John Chowning, The synthesis of complex audio spectra by means of frequency modulation, *J. Audio Engineering Society* **21** (7) (1973), 526–534. Reprinted as Chapter 1 of Roads and Strawn (1985), pages 6–29.

John Chowning, Frequency modulation synthesis of the singing voice, in Mathews and Pierce (1989), Chapter 6, pages 57–63.

John Chowning and David Bristow (1986), *FM Theory and Applications*.

L. Demany and K. I. McAnally, The perception of frequency peaks and troughs in wide frequency modulations, *J. Acoust. Soc. Amer.* **96** (2) (1994), 706–715.

L. Demany and S. Clément, The perception of frequency peaks and troughs in wide frequency modulations, II. Effects of frequency register, stimulus uncertainty, and intensity, *J. Acoust. Soc. Amer.* **97** (4) (1995), 2454–2459; III. Complex carriers, *J. Acoust. Soc. Amer.* **98** (5) (1995), 2515–2523; IV. Effect of modulation waveform, *J. Acoust. Soc. Amer.* **102** (5) (1997), 2935–2944.

A. Horner, Double-modulator FM matching of instrument tones, *Computer Music Journal* **20** (2) (1996), 57–71.

A. Horner, A comparison of wavetable and FM parameter spaces, *Computer Music Journal* **21** (4) (1997), 55–85.

A. Horner, J. Beauchamp and L. Haken, FM matching synthesis with genetic algorithms, *Computer Music Journal* **17** (4) (1993), 17–29.

M. LeBrun, A derivation of the spectrum of FM with a complex modulating wave, *Computer Music Journal* **1** (4) (1977), 51–52. Reprinted as Chapter 5 of Roads and Strawn (1985), pages 65–67.

F. Richard Moore (1990), *Elements of Computer Music*, pages 316–332.

D. Morrill, Trumpet algorithms for computer composition, *Computer Music Journal* **1** (1) (1977), 46–52. Reprinted as Chapter 2 of Roads and Strawn (1985), pages 30–44.

C. Roads (1996), *The Computer Music Tutorial*, pages 224–250.

S. Saunders, Improved FM audio synthesis methods for real-time digital music generation, *Computer Music Journal* **1** (1) (1977), 53–55. Reprinted as Chapter 3 of Roads and Strawn (1985), pages 45–53.

W. G. Schottstaedt, The simulation of natural instrument tones using frequency modulation with a complex modulating wave, *Computer Music Journal* **1** (4) (1977), 46–50. Reprinted as Chapter 4 of Roads and Strawn (1985), pages 54–64.

D. Slater, Chaotic sound synthesis, *Computer Music Journal* **22** (2) (1998), 12–19.

B. Truax, Organizational techniques for c : m ratios in frequency modulation, *Computer Music Journal* **1** (4) (1977), 39–45. Reprinted as Chapter 6 of Roads and Strawn (1985), pages 68–82.

8.10 CSound

CSound is a public domain synthesis programme written by Barry Vercoe at the Media Lab in MIT in the C programming language. It has been compiled for various platform, and both source code and executables are freely available.

The programme takes as input two files, called the *orchestra* file and the *score* file. The orchestra file contains the instrument definitions, or how to synthesize the desired sounds. It makes use of almost every known method of synthesis, including FM synthesis, the Karplus–Strong algorithm, phase vocoder, pitch envelopes, granular synthesis and so on, to define the instruments. The score file uses a language similar in conception to MIDI but different in execution, in order to describe the information for playing the instruments, such as amplitude, frequency, note durations and start times. The utility MIDI2CS provides a flexible way of turning MIDI files into CSound score files. The final output of the CSound programme is a file in some chosen sound format, for example a WAV file or an AIFF file, which can be played through a computer sound card, downloaded into a synthesizer with sampling features, or written onto a CD.

We limit ourselves to a brief description of some of the main features of CSound, with the objective of getting as far as describing how to realise FM synthesis. The examples are adapted from the CSound manual.

Getting it

The source code and executables for CSound5.01 for a number of platforms, including Linux, Mac, MS-DOS and Windows can be obtained from

sourceforge.net/projects/csound/

(files are at sourceforge.net/project/showfiles.php?group_id=81968)

as can the manual and some example files. The files you need are as follows:

For all systems, the manual

CSound5.01_manual_pdf.zip (US letter size)
CSound5.01_manual_pdf_AA.zip (A4 for the rest of the world)

Executables (you don't need the source code unless you're compiling the programme yourself):

CSound5.01_src.tar.gz (Source code in C)
CSound5.01_src.zip (Source code in C)
CSound5.01_OS9_src.smi.bin (Source for Mac OS 9)
CSound5.01_i686.rpm (Compiled for Linux)
CSound5.01_x86_64.rpm (Compiled for Linux)
CSound5.01_OSX10.4.tar.gz (Compiled for Mac OS 10.4)
CSound5.01_OSX10.3.tar.gz (Compiled for Mac OS 10.3)
CSound5.01_OSX10.2.tar.gz (Compiled for Mac OS 10.2)
CSound5.01_OS9.smi.bin (Compiled for Mac OS 9)
CSound5.01_win32.i686.zip (Compiled for Windows)
CSound5.01_win32.exe (Compiled for Windows with installer)

For Mac OS X, another way to obtain and install CSound is to download Mac-Csound from csounds.com/matt/MacCsound. This is a packaged complete installation, including a primitive GUI.

The orchestra file

This file has two main parts, namely the *header* section, which defines the sample rate, control rate and number of output channels, and the *instrument* section, which gives the instrument definitions. Each instrument is given its own number, which behaves like a patch number on a synthesizer.

The header section has the following format (everything after a semicolon is a comment):

```
sr = 44100 ; sample rate in samples per second
kr = 4410  ; control rate in control signals per second
ksmps = 10 ; ksmps = sr/kr must be an integer,
           ; samples per control period
nchnls = 1 ; number of channels
```
(8.10.1)

An instrument definition consists of a collection of statements which generate or modify a digital signal. For example, the statements

```
instr 1
   asig oscil 10000, 440, 1
      out asig
endin
```
(8.10.2)

generate a 440 Hz wave with amplitude 10 000, and send it to an output. The two lines of code representing the waveform generator are encased in a pair of statements which define this to be an instrument. For WAV file output, the possible range of amplitudes before clipping takes effect is from $-32\,768$ to $+32\,767$, for a total of 2^{15} possible values (see Section 7.3). The final argument 1 is a waveform number. This determines which waveform is taken from an f statement in the score file (see below). In our first example below, it will be a sine wave. The label `asig` is allowed to be any string beginning with `a` (for 'audio signal'). So, for example, `a1` would have worked just as well. The `oscil` statement is one of CSound's many signal generators, and its effect is to output periodic signals made by repeating the values passed to it, appropriately scaled in amplitude and frequency. There is also another version called `oscili`, with the same syntax, which performs linear interpolation rather than truncation to find values at points between the sample points. This is slower by approximately a factor of two, but in some situations it can lead to better sounding output. In general, it seems to be better to use `oscil` for sound waves and `oscili` for envelopes (see page 297).

As it stands, the instrument (8.10.2) isn't very useful, because it can only play one pitch. To pass a pitch, or other attributes, as parameters from the score file to the orchestra file, an instrument uses variables named `p1`, `p2`, `p3`, and so on. The first three have fixed meanings, and then `p4`, `p5`, ... can be given other meanings. If we replace 440 by `p5`,

```
asig oscil 10000, p5, 1
```

then the parameter `p5` will determine pitch.

The score file
Each line begins with a letter called an *opcode*, which determines how the line is to be interpreted. The rest of the line consists of numerical parameter fields `p1`, `p2`, `p3` and so on. The possible opcodes are:

`f` (function table generator),

`i` (instrument statement; i.e., play a note),

`t` (tempo),

`a` (advance score time; i.e., skip parts),

`b` (offset score time),

`v` (local textual time variation),

`s` (section statement),

`r` (repeat sections),

`m` and `n` (repeat named sections),

e (end of score),

c (comment; semicolon is preferred).

If a line of the score file does not begin with an opcode, it is treated as a continuation line.

Each parameter field consists of a floating point number with optional sign and optional decimal point. Expressions are not permitted.

An f statement calls a subroutine to generate a set of numerical values describing a function. The set of values is intended for passing to the orchestra file for use by an instrument definition. The available subroutines are called GEN01, GEN02, ... Each takes some number of numerical arguments. The parameter fields of an f statement are as follows.

p1 waveform number,

p2 when to begin the table, in beats,

p3 size of table; a power of 2, or one more, maximum 2^{24},

p4 number of GEN subroutine,

p5, p6, . . . parameters for GEN subroutine.

Beats are measured in seconds, unless there is an explicit t (tempo) statement; in our examples, t statements are omitted for simplicity.

So, for example, the statement

```
f1 0 8192 10 1
```

uses GEN10 to produce a sine wave, starting 'now', of size 8192, and assigns it to waveform 1. The subroutine GEN10 produces waveforms made up of weighted sums of sine waves, whose frequencies are integer multiples of the fundamental. So, for example,

```
f2 0 8192 10 1 0 0.5 0 0.333
```

produces the sum of the first five terms in the Fourier series for a square wave, and assigns it to waveform 2.

An i statement activates an instrument. This is the kind of statement used to 'play a note'. Its parameter fields are as follows.

p1 instrument number,

p2 starting time in beats,

p3 duration in beats,

p4, p5, . . . parameters used by the instrument.

An e statement denotes the end of a score. It consists of an e on a line on its own. Every score file must end in this way.

For example, if instrument 1 is given by (8.10.2), then the score file

```
f1 0 8192 10 1 ; use GEN10 to create a sine wave
i1 0 4          ; play instr 1 from time 0 for 4 secs
e
```

$$(8.10.3)$$

will play a 440Hz tone for 4 seconds.

Running CSound

The programme CSound was designed as a command line programme, and although various front ends have been designed for it, the command line remains the most convenient method. Having installed CSound according to the instructions that accompany the programme, the procedure is to create an orchestra file called `<filename>.orc` and a score file called `<filename>.sco` using your favourite (ascii) text processor.[2] The basic syntax for running CSound is

```
csound <flags> <filename>.orc <filename>.sco
```

For example, if your files are called `ditty.orc` and `ditty.sco`, and you want a WAV file output, then use the `-W` flag (this is case sensitive).

```
csound -W ditty.orc ditty.sco
```

This will produce as output a file called `test.wav`. If you want some other name, it must be specified with the `-o` flag.

```
csound -W -o ditty.wav ditty.orc ditty.sco
```
$$(8.10.4)$$

If you want to suppress the graphical displays of the waveforms, which CSound gives by default, this is achieved with the `-d` flag.

We are now ready to run our first example. Make two text files, one called `ditty.orc` containing the statements (8.10.1) followed by (8.10.2), and one called `ditty.sco` containing the statements (8.10.3). If the programme is properly installed, then typing the command (8.10.4) at the command line should produce a file `ditty.wav`. Playing this file through a sound card or other audio device should then sound a pure sine wave at 440Hz for 4 seconds.

---

[2] Word processors such as Word Perfect or Word by default save files with special formatting characters embedded in them. CSound will choke on these characters. In MS-DOS, the command

```
edit <filename>
```

will invoke a simple ascii text processor whose output will not choke CSound in this way. If you are running in an MS-DOS box inside Windows, the command

```
notepad <filename>
```

will start up the ascii text processor called `notepad` in a separate window, which is more convenient for switching between the editor and running CSound.

Warning

Both the orchestra and the score file are case sensitive. If you are having problems running CSound on the above orchestra and score files, check that you have typed everything in lower case.

There is also an annoying feature, which is that if the last line of text in the input file does not have a carriage return, then a wave file will be generated, but it will be unreadable. So it is best to leave a blank line at the end of each file.

Our 'ditty' wasn't really very interesting, so let's modify it a bit. In order to be able to vary the amplitude and pitch, let us modify the instrument (8.10.2) to read

```
instr 1
   asig oscil p4, p5, 1 ; p4 = amplitude, p5 = frequency
      out asig
endin
```

(8.10.5)

Now we can play the first ten notes of the harmonic series (see page 144) using the following score file.

```
f1 0 8192 10 1            ; sine wave
i1 0.0 0.4 32000   261.6 ; fundamental (C, to nearest
                               tenth of a Hz)
i1 0.5 0.4 24000   523.2 ; second harmonic, octave
i1 1.0 0.4 16000   784.8 ; third harmonic, perfect fifth
i1 1.5 0.4 12000  1046.4 ; fourth harmonic, octave
i1 2.0 0.4  8000  1308.0 ; fifth harmonic, just major third
i1 2.5 0.4  6000  1569.6 ; sixth harmonic, perfect fifth
i1 3.0 0.4  4000  1831.2 ; seventh harmonic, listen
                               carefully to this
i1 3.5 0.4  3000  2092.8 ; eighth harmonic, octave
i1 4.0 0.4  2000  2354.4 ; ninth harmonic, just major second
i1 4.5 0.4  1500  2616.0 ; tenth harmonic, just major third
e
```

(8.10.6)

This file plays a series of notes at half second intervals, each lasting 0.4 seconds, at successive integer multiples of 220Hz, and at steadily decreasing amplitudes. Make an orchestra file from (8.10.1) and (8.10.5), and a score file from (8.10.6), run CSound as before, and listen to the results.

Data rates

Recall from (8.10.1) that the header of the orchestra file defines two rates, namely the *sample rate* and the *control rate*. There are three different kinds of variables in CSound, which are distinguished by how often they get updated. a-rate variables, or audio rate variables, are updated at the sample rate, while the k-rate variables, or control rate variables, are updated at the control rate. Audio signals should be taken to be a-rate, while an envelope, for example, is usually assigned to a k-rate variable. It is possible to make use of audio rate signals for control, but this will increase the computational load. A third kind of variable, the i-rate variable, is updated just once when a note is played. These variables are used primarily for setting values to be used by the instrument. The first letter of the variable name (a, k or i) determines which kind of variable it is.

The variables discussed so far are all *local* variables. This means that they only have meaning within the given instrument. The same variable can be reused with a different meaning in a different instrument. There are also global versions of variables of each of these rates. These have names beginning with ga, gk and gi. Assignment of a global variable is done in the *header* section of the orchestra file.

Envelopes

One way to apply an envelope is to make an oscillator whose frequency is 1/p3, the reciprocal of the duration, so that exactly one copy of the waveform is used each time the note is played. It is better to use oscili rather than oscil for envelopes, because many sample points of the envelope will be used in the course of the one period. So, for example,

```
kenv oscili p4, 1/p3, 2
```

uses waveform 2 to make an envelope. The first letter k of the variable name kenv means that this is a control rate variable. It would work just as well to make it an audio rate variable by using a name like aenv, but it would demand greater computation time, and result in no audible improvement.

The subroutine GEN07, which performs linear interpolation, is ideal for an envelope made from straight lines. The arguments p4, p5, ... of this subroutine alternate between numbers of points and values. So, for example, the statement

```
f2 0 513 7 0 80 1 50 0.7 213 0.7 170 0 ; ADSR envelope
```

in the score file produces an envelope resembling the one in Figure 8.2 with ADSR sections of length 80, 50, 213, 170 samples, with heights varying linearly

$$0 \to 1 \to 0.7 \to 0.7 \to 0,$$

and assigns it to waveform 2. The numbers of sample points in the sections should always add up to the total length p3.

Recall that the total number of sample points must be either a power of two, or one more than a power of two. It is usual to use a power of two for repeating waveforms. For waveforms that will be used only once, such as an envelope, we use one more than a power of two so that the number of intervals between sample points is a power of two.

To apply the envelope to the instrument (8.10.5), we replace p4 with `kenv` to make

```
instr 1
   kenv oscili p4, 1/p3, 2 ; envelope from waveform 2
                            ; p4 = amplitude
   asig oscil kenv, p5, 1   ; p5 = frequency
      out asig
endin
```

It would also be possible to replace the waveform number 2 in the definition of `kenv` with another variable, say p6, to give a more general purpose shaped sine wave.

Exercises

1. Make orchestra and score files to generate two sine waves, one at just greater than twice the frequency of the other, and listen to the output. (See also Exercise **6** in Section 1.8.)
2. Make orchestra and score files to play a major scale using a sine wave with an ADSR envelope. Check that your files work by running CSound on them and listening to the result.

8.11 FM synthesis using CSound

Here is the most basic two-operator FM instrument:

```
instr 1
   amod oscil p6 * p7, p6, 1      ; modulating wave
                                  ; p6 = modulating frequency
                                  ; p7 = index of modulation
   kenv oscili p4, 1/p3, 2        ; envelope, p4 = amplitude
   asig oscil kenv, p5 + amod, 1  ; p5 = carrier frequency
      out asig
endin
```
$$(8.11.1)$$

The parameter p7 here represents the index of modulation; the reason why it is multiplied by p6 in the definition of the modulating wave `amod` is that the modulation is taking place directly on the frequency rather than on the phase. According to Equation (8.8.2), this means that the index of modulation must be multiplied by the frequency of the modulating wave before being applied. The argument p5 + `amod` in the definition of `asig` is the carrier frequency p5 plus the modulating wave `amod`. The wave has been given an envelope `kenv`.

For a score file to illustrate this simple instrument, we introduce some useful abbreviations available for repetitive scores. First, note that the `i` statements in a score do not have to be in order of time of execution. The score is sorted with respect to time before it is played. The *carry* feature works as follows. Within a group of consecutive i statements in the score file (not necessarily consecutive in time) whose p1 parameters are equal, empty parameter fields take their value from the previous statement. An empty parameter field is denoted by a dot, with spaces between consecutive fields. Intervening comments or blank lines do not affect the carry feature, but other non-i statements turn it off.

For the second parameter field p2 only, the symbol + gives the value of p2 + p3 from the previous i statement. This begins a note at the time the last one ended. The symbol + may also be carried using the carry feature described above. Liberal use of the carry and + features greatly simplify typing in and subsequent alteration of a score. Here, then, is a score illustrating simple FM synthesis with $f_m = f_c$, with gradually increasing index of modulation.

```
f1 0 8192 10 1                               ; sine wave
f2 0 513 7 0 80 1 50 0.7 213 0.7 170 0 ; ADSR
i1 1 1 10000 200 200 0                       ; index = 0
                                                (pure sine wave)
i1 + .    .      .    . 1            ; index = 1
i1 + .    .      .    . 2            ; index = 2
i1 + .    .      .    . 3            ; index = 3
i1 + .    .      .    . 4            ; index = 4
i1 + .    .      .    . 5            ; index = 5
e
```

Sections

An s statement consisting of a single s on a line by itself ends a section and starts a new one. Sorting of i and f statements (as well as a, which we haven't discussed) is done by section, and the timing starts again at the beginning for each section. Inactive instruments and data spaces are purged at the end of a section, and this frees up computer memory.

The following score, using the same instrument (8.11.1), has three sections with different ratios $f_m : f_c$ and with gradually increasing index of modulation.

```
f1 0 8192 10 1 ; sine wave
i1 1 1 10000 200 200 0 ; index = 0, fm:fc = 1:1
i1 + .    .      .    . 1 ; index = 1
i1 + .    .      .    . 2 ; index = 2
i1 + .    .      .    . 3 ; index = 3
i1 + .    .      .    . 4 ; index = 4
i1 + .    .      .    . 5 ; index = 5
s
i1 1 1 10000 200 400 0 ; index = 0, fm:fc = 1:2
i1 + .    .      .    . 1 ; index = 1
```

```
instr 1 ; FM bell
ifc  = cpspch(p5)              ; carrier frequency
ifm  = cpspch(p5) * 1.618   ; modulating frequency
   kenv  oscili  p4, 1/p3, 2 ; envelope, p3 = duration, exp
                                          decay f2
                              ; p4 = amplitude
   ktmb oscili ifm * 10, 1/p3, 3 ; timbre envelope,
                                           max = 10,
                              ; exp decay f3
   amod oscil ktmb, ifm, 1    ; modulator
   asig oscil kenv, ifc + amod, 1 ; carrier
      out asig
endin
```

Here is the score file to play notes E, C, D, G for a chime, using this instrument.

```
f1    0   8192     10   1
f2    0    513      5   1      513 .0001
f3    0    513      5   1      513 .001
i1    1     15   8000   8.04 ; 15 seconds at amplitude 8000
                                      at middle C
i1   2.5    .      .   8.00
i1    4     .      .   8.02
i1   5.5    .      .   7.07
e
```

A general purpose instrument

It is not hard to modify the instrument described above to make a general purpose two operator FM synthesis instrument.

```
instr 1                           ; Two operator FM instrument
ifc = cpspch(p5) * p6          ; p6 = carrier frequency
                                          multiplier

ifm = cpspch(p5) * p7          ; p7 = modulator frequency
                                          multiplier

   kenv oscili p4, 1/p3, p8    ; p3 = duration
                               ; p4 = amplitude
                               ; p8 = carrier envelope
   ktmb oscili ifm * p10, 1/p3, p9 ; p9 = modulator envelope
                               ; p10 = maximum index of
                                          modulation
   amod oscil ktmb, ifm, 1       ; modulator
   asig oscil kenv, ifc + amod, 1  ; carrier
      out asig
endin
```

The rest of the examples in this section are described in terms of this setup.

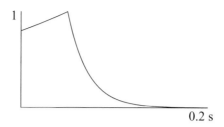

0.2 s

Figure 8.29 Amplitude envelope for a wood drum sound.

0.2 s

Figure 8.30 Index of modulation envelope for a wood drum sound.

The wood drum

To make a reasonably convincing wood drum, the amplitude envelope is made up of two exponential curves using GEN05 (Figure 8.29), while the envelope for the index of modulation is made up of two straight line segments, decreasing to zero and then staying there, using GEN07.

It turns out to be better to use a modulating frequency lower than the carrier frequency. So we use the reciprocal of the golden ratio, which is 0.618. We also use a large index of modulation, with a peak of 25, and a note duration of 0.2 seconds. This instrument works best in the octave going down from middle C. So the function table generators take the form

```
f1 0 8192   10  1                      ;  sine wave
f2 0  513    5  .8 128  1 385 .0001 ;  amplitude envelope
f3 0  513    7  1   64  0 449 0      ;  modulating index
                                           envelope
```

and the instrument statements take the form

```
i1 <time> 0.2 <amplitude> <pitch> 1.0 0.618 2 3 25
```

Brass

For a brass instrument, we use a harmonic spectrum containing all multiples of the fundamental. This is easily achieved by taking $f_c = f_m$. The relative amplitude of higher harmonics is greater when the overall amplitude is greater, so the timbre and amplitude are given the same envelope. This is chosen to look like the ADSR curve in Figure 8.2, to represent an overshoot in intensity during the attack. The index of modulation does not want to be as great as in the above examples. A maximum index of 5 gives a reasonable sound. The envelope given below is suitable for a note of duration around 0.6 seconds. It would need to be modified slightly for other durations.

```
f1 0 8192 10 1 ; sine wave
f2 0 513 7 0 85 1 86 0.75 256 0.7 86 0 ; envelope for brass
```

A typical note would then be represented by a statement of the form

```
i1 <time> 0.6 <amplitude> <pitch> 1.0 1.0 2 2 5
```

To improve the sound slightly on the brass tone presented here, we may wish to add a small deviation to the modulating frequency, so that there is a slight tremolo effect in the sound. If we replace the definition of the modulating frequency by the statement

```
ifm = cpspch(p5) * p7 + 0.5
```

then this will have the required effect.

Woodwind
For woodwind instruments, higher harmonics are present during the attack, and then the low frequencies enter. So we want the carrier frequency to be a multiple

of the modulating frequency, and use an envelope of the form ⟋‾‾⟍ for the

carrier and ⟋‾‾‾‾ for the modulator. So the function table generators take the

form

```
f1 0 8192 10 1                    ; sine wave
f2 0   513  7 0 50 1 443 1 20 0 ; amplitude envelope
f3 0   513  7 0 50 1 463 1        ; modulating index envelope
```

For a clarinet, where odd harmonics dominate, we take $f_c = 3 f_m$ and a maximum index of 2. A bassoon sound is produced by giving the odd harmonics a more irregular distribution. This can be achieved by taking $f_c = 5 f_m$ and a maximum index of 1.5.

8.13 Further techniques in CSound

The CSound language is vast. In this section, we cover just a few of the features which we have not touched on in the previous sections. For more information, see the CSound manual.

Tempo

The default tempo is 60 beats per minute, or one beat per second. To change this, a tempo statement is put in the score file. An example of the simplest form of tempo statement is

```
t 0 80
```

which sets the tempo to 80 beats per minute. The first argument (p1) of the tempo statement must always be zero. A tempo statement with more arguments causes accelerandos and ritardandos. The arguments are alternately times in beats (p1 = 0, p3, p5 ...) and tempi in beats per minute (p2, p4, p6, ...). The tempi between the specified times are calculate by making the durations of beats vary linearly. So, for example, the tempo statement

```
t 0 100 20 120 40 120
```

causes the initial tempo to be 100 beats per minute. By the twentieth beat, the tempo is 120 beats per minute. But the number of beats per minute is not linear between these values. Rather, the durations decrease linearly from 0.6 seconds to 0.5 seconds over the first twenty beats. The tempo is then constant from beat 20 until beat 40. By default, the tempo remains constant after the last beat where it is specified, so in this example the last two parameters are superfluous.

The tempo statement is only valid within the score section (cf. page 299) in which it is placed, and only one tempo statement may be used in each section. Its location within the section is irrelevant.

Stereo and Panning

For stereo output, we want to set nchnls = 2 in the header of the orchestra file (8.10.1). In the instrument definition, instead of using out, we use outs with two arguments. So, for example, to do a simple pan from left to right, we might want the following lines in the instrument definition.

```
kpanleft lineseg 0, p3, 1
kpanright = 1 - kpanleft
   outs asig * kpanleft, asig * kpanright
```

The problem with this method of panning is that the total sound energy is proportional to the square of amplitude, summed over the two channels. So in the middle of the pan, the total energy is only $1/\sqrt{2}$ times the total enery on the left or right. So it sounds like there's a hole in the middle. The easiest way to correct this is to take the square root of the straight line produced by the signal generator lineseg. So, for example, we could have the following lines.

```
kpan lineseg 0, p3, 1
kpanleft  = sqrt(kpan)
kpanright = sqrt(1-kpan)
```

Since $\sin^2 \theta + \cos^2 \theta = 1$, another way to keep uniform total sound energy is as follows.

```
kpan lineseg 0, p3, 1
ipibytwo = 1.5708
kpanleft = sin(kpan * ipibytwo)
kpanright = cos(kpan * ipibytwo)
```

A good trick for obtaining what sounds like a wider sweep for the pan, especially when using headphones to listen to the output, is to make the angle go from $-\pi/4$ to $3\pi/4$ instead of 0 to $\pi/2$. This can be achieved by replacing the definition of kpan above with the following line.

```
kpan lineseg -0.5, p3, 1.5
```

Display and spectral display
There is a facility for displaying either a waveform in an instrument file or its spectrum. So, for example, the instrument

```
instr 1
   asig oscil 10000 440 1
      out asig
      display asig p3
endin
```

is the same as (8.10.2), except that the extra line causes the graph of asig (of length p3) to be displayed. If the flag -d (see page 295) is set, this line makes no difference at all. Replacing the display line with

```
         dispfft asig p3, 1024
```

causes a fast Fourier transform of asig to be displayed, using an input window size of 1024 points. The number of points must be a power of two between 16 and 4096.

Arithmetic
In the orchestra file, variables represent signed floating point real numbers. The standard arithmetic operations +, -, * (times) and / (divide) can be used, as well as parentheses to any depth. Powers are denoted a^b, but b is not allowed

to be audio rate. The expression a % b returns a reduced modulo b. Among the available functions are

int (integer part),
frac (fractional part),
abs (absolute value),
exp (exponential function, raises e to the given power),
log and log10 (natural and base ten logarithm; argument must be positive),
sqrt (square root),
sin, cos and tan (sine, cosine and tangent, argument in radians),
sininv, cosinv, taninv (arcsine, arccos and arctan, answer in radians),
sinh, cosh and tanh (hyperbolic sine, cosine and tangent),
rnd (random number between zero and the argument),
birnd (random number bewteen plus and minus the argument).

Conditional values can also be used. For example,

(ka > kb ? 3 : 4)

has value 3 if ka is greater than kb, and 4 otherwise. Comparisons may be made using

> (greater than),
< (less than),
>= (greater than or equal to),
<= (less than or equal to),
== (equal to),
!= (not equal to).

Expressions, as well as variables, may be compared in this way, but audio rate variables and expressions are not permitted.

Automatic score generation
There are a number of methods of avoiding the tedious process of writing a score file for CSound. One method is to use the *score translator* programme scot. This takes a text file <filename>.sc written in a compressed score notation and writes out a score file <filename>.sco. Another is to use Cscore, which is a programme for making and manipulating score files. The user writes a control programme in the C language, which makes use of a set of function definitions contained in a header file cscore.h. Finally, there is MIDI2CS, a programme which takes a MIDI file as input, and outputs a score file. There is also a considerable amount of support for MIDI within the CSound language.

CSoundAV is a realtime version of CSound for the PC, and can be obtained from Gabriel Maldonado's home page at

web.tiscali.it/G-Maldonado/

I have not tried it out, so I cannot comment on how well it works, but it looks promising.

Further reading on CSound

Richard Boulanger (2000), *The CSound Book*.
Electronic Musician, February 1998 issue.
Keyboard, January 1997 issue.

8.14 Other methods of synthesis

Sampling is not really a form of synthesis at all, but is often used in digital synthesizers. It is usual to sample sounds at only a small collection of pitches, and then to pitch shift by stretching or compressing the waveform, in order to fill in the gaps. Pitch shifting a digital signal introduces high frequency noise, related to the fact that the sample rate is not being shifted at the same time. This is removed using a low pass filter.

Wavetable synthesis is a method related to sampling, in which digitally recorded wave files are used as raw material to produce sounds which are a sort of hybrid between synthesis and sampling. It is usual to use one wave file for the attack portion of the sound, and another for the sustain portion. In the case of the sustain portion, a whole number of periods of the sound are used to form a loop which is repeated. An envelope is then applied to shape the sound, and then finally the result is pitch shifted and put through a low pass filter. An exception to this general procedure is 'one shot' sounds such as short percussive sounds. These are usually just recorded as a single wavefile without looping.

Granular synthesis is a method where the sound comes in small packets called *grains*, whose duration is usually of the order of ten milliseconds. Thousands of these grains are used in each second, to create a sound texture. Usually, some algorithm is used for describing large quantities of grains at a time, so that each grain does not have to be described separately.

Further reading on granular synthesis

S. Cavaliere and A. Piccialli, Granular synthesis of musical signals, article 5 in Roads *et al.* (1997), pages 155–186.

John Duesenberry, Square one: a world in a grain of sound, *Electronic Musician*,
November 1999.
Curtis Roads, Granular synthesis of sound, *Computer Music Journal* **2** (2) (1978), 61–62.
A revised and updated version of this article appears as Chapter 10 of Roads and
Strawn (1985), pages 145–159.
Curtis Roads, Granular synthesis, *Keyboard*, June 1997.
Curtis Roads (2001), *Microsound*.

8.15 The phase vocoder

The phase vocoder is a method of sound analysis and manipulation. It is based
on the technique of applying a discrete Fourier transform to small windows of the
original sound. The transform may then be manipulated, and finally the sound may
reconstructed from the manipulated transform. For example, it is not hard to stretch
a sound without altering the pitch using this technique.

Further reading

Mark Dolson, The phase vocoder: a tutorial, *Computer Music Journal* **10** (4) (1986),
14–27.
Marie-Hélène Serra, Introducing the phase vocoder, article 2 in Roads *et al.* (1997), pages
31–90.

8.16 Chebyshev polynomials

Composition of functions in general is a good way of obtaining synthetic tones. For
example, if we take a basic cosine wave $\cos vt$ and compose it with the function
$f(x) = 2x^2 - 1$, then we obtain

$$2\cos^2 vt - 1 = \cos 2vt.$$

So, composing with this function has the effect of doubling frequency. The corresponding functions for arbitrary integer multiples of frequency are called the
Chebyshev[5] *polynomials of the first kind*, which we now investigate. Let $T_n(x)$ be
the polynomial defined inductively by $T_0(x) = 1$, $T_1(x) = x$, and for $n > 1$,

$$T_n(x) = 2x T_{n-1}(x) - T_{n-2}(x).$$

[5] Other spellings for this name include Tchebycheff and Chebichev. These are all just transliterations of the
Russian Чебышев.

Thus, for example, we have

$$T_0(x) = 1,$$
$$T_1(x) = x,$$
$$T_2(x) = 2x^2 - 1,$$
$$T_3(x) = 4x^3 - 3x,$$
$$T_4(x) = 8x^4 - 8x^2 + 1,$$
$$T_5(x) = 16x^5 - 20x^3 + 5x,$$
$$T_6(x) = 32x^6 - 48x^4 + 18x^2 - 1,$$
$$T_7(x) = 64x^7 - 112x^5 + 56x^3 - 7x.$$

Lemma 8.16.1 *For $n \geq 0$ we have $T_n(\cos vt) = \cos nvt$.*

Proof The proof is by induction on n. We begin by observing that

$$\cos vt \, \cos(n-1)vt - \sin vt \, \sin(n-1)vt = \cos nvt,$$
$$\cos vt \, \cos(n-1)vt + \sin vt \, \sin(n-1)vt = \cos(n-2)vt$$

(see Section 1.8), so that, adding and rearranging, we have

$$\cos nvt = 2 \cos vt \, \cos(n-1)vt - \cos(n-2)vt.$$

Now for $n = 0$ and $n = 1$, the statement of the lemma is obvious from the definition. For $n \geq 2$, assuming the statement to be true for smaller values of n, we have

$$T_n(\cos vt) = 2 \cos vt \, T_{n-1}(\cos vt) - T_{n-2}(\cos vt)$$
$$= 2 \cos vt \, \cos(n-1)vt - \cos(n-2)vt$$
$$= \cos nvt.$$

So, by induction, the lemma is true for all $n \geq 0$. □

Using a weighted sum of Chebyshev polynomials and composing, we can obtain a waveform with the corresponding weights for the harmonics. Changing the weighting with time will change the timbre of the resulting tone. So, for example, if we apply the operation

$$T_1 + \tfrac{1}{3}T_3 + \tfrac{1}{5}T_5 + \tfrac{1}{7}T_7 + \tfrac{1}{9}T_9 + \tfrac{1}{11}T_{11}$$

to a cosine wave, we obtain an approximation to a square wave (see Equation (2.2.10)). This operation will turn any mixture of cosine waves into the same mixture of square waves.

Exercises

1. Show that $y = T_n(x)$ satisfies *Chebyshev's differential equation*

$$(1 - x^2)\frac{d^2 y}{dx^2} - x\frac{dy}{dx} + n^2 y = 0.$$

2. Show that

$$T_n(x) = x^n - \binom{n}{2}x^{n-2}(1 - x^2) + \binom{n}{4}x^{n-4}(1 - x^2)^2 - \cdots$$

Hint: use de Moivre's theorem and the binomial theorem.

3. Draw a graph of $y = T_n(x)$ for $-1 \le x \le 1$ and $0 \le n \le 5$.

9

Symmetry in music

Limerick

First, let me explain that I'm cursed;
I'm a poet whose time gets reversed.
   Reversed gets time
   Whose poet a I'm;
Cursed I'm that explain me let, first.

9.1 Symmetries

Music contains many examples of symmetry. In this chapter, we investigate the symmetries that appear in music, and the mathematical language of group theory for describing symmetry.

We begin with some examples. *Translational symmetry* looks like Figure 9.1. In group theoretic language, which we explain in the next few sections, the symmetries form an *infinite cyclic group*. In music, this would just be represented by *repetition* of some rhythm, melody, or other pattern. Figure 9.2 shows the beginning of the right hand of Beethoven's *Moonlight Sonata*, Op. 27 No. 2.

Of course, any actual piece of music only has finite length, so it cannot really have true translational symmetry. Indeed, in music, approximate symmetry is much more common than perfect symmetry. The musical notion of a *sequence* is a good example of this. A sequence consists of a pattern that is repeated with a shift; but the shift is usually not exact. The intervals are not the same, but rather they are modified to fit the harmony. For example, the sequence shown in Figure 9.3 comes from J. S. Bach's *Toccata and Fugue in D*, BWV 565, for organ. Although the general motion is downwards, the numbers of semitones between the notes in the triplets is constantly varying in order to give the appropriate harmonic structure.

Reflectional symmetry appears in music in the form of *inversion* of a figure or phrase. For example, the bar from Béla Bartók's *Fifth String Quartet*

Figure 9.1 Translational symmetry.

Figure 9.2 Opening of Beethoven's *Moonlight Sonata*.

Figure 9.3 Part of Bach's *Toccata and Fugue in D*.

Figure 9.4 Bar from Bartók's *Fifth String Quartet*. With kind permission of Bossey & Hawkes.

shown in Figure 9.4 displays a reflectional symmetry whose horizontal axis is the note B♭. The lower line is obtained by inverting the upper line. The symmetry group here is cyclic of order two.

Such symmetry can also be more global in character. For example, in Richard Strauss' *Elektra* (1906–1908), although symmetry plays little or no role in the choice of individual notes, its influence is apparent in the choice of keys. The introduction starts with Agamemnon's motive in in D minor. Then Elektra's motive consists of B minor and F minor triads, symmetrically placed around D. Then in Elektra's monologue, Agamemnon is associated with B♭ and Klytemnestra with F♯, again symmetrical around D. The opera continues this way, working either side of the initial D. The ending is in C major, with a prominent major third E in the last four

Figure 9.5 Opening of Chopin's *Waltz*, Op. 34 No. 2, left hand.

bars. These observations are taken from pages 15–16 of Antokoletz, *The Music of Béla Bartók* (reference on page 321).

It is more common for horizontal reflection to be combined with a displacement in time. For example, the left hand of Chopin's *Waltz*, Op. 34 No. 2, begins as shown in Figure 9.5. Each bar of the upper line of the left hand is inverted to form the next bar. Because of the displacement in time, this is really a *glide reflection*; namely a translation followed by a reflection about a mirror parallel to the direction of translation. In group theoretic terms, this is another manifestation of the infinite cyclic group. See Figure 9.7.

The reason for the importance of symmetry in music is that regularity of pattern builds up expectations as to what is to come next. But it is important to break the expectations from time to time, to prevent boredom. Good music contains just the right balance of predictability and surprise.

In the above example, the mirror line for the reflectional symmetry was horizontal. It is also possible to have *temporal* reflectional symmetry with a vertical mirror line, so that the notes form a palindrome. For example, an ascending scale followed by a descending scale has this kind of reflectional symmetry, as in the elementary vocal exercise shown in Figure 9.8. The symmetry group here is cyclic of order two.

This is the musical equivalent of the *palindrome*. One example of a musical form involving this kind of symmetry is the *retrograde canon* or *crab canon* (Cancrizans). This term denotes a work in the form of a canon and exhibiting temporal reflectional symmetry by means of playing the melody forwards and backwards at the same time. For example, the first canon of J. S. Bach's *Musical Offering* (BWV 1079) is a retrograde canon formed by playing Frederick the Great's royal theme, consisting of the 18 bars shown in Figure 9.9, simultaneously forwards and backwards in this way. The first voice starts at the beginning of the first bar and works forward to the end, while the second voice starts at the end of the last bar and works backwards to the beginning. Other examples can be found at the end of this section, under 'further listening'. The other parts of Bach's *Musical Offering* exhibit various other tricky ways of playing with symmetry and form.

Examples of *rotational* symmetry can also be found in music. For example, the four note phrase in Figure 9.10 has perfect rotational symmetry, whose centre is at the end of the second beat, at the pitch D♯.

Der Spiegel (The Mirror) Duet

Public Domain. Sequenced by Fred Nachbaur using NoteWorthy
Confused? Try playing this from opposite sides of a table.

(Note: the attribution to Mozart is dubious)

Figure 9.6.

Figure 9.11 Translation and rotation.

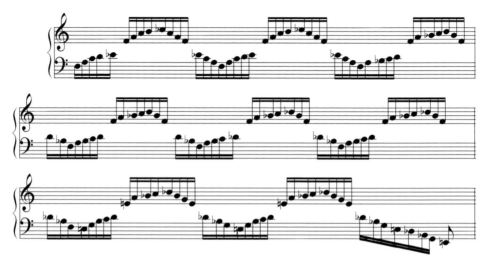

Figure 9.12 Part of Mozart's *Capriccio*, K. 395.

Figure 9.11. In group theoretic language, the symmetries form an *infinite dihedral group*.

In Figure 9.12, from the middle of Mozart's *Capriccio*, K. 395 for piano, the symmetry is approximate. It is easy to observe that each beamed set of notes for the right hand has a gradual rise followed by a steeper descent, while those for the left hand have a steep descent followed by a more gradual rise. Each pair of beams is slightly different from the previous, so we do not get bored. Our expectations are finally thwarted in the last beam, where the descent continues all the way down to a low E♮.

Horizontally repeated patterns are sometimes known as *frieze patterns*, and they are classified into seven types. The numbering scheme shown in Figure 9.13 is the international one usually used by mathematicians and crystallographers, for reasons which are not likely to become clear any time soon (see, for example, pages 39 and 44 of Grünbaum and Shephard, reference on page 321). The abstract groups are explained later on in this chapter.

Example	name	abstract group
	p111	$\mathbb{Z}$
	p1a1	$\mathbb{Z}$
	p1m1	$\mathbb{Z} \times \mathbb{Z}/2$
	pm11	D_∞
	p112	D_∞
	pma2	D_∞
	pmm2	$D_\infty \times \mathbb{Z}/2$

Figure 9.13 The seven frieze types.

For example, the upper line of the left hand of the Chopin Waltz example in Figure 9.5 belongs to frieze type p1a1, while the Ravel example in Figure 9.10 belongs to frieze type p112.

Exercises

1. What symmetry is present in the following extract (with kind permission of Bossey & Hawkes) from Béla Bartók's *Music for Strings, Percussion and Celeste*? Is it exact or approximate?

2. Find the symmetries in the following two bars from *The Lamb*. (Music by John Tavener; words by William Blake. Copyright 1976, 1989 Chester Music Limited. All rights reserved. International copyright secured. Reprinted by permission.) Are the symmetries exact or approximate?

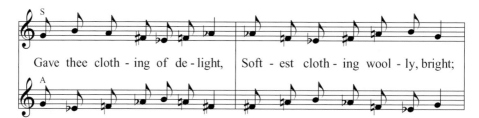

Gave thee cloth - ing of de - light,   Soft - est cloth - ing wool - ly, bright;

Gave thee cloth - ing of de - light,   Soft - est cloth - ing wool - ly, bright;

3. The symmetry in the first two bars of Schoenberg's *Klavierstück* Op. 33a is somewhat harder to see.

You may find it helpful to draw the chords on a circle; the first chord will come out as follows.

4. Which frieze pattern appears in the first few bars of Debussy's *Rêverie*, which are as follows?

5. (Perle, 1977, page 20) Find the symmetries in the following three bars from the beginning of Berg's *Lyric Suite* (bars 2–4).

You may find it helpful to draw the notes on a circle, as in question **3**, and break them up into two sets of six.

Figure 9.14 Piano symmetry.

Further reading

Elliott Antokoletz, *The Music of Béla Bartók*, University of California Press, 1984.
Bruce Archibald, Some thoughts on symmetry in early Webern, *Perspectives in New Music* **10** (1972), 159–163.
K. Bailey, Symmetry as nemesis: Webern and the first movement of the Concerto, Opus 24, *J. Music Theory* **40** (2) (1996), 245–310.
J. W. Bernard, Space and symmetry in Bartók, *J. Music Theory* **30** (2) (1986), 185–201.
F. J. Budden, *The Fascination of Groups*, Cambridge University Press, 1972. Chapter 23 is titled Groups and music.
Branko Grünbaum and G. C. Shephard, *Tilings and Patterns, an Introduction*. W. H. Freeman and Company, 1989.
E. Lendvai (1993), *Symmetries of Music*.
R. P. Morgan, Symmetrical form and common-practice tonality, *Music Theory Spectrum* **20** (1) (1998), 1–47.
D. Muzzulini, Musical modulation by symmetries, *J. Music Theory* **39** (2) (1995), 311–327.
L. J. Solomon, New symmetric transformations, *Perspectives in New Music* **11** (2) (1973), 257–264.
George Perle, Symmetric formations in the string quartets of Béla Bartók, *Music Review* **16** (1955), 300–312.

Further listening (see Appendix G)

William Byrd, *Diliges Dominum* exhibits temporal reflectional symmetry, making it a perfect palindrome.
In Joseph Haydn's *Sonata 41* in A, the movement *Menuetto al rovescio* is also a perfect palindrome.
The first and last of the 25 pieces making up Paul Hindemith's *Ludus Tonalis* are the *Praeludium* and the *Postludium*; the latter is obtained from the former by a perfect rotation, but with the addition of one final bar.

Figure 9.15 An unfinished keyboard piece employing invertible counterpoint. From Prof. Peter Schickele, *The definitive biography of P.D.Q. Bach* (1807–1742)?, Random House, 1976.

Guillaume de Machaut, *Ma fin est mon commencement* (My end is my beginning) is a retrograde canon in three voices, with a palindromic tenor line. The other two lines are exact temporal reflections of each other.

9.2 The harp of the Nzakara

In this section, we take a look at an example taken from the article of Chemillier in (2002). The Nzakara and Zande people of the Central African Republic, Congo and Sudan have a musical tradition of the court which is now in a state of neglect. The music consists of poetry sung to the accompaniment of a five string harp. The harpist plays a formulaic repeating pattern of pairs of notes.

The five strings of the harp are tuned to notes which can be transcribed roughly as C, D, E, G, B♭. These five strings are regarded as having a cyclic order rather than a linear order, so that the lowest string is regarded as adjacent to the highest string.

$$
\begin{array}{ccc}
 & 0 & \\
4 & & 1 \\
 & & \\
3 & & 2
\end{array}
$$

The strings are plucked in pairs, and the two strings of a pair are *never* adjacent in the cycle. So there are only five possible pairs. The strings in the pair have a unique common neighbor, and we can label the pair using this common neighbor. So the five pairs are as follows.

label	strings	
0	1	4
1	0	2
2	1	3
3	2	4
4	0	3

The repeating harp patterns are divided into categories with names such as *ngbàkià*, *limanza* and *gitangi*. An example of a *limanza* line is given by repeating the following sequence of pairs.

Transcribing this using our labels, we obtain the sequence

$$120141403424231202014030 3422313.$$

At first sight, it is hard to see any pattern. But we divide it into groups of six as follows.

$$12\ 014140\ 342423\ 120201\ 403034\ 2313.$$

Since the pattern is supposed to repeat, the initial pair can be thought of as being at the end of the last group of four to make a group of six,

$$014140\ 342423\ 120201\ 403034\ 231312.$$

Now we can see that each group of six is obtained from the previous group by moving two places down the cycle of five strings. This forms a sort of twisted translational symmetry.

There is also a kind of rotational symmetry (this explains why we chose to move two time slots from the beginning to the end). We can reverse time, giving

$$213132\ 430304\ 102021\ 324243\ 041410$$

and then reverse the cyclic ordering of the five strings, by replacing string x by string $2 - x$ (mod 5). This gives the sequence

014140 342423 120201 403034 231312,

which is the same as the original sequence.

Exercise

Here is a repeating *ngbàkià* harp line taken from the same article of Chemillier.

Find the symmetries in this pattern.

Further reading

Marc Chemillier, Mathématiques et musiques de tradition orale, pages 133–143 of
      Genevois and Orlarey (1997).
Marc Chemillier, Ethnomusicology, ethnomathematics. The logic underlying orally
      transmitted artistic practices, pages 161–183 of Assayag *et al.* (2002).

Further listening (see Appendix G)

Marc Chemillier, *Central African Republic. Music of the former Bandia courts.*

9.3 Sets and groups

The mathematical structure which captures the notion of symmetry is the notion of a *group*. In this section, we give the basic axioms of group theory, and we describe how these axioms capture the notion of symmetry.

A *set* is just a collection of objects. The objects in the set are called the *elements* of the set. We write $x \in X$ to mean that an object x is an element of a set X, and we write $x \notin X$ to mean that x is not an element of X.

Strictly speaking, a set shouldn't be too big. For example, the collection of all sets is too big to be a set, and if we allow it to be a set then we run into Russell's paradox, which goes as follows. If the collection of all sets is regarded as a set, then it is possible for a set to be an element of itself: $X \in X$. Now form the set S consisting of all sets X such that $X \notin X$. If $S \notin S$ then S is one of the sets X satisfying the condition for being in S, and so $S \in S$. On the other hand, if $S \in S$ then S is not one of these sets X, and so $S \notin S$. This contradictary conclusion is Russel's paradox. Fortunately, finite and countably infinite collections are small enough to be sets, and we are mostly interested in such sets.[1] If a set X is finite, we write $|X|$ for the number of elements in X.

A *group* is a set G together with an operation which takes any two elements g and h of G and multiplies them to give again an element of G, written gh. For G to be a group, this multiplication must be defined for all pairs of elements g and h in G, and it must satisfy three axioms:

(i) (Associative law) Given any elements g, h and k in G (not necessarily different from each other), if we multiply gh by k we get the same answer as if we multiply g by hk:

$$(gh)k = g(hk).$$

(ii) (Identity) There is an element $e \in G$ called the *identity element*, which has the following property. For every element g in G, we have $eg = g$ and $ge = g$.

(iii) (Inverses) For each element $g \in G$, there is an *inverse* element, written g^{-1}, with the property that $gg^{-1} = e$ and $g^{-1}g = e$.

It is worth noticing that a group does *not* necessarily satisfy the commutative law. An *abelian group* is a group satisfying the following axiom in addition to axioms (i)–(iii):

(iv) (Commutative law)[2] Given any elements g and h in G, we have $gh = hg$.

We can give a group by writing down a *multiplication table*. For example, here is the multiplication table for a group with three elements.

	e	a	b
e	e	a	b
a	a	b	e
b	b	e	a

To multiply elements g and h of a group using a multiplication table, we look in row g and column h, and the entry is gh. So for example, looking in the above table, we see that $ab = e$. The above example is an abelian group, because the

---

[1] For a reasonably modern and sophisticated introduction to set theory, I recommend W. Just and M. Weese, *Discovering Modern Set Theory*, two volumes, published by the American Mathematical Society, 1995. None of the sophistication of modern set theory is necessary for music theory.

[2] In real life, as in group theory, operations seldom satisfy the commutative law. For example, if we put on our socks and then put on our shoes, we get a very different effect from doing it the other way round. The associative law is much more commonly satisfied.

table is symmetric about its diagonal. The following multiplication table describes
a nonabelian group G with six elements.

	e	v	w	x	y	z
e	e	v	w	x	y	z
v	v	w	e	y	z	x
w	w	e	v	z	x	y
x	x	y	z	e	v	w
y	y	z	x	w	e	v
z	z	x	y	v	w	e

In this group, we have $xy = v$ but $yx = w$, which shows that the group is not
abelian. We write $|G| = 6$ to indicate that the group G has six elements.

Groups don't have to be finite of course. For example, the set $\mathbb{Z}$ of integers with
operation of *addition* forms an abelian group. Usually, a group operation is only
written additively if the group is abelian. The identity element for the operation of
addition is 0, and the additive inverse of an integer n is $-n$.

It should by now be apparent that multiplication tables aren't a very good way of
describing a group. Suppose we want to check that the above multiplication table
satisfies the axioms (i)–(iii). We would have to make $6 \times 6 \times 6 = 216$ checks just
for the associative law. Now try to imagine making the checks for a group with
thousands of elements, or even millions.

Fortunately, there is a better way, based on permutation groups. A *permutation*
of a set X is a function f from X to X such that each element y of X can be written
as $f(x)$ for a unique $x \in X$. See also page 332, where this is described as a *bijective*
function from X to itself. This ensures that f has an *inverse function*, f^{-1}, which
takes y back to x. So we have $f^{-1}(f(x)) = f^{-1}(y) = x$, and $f(f^{-1}(y)) = f(x) =$
y.

For example, if $X = \{1, 2, 3, 4, 5\}$, the function f defined by

$$f(1) = 3, \quad f(2) = 5, \quad f(3) = 4, \quad f(4) = 1, \quad f(5) = 2$$

is a permutation of X, whose inverse is given by

$$f^{-1}(1) = 4, \quad f^{-1}(2) = 5, \quad f^{-1}(3) = 1, \quad f^{-1}(4) = 3, \quad f^{-1}(5) = 2.$$

There are two common notations for writing permutations on finite sets, both
of which are useful. The first notation lists the elements of X and where they
go. In this notation, the permutation f described above would be written as
follows.

$$\begin{pmatrix} 1 & 2 & 3 & 4 & 5 \\ 3 & 5 & 4 & 1 & 2 \end{pmatrix}$$

The other notation is called *cycle notation*. For the above example, we notice that 1 goes to 3 goes to 4 goes back to 1 again, and 2 goes to 5 goes back to 2. So we write the permutation as

$$f = (1, 3, 4)(2, 5).$$

This notation is based on the fact that if we apply a permutation repeatedly to an element of a finite set, it will eventually cycle back round to where it started. The entire set can be split up into disjoint cycles in this way, so that each element appears in one and only one cycle. If a permutation is written in cycle notation, to see its effect on an element, we locate the cycle containing the element. If the element is not at the end of the cycle, the permutation takes it to the next one in the cycle. If it is at the end, it takes it back to the beginning. The *length* of a cycle is the number of elements appearing in it. If a cycle has length one, then the element appearing in it is a *fixed point* of the permutation. Fixed points are often omitted when writing a permutation in cycle notation.

To multiply permutations, we compose functions. In the above example, suppose we have another permutation g of the same set X, given by

$$g = \begin{pmatrix} 1 & 2 & 3 & 4 & 5 \\ 2 & 5 & 1 & 4 & 3 \end{pmatrix}$$

or in cycle notation,

$$g = (1, 2, 5, 3)(4).$$

If we omit the fixed point 4 from the notation, this element is written $g = (1, 2, 5, 3)$. Then $f(g(1)) = f(2) = 5$. Continuing this way, fg is the following permutation,

$$fg = \begin{pmatrix} 1 & 2 & 3 & 4 & 5 \\ 5 & 2 & 3 & 1 & 4 \end{pmatrix} = (1, 5, 4)$$

whereas gf is given by

$$gf = \begin{pmatrix} 1 & 2 & 3 & 4 & 5 \\ 1 & 3 & 4 & 2 & 5 \end{pmatrix} = (2, 3, 4).$$

The *identity permutation* takes each element of X to itself. In the above example, the identity permutation is

$$e = \begin{pmatrix} 1 & 2 & 3 & 4 & 5 \\ 1 & 2 & 3 & 4 & 5 \end{pmatrix} = (1)(2)(3)(4)(5).$$

Omitting fixed points from the identity permutation leaves us with a rather embarrassing empty space, which we fill with the sign e denoting the identity element.

The *order* of a permutation is the number of times it has to be applied to get back to the identity permutation. In the above example, f has order six, g has order four, and both fg and gf have order three. The order of an element g of any group is defined in the same way, as the least positive value of n such that $g^n = 1$. If there is no such n, then g is said to have *infinite order*. For example, the translation which began the chapter is a transformation of infinite order, whereas a reflection is a transformation of order two.

Notice how the commutative law is not at all built into the world of permutations, but the associative law certainly is. The inverse of a permutation is a permutation, and the composite of two permutations is also a permutation. So it is easy to check whether a collection of permutations forms a group. We just have to check that the identity is in the collection, and that the inverses and composites of permutations in the collection are still in the collection.

The set of *all* permutations of a set X forms a group which is called the *symmetric group* on the set X, with the multiplication given by composing permutations as above. We write the symmetric group on X as $\mathsf{Symm}(X)$. If $X = \{1, 2, \ldots, n\}$ is the set of integers from 1 to n, then we write S_n for $\mathsf{Symm}(X)$. Notice that the sets X and $\mathsf{Symm}(X)$ are quite different in size. If $X = \{1, 2, \ldots, n\}$ then X has n elements, but $\mathsf{Symm}(X)$ has $n!$ elements. To see this, if $f \in \mathsf{Symm}(X)$ then there are n possibilities for $f(1)$. Having chosen the value of $f(1)$, there are $n - 1$ possibilities left for $f(2)$. Continuing this way, the total number of possibilities for f is $n(n - 1)(n - 2) \cdots 1 = n!$.

The definition of a *permutation group* is that it is a subgroup of $\mathsf{Symm}(X)$ for some set X. In general, a *subgroup* H of a group G is a subset of G which is a group in its own right, with multiplication inherited from G. This is the same as saying that the identity element belongs to H, inverses of elements of H are also in H, and products of elements of H are in H. So to check that a set H of permutations of X is a group, we check these three properties so that H is a subgroup of $\mathsf{Symm}(X)$. Notice that the associative law is automatic for permutations, and does not need to be checked.

Exercises

1. If g and h are elements of a group, explain why gh and hg always have the same order.
2. Show that composition of functions always satisfies the associative law.

Further reading

Hans J. Zassenhaus, *The Theory of Groups*. Dover reprint, 1999. This is a solid introduction to group theory, originally published in 1949 by Chelsea.

9.4 Change ringing

The art of change ringing is peculiar to the English, and, like most English peculiarities, unintelligible to the rest of the world. To the musical Belgian, for example, it appears that the proper thing to do with a carefully tuned ring of bells is to play a tune upon it. By the English campanologist, the playing of tunes is considered to be a childish game, only fit for foreigners; the proper use of the bells is to work out mathematical permutations and combinations. When he speaks of the music of his bells, he does not mean musicians' music—still less what the ordinary man calls music. To the ordinary man, in fact, the pealing of bells is a monotonous jangle and a nuisance, tolerable only when mitigated by remote distance and sentimental association. The change-ringer does, indeed, distinguish musical differences between one method of producing his permutations and another; he avers, for instance, that where the hinder bells run 7, 5, 6, or 5, 6, 7, or 5, 7, 6, the music is always prettier, and can detect and approve, where they occur, the consecutive fifths of Tittums and the cascading thirds of the Queen's change. But what he really means is, that by the English method of ringing with rope and wheel, each several bell gives forth her fullest and her noblest note. His passion—and it is a passion—finds its satisfaction in mathematical completeness and mechanical perfection, and as his bell weaves her way rhythmically up from lead to hinder place and down again, he is filled with the solemn intoxication that comes of intricate ritual faultlessly performed.

(Dorothy L. Sayers, The Nine Tailors, *1934)*

The symmetric group, described at the end of the last section, is essential to the understanding of *change ringing*, or *campanology*. This art began in England in the tenth century, and continues in thousands of English churches to this day. A set of swinging bells in the church tower is operated by pulling ropes. There are generally somewhere between six and twelve bells. The problem is that the bells are heavy, and so the timing of the peals of the bells is not easy to change. So, for example, if there were eight bells, played in sequence as

$$1, 2, 3, 4, 5, 6, 7, 8,$$

then in the next round we might be able to change the timings of some adjacent bells in the sequence to produce

$$1, 3, 2, 4, 5, 7, 6, 8,$$

but we would not be able to move the timing of a bell in the sequence by more than one position. So the general rules for change ringing state that a change ringing composition consists of a sequence of rows. Each row is an order for the set of bells, and the position of a bell in the row can differ by at most one from its previous position. It is also stipulated that a row is not repeated in a composition, except that

```
1  2  3  4
2  1  4  3
2  4  1  3
4  2  3  1
4  3  2  1
3  4  1  2
3  1  4  2
1  3  2  4
1  3  4  2
3  1  2  4
3  2  1  4
2  3  4  1
2  4  3  1
4  2  1  3
4  1  2  3
1  4  3  2
1  4  2  3
4  1  3  2
4  3  1  2
3  4  2  1
3  2  4  1
2  3  1  4
2  1  3  4
1  2  4  3
1  2  3  4
```

Figure 9.16 Plain Bob.

the last row returns to the beginning. So, for example, Plain Bob on four bells goes as shown in Figure 9.16.

This sequence of rows is really a walk around the symmetric group S_4. So the image of the first row under each of the $4! = 24$ elements of S_4 appears exactly once in the list, except that the first is repeated as the last.

In order to fix the notation, we think of a row as a function from the bells to the time slots. To go from one row to the next, we compose with a permutation of the set of time slots. The permutation is only allowed to fix a time slot, or to swap it with an adjacent time slot. So in the above example, the first few steps involve alternately applying the permutations $(1, 2)(3, 4)$ and $(1)(2, 3)(4)$. Then when we reach the row 1 3 2 4, this prescription would take us back to the beginning. In order to avoid this, the permutation $(1)(2)(3, 4)$ is applied instead of $(1)(2, 3)(4)$, and then we may continue as before. At the line 1 4 3 2 we again have the problem that we would be taken to a previously used row, and we avert this by the same method. When we have exhausted all the permutations in S_4, we return to the beginning.

Exercise

The *Plain Hunt* consists of alternately applying the permutations

$$a = (1, 2)(3, 4)(5, 6)\ldots$$
$$b = (1)(2, 3)(4, 5)\ldots$$

If the number of bells is n, how many rows are there before the return to the initial order?

(Hint: treat separately the cases n even and n odd.)

Further reading

F. J. Budden, *The Fascination of Groups*, Cambridge University Press, 1972. Chapter 24 is titled Ringing the changes: groups and campanology.

D. J. Dickinson, On Fletcher's paper "Campanological groups", *Amer. Math. Monthly* **64** (5) (1957), 331–332.

T. J. Fletcher, Campanological groups, *Amer. Math. Monthly* **63** (9) (1956), 619–626.

B. D. Price, Mathematical groups in campanology, *Math. Gaz.* **53** (1969), 129–133.

R. A. Rankin, A campanological problem in group theory, *Math. Proc. Camb. Phil. Soc.* **44** (1948), 17–25.

R. A. Rankin, A campanological problem in group theory, II, *Math. Proc. Camb. Phil. Soc.* **62** (1966), 11–18.

J. F. R. Stainer, Change-ringing, *Proc. Musical Assoc., 46th Sess.* (1919–20), 59–71.

Ian Stewart, *Another Fine Math You've Got Me Into...*, W. H. Freeman & Co., 1992. Chapter 13 of this book, The group-theorist of Notre Dame, is about change ringing.

Richard G. Swan, A simple proof of Rankin's campanological theorem, *Amer. Math. Monthly* **106** (2) (1999), 159–161.

Arthur T. White, Ringing the changes, *Math. Proc. Camb. Phil. Soc.* **94** (1983), 203–215.

Arthur T. White, Ringing the changes II, *Ars Combinatorica* **20–A** (1985), 65–75.

Arthur T. White, Ringing the cosets, *Amer. Math. Monthly* **94** (8) (1987), 721–746.

Arthur T. White, Ringing the cosets II, *Math. Proc. Camb. Phil. Soc.* **105** (1989), 53–65.

Arthur T. White, Fabian Stedman: the first group theorist? *Amer. Math. Monthly* **103** (9) (1996), 771–778.

Arthur T. White and Robin Wilson, The hunting group, *Mathematical Gazette* **79** (1995), 5–16.

Wilfred G. Wilson, *Change Ringing*, October House Inc., 1965.

9.5 Cayley's theorem

Cayley's theorem explains why the axioms of group theory exactly capture the physical notion of symmetry. It says that any abstract group, in other words, any set with a multiplication satisfying the axioms described in Section 9.3, can be realised as a group of permutations of some set.

There is something mildly puzzling about this theorem. Where are we going to produce a set from? We're just given a group, and nothing else. So we do the obvious thing, and use the set of elements of the group itself as the set on which it will act as permutations. So, before reading this, make very sure you have separated in your mind the set of elements of a permutation group and the set on which it acts by permutations. Because otherwise what follows will be very confusing.

the addition table for this clock arithmetic.

+	0	1	2	3	4	5	6	7	8	9	10	11
0	0	1	2	3	4	5	6	7	8	9	10	11
1	1	2	3	4	5	6	7	8	9	10	11	0
2	2	3	4	5	6	7	8	9	10	11	0	1
3	3	4	5	6	7	8	9	10	11	0	1	2
4	4	5	6	7	8	9	10	11	0	1	2	3
5	5	6	7	8	9	10	11	0	1	2	3	4
6	6	7	8	9	10	11	0	1	2	3	4	5
7	7	8	9	10	11	0	1	2	3	4	5	6
8	8	9	10	11	0	1	2	3	4	5	6	7
9	9	10	11	0	1	2	3	4	5	6	7	8
10	10	11	0	1	2	3	4	5	6	7	8	9
11	11	0	1	2	3	4	5	6	7	8	9	10

To emphasize that an addition is being done in clock arithmetic rather than ordinary arithmetic, it is often written using the *congruence* symbol '$\equiv$' rather than the equals sign, as in

$$6 + 8 \equiv 2 \pmod{12}.$$

More generally, $a \equiv b \pmod{n}$ means that $a - b$ is a multiple of n.

In terms of group theory, the above addition table makes the set $\{0, 1, 2, 3, 4, 5, 6, 7, 8, 9, 10, 11\}$ into a group. The operation is written as addition; of course, clock arithmetic is abelian. The identity element is 0, and the inverse of i is either $-i$ or $12 - i$, depending which is in the range from 0 to 11. This group is written as $\mathbb{Z}/12$.

There is an obvious homomorphism from the group $\mathbb{Z}$ to $\mathbb{Z}/12$. It takes an integer to the unique integer in the range from 0 to 11 which differs from it by a multiple of 12.

In musical terms, we could think of the numbers from 0 to 11 as representing musical intervals in multiples of semitones, in the twelve tone equal tempered octave. So, for example, 1 is represented by the permutation which increases each note by one semitone, namely the permutation

$$\begin{pmatrix} C & C\sharp & D & E\flat & E & F & F\sharp & G & G\sharp & A & B\flat & B \\ C\sharp & D & E\flat & E & F & F\sharp & G & G\sharp & A & B\flat & B & C \end{pmatrix}$$

The circulating nature of clock arithmetic then becomes *octave equivalence* in the musical scale, where two notes belong to the same *pitch class* if they differ by a whole number of octaves. Each element of $\mathbb{Z}/12$ is then represented by a different permutation of the twelve pitch classes, with the number i representing an increase of i semitones. So, for example, the number 7 represents the permuation which makes each note higher by a fifth. Then addition has an obvious interpretation as addition of musical intervals.

This permutation representation looks like Cayley's theorem. But making this precise involves choosing a starting point somewhere in the octave. We choose to start by representing C as 0, so that the correspondence becomes

C	C♯	D	E♭	E	F	F♯	G	G♯	A	B♭	B
0	1	2	3	4	5	6	7	8	9	10	11

Under this correspondence, each element of $\mathbb{Z}/12$ is being represented by the permutation of the twelve notes of the octave given by Cayley's theorem.

Of course, there is nothing special about the number 12 in clock arithmetic. If n is any positive integer, we may form the group $\mathbb{Z}/n$ whose elements are the integers in the range from 0 to $n - 1$. Addition is described by adding as integers, and then subtracting n if necessary to put the answer back in the right range. So, for example, if we were interested in 31 tone equal temperament, which gives such a good approximation to quarter comma meantone (see Section 6.5), then we would use the group $\mathbb{Z}/31$.

Further reading

Gerald J. Balzano, The group-theoretic description of 12-fold and microtonal pitch systems, *Computer Music Journal* **4** (4) (1980), 66–84.

Paul Isihara and M. Knapp, Basic $\mathbb{Z}_{12}$ analysis of musical chords. With loose erratum, *UMAP J.* **14** (1993), 319–348.

D. Lewin, A label-free development for 12-pitch-class systems, *J. Music Theory* **21** (1) (1977), 29–48.

Paul F. Zweifel, Generalized diatonic and pentatonic scales: a group-theoretic approach. *Perspectives of New Music* **34** (1) (1996), 140–161.

9.7 Generators

If G is a group, a subset S of the set of elements of G is said to *generate* G if every element of G can be written as a product of elements of S and their inverses.[3] We say that G is *cyclic* if it can be generated by a single element g. In this case, the elements of the group can all be written in the form g^n with $n \in \mathbb{Z}$. The case $n = 0$ corresponds to the identity element, while negative values of n are interpreted to give powers of the inverse of g.

There are two kinds of cyclic groups. If there is no nonzero value of n for which g^n is the identity element, then the elements g^n multiply the same way that the integers n add. In this case, the group is isomorphic to the additive group $\mathbb{Z}$ of integers. If there is a nonzero value of n for which g^n is the identity element, then,

---

[3] To clarify, an empty product is considered to be the identity element. So if S is empty and G is the group with one element, then S does generate G.

by inverting if necessary, we can assume that n is positive. Then letting n be the smallest positive number with this property, it is easy to see that G is isomorphic to the group $\mathbb{Z}/n$ described in the last section.

How many generators does $\mathbb{Z}/n$ have? We can find out whether an integer i generates $\mathbb{Z}/n$ with the help of some elementary number theory.

Lemma 9.7.1 *Let d be the greatest common divisor of n and i. Then there are integers r and s such that $d = rn + si$.*

Proof This follows from Euclid's algorithm for finding the greatest common divisor of two integers.

Let's just recall how Euclid's algorithm goes, and then we'll see how it enables you to write the greatest common divisor in this form. If we're given two integers, let's assume that they're positive (otherwise, just negate them) and that the second is bigger than the first (otherwise, swap them round). If the first is an exact divisor of the second, then it is the greatest common divisor. If it isn't, subtract as many of the first as you can from the second without going negative, and then swap them round. Now repeat.

For example, suppose we're given the integers 24 and 34. Since 24 is smaller than 34, we subtract 24 from 34 and swap them round, so our new numbers are 10 and 24. We can now subtract two 10s from 24 and swap them round to get 4 and 10. We subtract two 4s from 10 and swap to get 2 and 4. Now 2 is an exact divisor of 4, so 2 is the greatest common divisor.

If we keep track of the operations, it enables us to write 2 as $r \times 24 + s \times 34$:

$$10 = -24 + 34$$
$$4 = 24 - 2 \times 10 = 24 - 2 \times (34 - 24) = 3 \times 24 - 2 \times 34$$
$$2 = 10 - 2 \times 4 = (-24 + 34) - 2(3 \times 24 - 2 \times 34) = -7 \times 24 + 5 \times 34.$$

So we have $r = -7$ and $s = 5$.   □

If i has no common factor with n, then $d = 1$, and the above equation says that s times i, considered as the sth power of i in the additive group $\mathbb{Z}/n$, is equal to 1. Since the element 1 is a generator of $\mathbb{Z}/n$, it follows that i is also a generator.

On the other hand, if n and i have a common factor $d > 1$, then all powers of i in $\mathbb{Z}/n$ (i.e., all multiples of i when thinking additively) give numbers divisible by d, so the number 1 is not a power of i. So we have the following.

Theorem 9.7.2 *The generators for $\mathbb{Z}/n$ are precisely the numbers i in the range $0 < i < n$ with the property that n and i have no common factor.* □

The number of possibilities for i in the above theorem is written $\phi(n)$, and called the *Euler phi function* of n.

For example, if $n = 12$, then the possibilities for i are 1, 5, 7 and 11, and so $\phi(12) = 4$. In terms of musical intervals, the fact that 7 is a generator for $\mathbb{Z}/12$ corresponds to the fact that all notes can be obtained from a given notes by repeatedly going up by a fifth. This is the *circle of fifths*. So it can be seen that, apart from the circle of semitones upwards and downwards, the only other ways of generating all the musical intervals is via the circle of fifths, again upwards or downwards. This, together with the consonant nature of the fifth, goes some way toward explaining the importance of the circle of fifths in music.

It is interesting to see that if $n = p$ happens to be a prime number, for example $p = 31$, then *every* element of $\mathbb{Z}/p$ apart from zero is a generator. So $\phi(p) = p - 1$.

In fact, there is a recipe for finding $\phi(n)$ in general, which goes as follows. If $n = p^a$ is a power of a prime then $\phi(n) = p^{a-1}(p-1)$. If m and n are relatively prime (i.e., have no common factors greater than one), then $\phi(mn) = \phi(m)\phi(n)$. Any positive integer can be written as a product of prime powers for different primes, so this gives a recipe for calculating $\phi(n)$. For example,

$$\phi(72) = \phi(2^3.3^2) = \phi(2^3)\phi(3^2) = 2^2(2-1)3(3-1) = 24.$$

Here are the values of $\phi(n)$ for small values of n.

n	1	2	3	4	5	6	7	8	9	10	11	12	13	14	15	16
$\phi(n)$	0	1	2	2	4	2	6	4	6	4	10	4	12	6	8	8

Exercises

1. Write down the generators for $\mathbb{Z}/24$. What is $\phi(24)$?
2. Show that each generator x of $\mathbb{Z}/n$ satisfies $x^2 \equiv 1 \pmod{n}$ if and only if n is a divisor of 24.
3. Find (a) $\phi(49)$, (b) $\phi(60)$, (c) $\phi(142)$, (d) $\phi(10\,000)$.
4. Let $\mathbb{C}^\times$ be the group of nonzero complex numbers under multiplication. Show that there are exactly n different homomorphisms from $\mathbb{Z}/n$ to $\mathbb{C}^\times$ (these are called the *characters* of $\mathbb{Z}/n$, and they play an important role in number theory and many other parts of mathematics). How many of these homomorphisms are injective? What do these homomorphisms have to do with the discrete Fourier transform of Section 7.9?

9.8 Tone rows

In twelve tone music, one begins with a *twelve tone row*, which consists of a sequence of twelve pitch classes in order, so that each of the twelve possible pitch classes appears just once.

If we want to be able to look at music which is not formally described as twelve tone as well, we should consider sequences of pitch classes of any length, and with possible repetitions.

A *transposition*[4] of a sequence $\mathbf{x}$ of pitch classes by n semitones is the sequence $\mathbf{T}^n(\mathbf{x})$ in which each of the pitch classes in $\mathbf{x}$ has been increased by n semitones. So, for example, if

$$\mathbf{x} = 3\ 0\ 8$$

then

$$\mathbf{T}^4(\mathbf{x}) = 7\ 4\ 0.$$

As another example, the first two bars of Chopin's *Étude*, Op. 25 No. 10, consist of the pitches

6–5–6 7–8–9 8–7–8 9–10–11 | 10–9–10 11–0–1 0–11–0 1–2–3

played as triplets, octave doubled, in both hands simultanously. The second half of the first bar is obtained by applying the transformation $\mathbf{T}^2$ to the first half. The transformation $\mathbf{T}^2$ is applied again to obtain the first half of the second bar, and again for the second half. So, if $\mathbf{x}$ is the sequence 6 5 6 7 8 9, then these two bars can be written

$$\mathbf{x}\quad \mathbf{T}^2(\mathbf{x}) \mid \mathbf{T}^4(\mathbf{x})\ \mathbf{T}^6(\mathbf{x}).$$

Bars 3 and 4 of this piece go as follows.

2–3–4 3–4–5 4–5–6 5–6–7 | 6–7–8 7–8–9 7–8–9 8–9–10.

Writing $\mathbf{y}$ for the sequence 2 3 4, we see that the last group in bar 2 is $\mathbf{T}^{-1}(\mathbf{y})$, while bars 3 and 4 can be written

$$\mathbf{y}\quad \mathbf{T}(\mathbf{y})\ \mathbf{T}^2(\mathbf{y})\ \mathbf{T}^3(\mathbf{y}) \mid \mathbf{T}^4(\mathbf{y})\ \mathbf{T}^5(\mathbf{y})\ \mathbf{T}^5(\mathbf{y})\ \mathbf{T}^6(\mathbf{y}).$$

Turning to the next operation, *inversion* $\mathbf{I}(\mathbf{x})$ of a sequence $\mathbf{x}$ just replaces each pitch class by its negative (in clock arithmetic). So, in the first example above with $\mathbf{x} = 3\ 0\ 8$, we have

$$\mathbf{I}(\mathbf{x}) = 9\ 0\ 4.$$

---

[4] Unfortunately, in group theory the word *transposition* is used to refer to a permutation which leaves all but two points fixed, and swaps those two points. These two usages from music and mathematics are not related, and this can be a source of confusion.

   Music theorists generally write $\mathbf{T}_n$ instead of $\mathbf{T}^n$; we shall stick with $\mathbf{T}^n$, as it conforms better to group theoretic notation.

The sequences $\mathbf{T}^n\mathbf{I}(\mathbf{x})$ are also regarded as inversions of $\mathbf{x}$. So, for example,

$$\mathbf{T}^6\mathbf{I}(\mathbf{x}) = 3\ 6\ 10$$

is an inversion of the above sequence $\mathbf{x}$.

The *retrograde* $\mathbf{R}(\mathbf{x})$ of $\mathbf{x}$ is just the same sequence in reverse order. So, in the above example,

$$\mathbf{R}(\mathbf{x}) = 8\ 0\ 3.$$

We have the following relations among the operations $\mathbf{T}$, $\mathbf{I}$ and $\mathbf{R}$:

$$\mathbf{T}^{12} = e, \quad \mathbf{T}^n\mathbf{R} = \mathbf{R}\mathbf{T}^n, \quad \mathbf{T}^n\mathbf{I} = \mathbf{I}\mathbf{T}^{-n}, \quad \mathbf{R}\mathbf{I} = \mathbf{I}\mathbf{R},$$

where e represents the identity operation, which does nothing (another name for this operation is $\mathbf{T}^0$). All relations between the operations $\mathbf{T}$, $\mathbf{I}$ and $\mathbf{R}$ follow from these.

There are four forms of a tone row $\mathbf{x}$. The *prime* form is the original form $\mathbf{x}$ of the row, or any of its transpositions $\mathbf{T}^n(\mathbf{x})$. The *inversion* form is any one of the rows $\mathbf{T}^n\mathbf{I}(\mathbf{x})$. The *retrograde* form is any one of the rows $\mathbf{T}^n\mathbf{R}(\mathbf{x})$. Finally, the *retrograde inversion* form of the row is any one of the rows $\mathbf{T}^n\mathbf{R}\mathbf{I}(\mathbf{x})$.

In group-theoretic terms, the operations $\mathbf{T}^n$ ($0 \le n \le 11$) form a cyclic group $\mathbb{Z}/12$. The operation $\mathbf{R}$ together with the identity operation form a cyclic group $\mathbb{Z}/2$. The operations $\mathbf{T}$ and $\mathbf{R}$ commute. The group theoretic way of describing a group with two types of operations which commute with each other is a Cartesian product, which we describe in Section 9.9. The relationship between $\mathbf{T}$ and $\mathbf{I}$ is more complicated, and is discussed in Section 9.10.

Exercise

Spot the retrograde tone row near the end of Spike Jones' *Liebestraum*.

Further reading

Allen Forte (1973), *The Structure of Atonal Music*.
George Perle (1977), *Twelve-tone Tonality*.
John Rahn, *Basic Atonal Theory*, Schirmer Books, 1980.

9.9 Cartesian products

If G and H are groups, then the *Cartesian product*, or *direct product*, $G \times H$ is the group whose elements are the ordered pairs (g, h) with $g \in G$ and $h \in H$. The multiplication is defined by

$$(g_1, h_1)(g_2, h_2) = (g_1 g_2, h_1 h_2).$$

The identity element is formed from the identity elements of G and H. The inverse of (g, h) is (g^{-1}, h^{-1}). The axioms of a group are easily verified, so that $G \times H$ with this multiplication does form a group.

Suppose that G and H are subgroups of a bigger group K, with the properties that each element of G commutes with each element of H, the only element which G and H have in common is the identity element (written $G \cap H = \{1\}$), and every element of K can be written as a product of an element of G and an element of H (written $K = GH$). Then there is an isomorphism from $G \times H$ to K given by sending (g, h) to gh. In this case, K is said to be an *internal direct product* of G and H.

For example, the group whose elements are the operations $\mathbf{T}^n$ and $\mathbf{T}^n \mathbf{R}$ of Section 9.8 is an internal direct product of the subgroup consisting of the operations $\mathbf{T}^n$ and the subgroup consisting of the identity and $\mathbf{R}$. So this group is isomorphic to $\mathbb{Z}/12 \times \mathbb{Z}/2$.

As another example, the lattice $\mathbb{Z}^2$ which we used in order to describe just intonation in Section 6.8 is really a direct product $\mathbb{Z} \times \mathbb{Z}$, where $\mathbb{Z}$ is the group of integers under addition, as usual. This can be viewed as an internal direct product, where the two copies of $\mathbb{Z}$ consist of the elements $(n, 0)$ and the elements $(0, n)$ for $n \in \mathbb{Z}$. Similarly, the lattice $\mathbb{Z}^3$ of Section 6.9 is $\mathbb{Z} \times \mathbb{Z} \times \mathbb{Z}$. This can be viewed as an internal direct product of three copies of $\mathbb{Z}$ consisting of the elements $(n, 0, 0)$, the elements $(0, n, 0)$ and the elements $(0, 0, n)$ with $n \in \mathbb{Z}$.

Exercises

1. Find an isomorphism between $\mathbb{Z}/3 \times \mathbb{Z}/4$ and $\mathbb{Z}/12$. Interpret this in terms of transpositions by major and minor thirds.

2. Show that there is no isomorphism between $\mathbb{Z}/12 \times \mathbb{Z}/2$ and $\mathbb{Z}/24$.
   (Hint: how many elements of order two are there?)

3. The group $\mathbb{Z}/2 \times \mathbb{Z}/2$ is called the *Klein four group*. Go back to Exercise **1** in Section 9.1 and explain what the Klein four group has to do with this example.

9.10 Dihedral groups

The operations $\mathbf{T}$ and $\mathbf{I}$ of Section 9.8 do not commute, but rather satisfy the relations $\mathbf{T}^n \mathbf{I} = \mathbf{I} \mathbf{T}^{-n}$. So we do not obtain a direct product in this case, but rather a more complicated construction, which in this case describes a *dihedral group*.

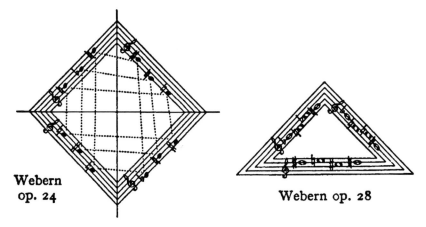

Webern op. 24

Webern op. 28

Figure 9.17.

A dihedral group has two elements g and h such that $h^2 = 1$ and $gh = hg^{-1}$. Every element is either of the form g^i or of the form $g^i h$. The powers of g form a cyclic subgroup which is either $\mathbb{Z}/n$ or $\mathbb{Z}$. In the former case, the group has $2n$ elements and is written[5] D_{2n}. In the latter case, the group has infinitely many elements, and is written D_∞ and called the *infinite dihedral group*. This is one of the groups which appeared in Section 9.1.

So the operations $\mathbf{T}^n$ and $\mathbf{T}^n\mathbf{I}$ form a group isomorphic to the dihedral group D_{24}. Finally, putting all this together, the group whose operations are

$$\mathbf{T}^n, \quad \mathbf{T}^n\mathbf{R}, \quad \mathbf{T}^n\mathbf{I}, \quad \mathbf{T}^n\mathbf{RI}$$

form a group which is isomorphic to $D_{24} \times \mathbb{Z}/2$.

The dihedral group D_{2n} has an obvious interpretation as the group of rigid symmetries of a regular polygon with n sides.

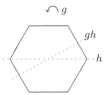

The element g corresponds to counterclockwise rotation through $1/n$ of a circle, while h corresponds to reflection about a horizontal axis. Then $g^i h$ corresponds to a reflection about an axis of symmetry which is rotated from the horizontal by i/n of a semicircle. The above diagram is for the case $n = 6$.

---

[5] Some authors write D_n for the dihedral group of order $2n$, just to confuse matters. Presumably these authors think that I'm confusing matters.

Exercises

1. Find an isomorphism between the dihedral group D_6 and the symmetric group S_3.

2. Find an isomorphism between D_{12} and $S_3 \times \mathbb{Z}/2$.

3. Show that D_{24} is not isomorphic to $S_3 \times \mathbb{Z}/4$.

4. Consider the group D_{24} generated by **T** and **I**. Which elements fix the following *diminished seventh* chord setwise? What sort of a group do they form?

5. Repeat Exercise **4** with the following 'augmented triad'.

6. Discuss the Nzakara harp example of Section 9.2 in terms of the Cartesian product of dihedral groups $D_{10} \times D_\infty$.

9.11 Orbits and cosets

If a group G acts as permutations on a set X, then we say that two elements x and x' of X are in the same *orbit* if there is an element $g \in G$ such that $g(x) = x'$. This partitions X into disjoint subsets, each consisting of elements related this way. These subsets are the orbits of G on X.

So, for example, if G is a cyclic group generated by an element g, then the cycles of g as described in Section 9.3 are the orbits of G on X.

As another example, the group $\mathbb{Z}/12$ acts on the set of tone rows of a given length, via the operations $\mathbf{T}^n$. Two tone rows are in the same orbit exactly when one is a transposition of the other.

If there is only one orbit for the action of G on X, we say that G acts *transitively* on X. So, for example, $\mathbb{Z}/12$ acts transitively on the set of twelve pitch classes, but not on the set of tone rows of a given length bigger than one.

We discussed the related concept of cosets briefly in Section 6.8. Here we make the discussion more precise, and show how this concept is connected with permutations. If H is a subgroup of a group G, we can partition the elements of G into *left cosets* of H as follows. Two elements g and g' are in the same left coset of H in G if there exists some element $h \in H$ such that $gh = g'$. This partitions the group G into disjoint subsets, each consisting of elements related in this way. These subsets are the left cosets of H in G. The notation for the left coset containing g is gH. So gH and $g'H$ are equal precisely when there exists an element $h \in H$ such that $gh = g'$; in other words, when $g^{-1}g'$ is an element of H. The coset gH consists of all the elements gh as h runs through the elements of H. The way of writing this is

$$gH = \{gh \mid h \in H\}.$$

The left cosets of H in G all have the same size as H does. So the number of left cosets, written $|G : H|$, is equal to $|G|/|H|$.

The example in Section 6.8 goes as follows. The group G is $\mathbb{Z}^2 = \mathbb{Z} \times \mathbb{Z}$. The subgroup H is the unison sublattice. Each coset consists of a set of vectors related by translation by the unison sublattice. The group theoretic notion corresponding to a periodicity block is a *set of coset representatives*. A set of left coset representatives for a subgroup H in a group G just consists of a choice of one element from each left coset.

If G acts as permutations on a set X, then there is a close connection between orbits and cosets of subgroups, which can be described in terms of *stabilizers*. If x is an element of X, then the stabilizer in G of x, written $\text{Stab}_G(x)$, is the subgroup of G consisting of the elements h satisfying $h(x) = x$.

Theorem 9.11.1 *Let $H = \text{Stab}_G(x)$. Then the map sending the coset gH to the element $g(x) \in X$ is well defined, and establishes a bijective correspondence between the left cosets of H in G and the elements of X in the orbit containing x.*

Proof To say that the map is well defined is to say that if we are given another element g' such that $gH = g'H$, then $g(x) = g'(x)$. The reason why this is true is that there is an element $h \in H$ such that $gh = g'$, and then $g'(x) = gh(x) = g(h(x)) = g(x)$.

To see that the map is injective, if $g(x) = g'(x)$ then $x = g^{-1}g'(x)$ and so $g^{-1}g' \in H$, and $gH = g'H$. It is obviously surjective, by the definition of an orbit. ☐

A consequence of this theorem is that the size of an orbit is equal to the index of the stabilizer of one of its elements,

$$|\text{Orbit}(x)| = |G : \text{Stab}_G(x)|. \tag{9.11.1}$$

9.12 Normal subgroups and quotients

In the last section, we discussed left cosets of a subgroup. Of course, right cosets make just as much sense; the reason why left rather than right cosets made their appearance in understanding orbits was that we write functions on the left of their arguments. We write Hg for the right coset containing g, so that

$$Hg = \{hg \mid h \in H\}.$$

It does not always happen that the left and right cosets of H are the same. For example, if G is the symmetric group S_3, and H is the subgroup consisting of the identity and the permutation (12), then the left cosets are

$$\{e, (12)\}, \quad \{(123), (13)\}, \quad \{(132), (23)\},$$

while the right cosets are

$$\{e, (12)\}, \quad \{(123), (23)\}, \quad \{(132), (13)\}.$$

This is because $(123)(12) = (13)$ while $(12)(123) = (23)$.

A subgroup N of G is said to be *normal* if the left cosets and the right cosets agree. For example, if G is abelian, then every subgroup is normal.

Theorem 9.12.1 *A subgroup N of G is normal if and only if, for each $g \in G$, we have $gNg^{-1} = N$.*

Proof To say that the subgroup N is normal means that for each $g \in G$ we have $gN = Ng$. Multiplying on the right by g^{-1}, and noticing that this can be undone by multiplication on the right by g, we see that this is equivalent to the condition that for each $g \in G$ we have $gNg^{-1} = N$. □

If N is normal in G, then the cosets of N in G can be made into a group called the *quotient group* of G by N, and denoted G/N, as follows. If $g_1 N$ and $g_2 N$ are cosets then we multiply them to form the coset $g_1 g_2 N$. To check that this is well defined, we must check that if $g_1 N = g_1' N$ then $g_1 g_2 N = g_1' g_2 N$, and that if $g_2 N = g_2' N$ then $g_1 g_2 N = g_1 g_2' N$. The second of these checks is easy enough, and just uses the associativity of multiplication. But for the first we must use normality. The easiest way to do this is to switch to right cosets, where we are checking that if $Ng_1 = Ng_1'$ then $Ng_1 g_2 = Ng_1' g_2$. This is like the second check for left cosets, and just uses the associativity of multiplication. Without normality, the multiplication of left cosets is not well defined.

To check that the axioms for a group are satisfied by this multiplication of cosets, we need an identity element, which is provided by the coset $eN = N$ containing the identity element e of G. The inverse of the coset gN is the coset $g^{-1}N$. It is an easy exercise to check the axioms with these definitions.

Clock arithmetic is a good example of a quotient group. Inside the additive group $\mathbb{Z}$ of integers, we have a (normal) subgroup $n\mathbb{Z}$ consisting of the integers divisible by n. The quotient group $\mathbb{Z}/n\mathbb{Z}$ is the clock arithmetic group, which we have been writing in the more usual notation $\mathbb{Z}/n$.

Another example is given by the unison vectors and periodicity blocks of Section 6.8. The quotient of $\mathbb{Z}^2$ (or more generally $\mathbb{Z}^n$) by the unison sublattice is a finite abelian group whose order is equal to the absolute value of the determinant of the matrix formed from the unison vectors.

There is a standard theorem of abstract algebra which says that every finite abelian group can be written in the form

$$\mathbb{Z}/n_1 \times \mathbb{Z}/n_2 \times \cdots \times \mathbb{Z}/n_r.$$

The positive integers $n_1, \ldots, n_r$ are not uniquely determined; for example $\mathbb{Z}/12$ is isomorphic to $\mathbb{Z}/3 \times \mathbb{Z}/4$. However, they can be chosen in such a way that each one is a divisor of the next one. If they are chosen in this way, then they are uniquely determined, and then they are called the *elementary divisors* of the finite abelian group. There is a standard algorithm for finding the elementary divisors, which can be found in many books on abstract algebra. From the point of view of scales, it seems relevant to try to choose the unison sublattice so that the quotient group is cyclic, which corresponds to the case where there is just one elementary divisor.

There is an intimate relationship between normal subgroups and homomorphisms. If f is a homomorphism from G to H, then the *kernel* of f is defined to be the set of elements $g \in G$ for which $f(g)$ is equal to the identity element of H. Writing N for the kernel of f, it is not hard to check that N is a normal subgroup of G.

Theorem 9.12.2 (First Isomorphism Theorem) *Let f be a homomorphism from G to H, with kernel N. Then there is an isomorphism between the quotient group G/N and the subgroup of H consisting of the image of the homomorphism f. This isomorphism takes a coset gN to $f(g)$.*

Proof There are a number of things to check here. We need to check that the function from G/N to the image of f which takes gN to $f(g)$ is well defined, that it is a group homomorphism, that it is injective, and that its image is the same as the image of f. These checks are all straightforward, and are left for the reader to fill in. □

There are actually three isomorphism theorems in elementary group theory, but we shall not mention the second or third.

An example of the first isomorphism theorem is again provided by clock arithmetic. The homomorphism from $\mathbb{Z}$ to $\mathbb{Z}/12$ is surjective and has kernel $12\mathbb{Z}$, and so $\mathbb{Z}/12$ is isomorphic to the quotient of $\mathbb{Z}$ by $12\mathbb{Z}$, as we already knew.

9.13 Burnside's lemma

This section and the next are concerned with problems of counting. A typical example of the kind of problem we are interested in is as follows. Recall that a tone

row consists of the twelve possible pitch classes in some order. The total number
of tone rows is

$$12 \times 11 \times 10 \times 9 \times \cdots \times 3 \times 2 \times 1 = 12!$$

or 479 001 600.

We might wish to count the number of possible twelve tone rows, where two
tone rows are considered to be the same if one can be obtained from the other by
applying an operation of the form $\mathbf{T}^n$. In this case, each tone row has twelve distinct
images under these operations. So the total number of tone rows up to this notion
of equivalence is $1/12$ of the number of tone rows, or $11! = 39\,916\,800$.

If we want to complicate the situation further, we might consider two tone rows
to be equivalent if one can be obtained from the other using the operations $\mathbf{T}^n$, $\mathbf{I}$
and $\mathbf{R}$. Now the problem is that some of the tone rows are fixed by some of the
elements of the group. So the counting problem degenerates into a lot of special
cases, unless we find a more clever way of counting. This is the kind of problem
that can be solved using Burnside's counting lemma.

The abstract formulation of the problem is that we have a finite group acting as
permutations on a finite set, and we want to know the number of orbits.

Burnside's lemma allows us to count the number of orbits of a finite group G
on a finite set X, provided we know the number of fixed points of each element
$g \in G$. It says that the number of orbits is the average number of fixed points.

Lemma 9.13.1 (Burnside) *Let G be a finite group acting by permutations on a
finite set X. For an element $g \in G$, write $n(g)$ for the number of fixed points of g
on X. Then the number of orbits of G on X is equal to*

$$\frac{1}{|G|} \sum_{g \in G} n(g).$$

Proof We count in two different ways the number of pairs (g, x) consisting of an
element $g \in G$ and a point $x \in X$ such that $g(x) = x$. If we count the elements of
the group first, then for each element of the group we have to count the number of
fixed points, and we get $\sum_{g \in G} n(g)$. On the other hand, if we count the elements
of X first, then for each x, Equation (9.11.1) shows that the number of elements
$g \in G$ stabilizing it is equal to $|G|$ divided by the length of the orbit in which x
lies. So each orbit contributes $|G|$ to the count.   □

So let us return to the problem of counting tone rows. Suppose that we wish to
count the number of tone rows, and we wish to regard one tone row as equivalent to
another if the first can be manipulated to the second using the operations $\mathbf{T}$, $\mathbf{I}$ and $\mathbf{R}$.
In other words, we wish to count the number of orbits of the group $G = D_{24} \times \mathbb{Z}/2$
generated by $\mathbf{T}$, $\mathbf{I}$ and $\mathbf{R}$ on the set X of tone rows.

In order to apply Burnside's lemma, we should find the number of tone rows fixed by each operation in the group. The identity operation fixes all tone rows, so that one is easy. The operations $\mathbf{T}^n$ with $1 \leq n \leq 11$ don't fix any tone rows, so that's also easy. The operation $\mathbf{R}$ fixes the tone rows whose last six entries are the reverse of the first six; but then there are repetitions so these aren't allowed as tone rows. For the operation $\mathbf{T}^6\mathbf{R}$, the fixed tone rows are the ones where the last six entries are the reverse of the first six, but transposed by a tritone (half an octave). So the first six have to be chosen in a way that uses just one of each pair related by a tritone. The number of ways of doing this is

$$12 \times 10 \times 8 \times 6 \times 4 \times 2 = 46\,080.$$

For values of n other than zero or six, $\mathbf{T}^n\mathbf{R}$ does not fix any tone rows, because doing this operation twice gives $\mathbf{T}^{2n}$, which doesn't fix any tone rows.

Next, we need to consider inversions. The operation $\mathbf{I}$ fixes only those tone rows comprised of the entries 0 and 6; but then there must be repetitions, so these aren't tone rows. The same goes for any operation of the form $\mathbf{T}^n\mathbf{I}$; the entries come from a subset of size at most two, so we can't form a tone row this way.

Finally, for an operation $\mathbf{T}^n\mathbf{IR}$, the entries in a fixed tone row are again determined by the first six entries. So the tone row has the form

$$a_1,\ a_2,\ a_3,\ a_4,\ a_5,\ a_6,\ n - a_6,\ n - a_5,\ n - a_4,\ n - a_3,\ n - a_2,\ n - a_1.$$

If n is even, there is some tone fixed by $\mathbf{T}^n\mathbf{I}$, which forces us to repeat a tone, so there are no fixed tone rows. If n is odd, however, there are fixed tone rows, and there are

$$12 \times 10 \times 8 \times 6 \times 4 \times 2 = 46\,080$$

of them.

We summarize this information in the following table.

operation	how many in G	fixed points
identity	1	479001600
$\mathbf{T}^n$ ($1 \leq n \leq 11$)	11	0
$\mathbf{T}^6\mathbf{R}$	1	46080
$\mathbf{T}^n\mathbf{R}$ ($n \neq 6$)	11	0
$\mathbf{T}^n\mathbf{I}$	12	0
$\mathbf{T}^n\mathbf{IR}$ (n even)	6	0
$\mathbf{T}^n\mathbf{IR}$ (n odd)	6	46080

So the sum over $g \in G$ of the number of fixed points of g on X is

$$479\,001\,600 + 7 \times 46\,080 = 479\,324\,160.$$

Dividing by $|G| = 48$, the total number of orbits of G on tone rows is equal to 9985 920. This proves the following theorem.

Theorem 9.13.2 (David Reiner) *If two twelve tone rows are considered the same when one may be obtained from the other using the operations* **T**, **I** *and* **R**, *then the total number of tone rows is* 9985 920. $\qquad\square$

Further reading

James A. Fill and Alan J. Izenman, Invariance properties of Schoenberg's tone row system, *J. Austral. Math. Soc.* B **21** (1979/80), 268–282.

James A. Fill and Alan J. Izenman, The structure of RI-invariant twelve-tone rows, *J. Austral. Math. Soc.* B **21** (1979/80), 402–417.

Colin D. Fox, Alban Berg the mathematician, *Math. Sci.* **4** (1979), 105–107.

David Reiner, Enumeration in music theory, *Amer. Math. Monthly* **92** (1) (1985), 51–54.

9.14 Pitch class sets

A *pitch class set* is defined to be a subset of the set of twelve pitch classes. For convenience, we number the pitch classes $\{0, 1, \ldots, 11\}$ as in Section 9.6.

Atonal theorists and composers such as Milton Babbitt, Allen Forte and Elliott Carter put an equivalence relation on pitch class sets. They say that two pitch class sets are *equivalent* if one can be obtained from the other using only transpositions $\mathbf{T}^n$ and inversion $\mathbf{I}$. In other words, the equivalence classes are the orbits of the dihedral group D_{24} generated by $\mathbf{T}$ and $\mathbf{I}$ on the collection of subsets of $\{0, 1, \ldots, 11\}$.

We can use Burnside's lemma 9.13.1 to count how many equivalence classes there are of each size. For this purpose, we need to count the fixed points of the elements of D_{24} on the collection of sets of a given size. It is easy to verify the following table.

Group element	\multicolumn{13}{c}{size of subset}

Group element	0	1	2	3	4	5	6	7	8	9	10	11	12
Identity	1	12	66	220	495	792	924	792	495	220	66	12	1
$\mathbf{T}, \mathbf{T}^5, \mathbf{T}^7, \mathbf{T}^{11}$	1	0	0	0	0	0	0	0	0	0	0	0	1
$\mathbf{T}^2, \mathbf{T}^{10}$	1	0	0	0	0	0	2	0	0	0	0	0	1
$\mathbf{T}^3, \mathbf{T}^9$	1	0	0	0	3	0	0	0	3	0	0	0	1
$\mathbf{T}^4, \mathbf{T}^8$	1	0	0	4	0	0	6	0	0	4	0	0	1
$\mathbf{T}^6$	1	0	6	0	15	0	20	0	15	0	6	0	1
$\mathbf{T}^{2m}\mathbf{I}$	1	2	6	10	15	20	20	20	15	10	6	2	1
$\mathbf{T}^{2m+1}\mathbf{I}$	1	0	6	0	15	0	20	0	15	0	6	0	1

For example, the first row just consists of the binomial coefficients $\binom{12}{j}$, where j is the size of the subset. The remaining rows of the table for powers of **T** are also just binomial coefficients, but interspersed with zeros. The inversions $\mathbf{T}^n\mathbf{I}$ come in two varieties. If $n = 2m + 1$ is odd, then there are no fixed pitch classes. So the fixed subsets have even size and the numbers are again binomial coefficients $\binom{6}{j}$, where $2j$ is the size of the subset. If $n = 2m$ is even, then there are two fixed pitch classes, so there are $2\binom{5}{j}$ fixed subsets of odd size $2j + 1$.

We can now apply Burnside's lemma 9.13.1 to find how many orbits of D_{24} there are on the subsets of various sizes. The answers are as follows.

size of subset	0	1	2	3	4	5	6	7	8	9	10	11	12
number of orbits	1	1	6	12	29	38	50	38	29	12	6	1	1

For example, to compute how many subsets there are of size 5, we compute

$$\tfrac{1}{24}(792 + 6 \times 20) = \tfrac{912}{24} = 38.$$

For reference, we can also compute the number of orbits under the group $\mathbb{Z}/12$ consisting of powers of **T** using the same data. The answers are as follows.

size of subset	0	1	2	3	4	5	6	7	8	9	10	11	12
number of orbits	1	1	6	19	43	66	80	66	43	19	6	1	1

Incidentally, the reason for the symmetry in the above tables is that complementation gives a one-to-one correspondence between subsets of size j and subsets of size $12 - j$, and this correspondence is preserved by the action of the group D_{24}.

Allen Forte describes the following method for choosing a preferred representative from each orbit, called the *prime form*.[6] When the elements of the subset are listed in increasing order, the first should be zero, and the last should be as small as possible. If there is more than one representative with the same last term, then the second should be as small as possible, then the third, and so on up to the next to last. In other words, the prime form is the earliest in the lexicographic order with respect to (first, last, second, third, ..., next to last).

For example, take the set $\{1, 7, 9\}$. We can use $\mathbf{T}^{11}$ to take it to a set containing zero, namely $\{0, 6, 8\}$. Or we could use $\mathbf{T}^5$ to take it to $\{0, 2, 6\}$, or $\mathbf{T}^3$ to take it to $\{0, 4, 10\}$. We also need to use **I** to get $\{3, 5, 11\}$, and then use powers of **T** to get $\{0, 2, 8\}$, $\{0, 4, 6\}$ and $\{0, 6, 10\}$. Of the six possibilities, the ones with the smallest last term are $\{0, 2, 6\}$ and $\{0, 4, 6\}$. To break the tie, we compare second terms, and we see that $\{0, 2, 6\}$ is the prime form.

---

[6] This should not be confused with the prime form of a tone row, described in Section 9.8.

There is an easy way to attach an invariant to each orbit, called the *interval vector*. This is computed as follows. To an unordered pair of distinct pitch classes, we can assign a difference, in the range from 1 to 6, by going around the circle of pitch classes in the shorter of the two possible directions. Take all unordered pairs in the set, and to each pair find the difference in this way. Then record how many times one, two, up to six occur in a row vector of length six. For example, for the set {1, 7, 9} the three differences are 2, 4 and 6. So the interval vector for this pitch class set is (0,1,0,1,0,1). It is clear that equivalent pitch class sets yield the same interval vectors. The converse is false; for example the sets {0, 1, 4, 6} and {0, 1, 3, 7} both have interval vector (1,1,1,1,1,1).

Here is a list of the prime forms of pitch class sets of size three, together with Allen Forte's name and Elliott Carter's numbering for them, and the interval vector.

Set	Forte	Carter	Vector
{0,1,2}	3-1(12)	4	(2,1,0,0,0,0)
{0,1,3}	3-2	12	(1,1,1,0,0,0)
{0,1,4}	3-3	11	(1,0,1,1,0,0)
{0,1,5}	3-4	9	(1,0,0,1,1,0)
{0,1,6}	3-5	7	(1,0,0,0,1,1)
{0,2,4}	3-6(12)	3	(0,2,0,1,0,0)
{0,2,5}	3-7	10	(0,1,1,0,1,0)
{0,2,6}	3-8	8	(0,1,0,1,0,1)
{0,2,7}	3-9(12)	5	(0,1,0,0,2,0)
{0,3,6}	3-10(12)	2	(0,0,2,0,0,1)
{0,3,7}	3-11	6	(0,0,1,1,1,0)
{0,4,8}	3-12(4)	1	(0,0,0,3,0,0)

Forte's number consists of the set size followed by a number indicating the placement with respect to lexicographical ordering on the interval vector, in backward order. The numbers in parentheses give the orbit size under the action of D_{24}, in case this is not 24. For reference, we give the corresponding information for sets of size four, five and six below. Sets of size greater than six are not named by Carter, and Forte uses the names of the complementary set, but with the initial number changed. So, for example, 9-3 is obtained by complementing 3-3 to obtain {2, 3, 5, 6, 7, 8, 9, 10, 11}, which is then put into prime form as {0, 1, 2, 3, 4, 5, 6, 8, 9}.

There is an easy way to obtain the interval vector for the complement of a set. For size three, add the vector (6,6,6,6,6,3); for size four add (4,4,4,4,4,2); and for size five add (2,2,2,2,2,1). The interval vector for the above three element set is (1,0,1,1,0,0), so for its nine element complement we get (7,6,7,7,6,3).

Set	Forte	Carter	Vector	Set	Forte	Carter	Vector
{0,1,2,3}	4-1(12)	1	(3,2,1,0,0,0)	{0,1,5,7}	4-16	19	(1,1,0,1,2,1)
{0,1,2,4}	4-2	17	(2,2,1,1,0,0)	{0,3,4,7}	4-17(12)	13	(1,0,2,2,1,0)
{0,1,3,4}	4-3(12)	9	(2,1,2,1,0,0)	{0,1,4,7}	4-18	21	(1,0,2,1,1,1)
{0,1,2,5}	4-4	20	(2,1,1,1,1,0)	{0,1,4,8}	4-19	24	(1,0,1,3,1,0)
{0,1,2,6}	4-5	22	(2,1,0,1,1,1)	{0,1,5,8}	4-20(12)	15	(1,0,1,2,2,0)
{0,1,2,7}	4-6(12)	6	(2,1,0,0,2,1)	{0,2,4,6}	4-21(12)	11	(0,3,0,2,0,1)
{0,1,4,5}	4-7(12)	8	(2,0,1,2,1,0)	{0,2,4,7}	4-22	27	(0,2,1,1,2,0)
{0,1,5,6}	4-8(12)	10	(2,0,0,1,2,1)	{0,2,5,7}	4-23(12)	4	(0,2,1,0,3,0)
{0,1,6,7}	4-9(6)	2	(2,0,0,0,2,2)	{0,2,4,8}	4-24(12)	16	(0,2,0,3,0,1)
{0,2,3,5}	4-10(12)	3	(1,2,2,0,1,0)	{0,2,6,8}	4-25(6)	12	(0,2,0,2,0,2)
{0,1,3,5}	4-11	26	(1,2,1,1,1,0)	{0,3,5,8}	4-26(12)	14	(0,1,2,1,2,0)
{0,2,3,6}	4-12	28	(1,1,2,1,0,1)	{0,2,5,8}	4-27	29	(0,1,2,1,1,1)
{0,1,3,6}	4-13	7	(1,1,2,0,1,1)	{0,3,6,9}	4-28(3)	5	(0,0,4,0,0,2)
{0,2,3,7}	4-14	25	(1,1,1,1,2,0)	{0,1,3,7}	4-Z29	23	(1,1,1,1,1,1)
{0,1,4,6}	4-Z15	18	(1,1,1,1,1,1)				

The only extra thing to describe here is the meaning of the symbol Z in the Forte naming system. This indicates that there are two orbits with the same interval vector; the second one is listed at the end for some reason which he never explains. The same happens for sets of size five and six, but more often.

Set	Forte	Carter	Vector	Set	Forte	Carter	Vector
{0,1,2,3,4}	5-1(12)	1	(4,3,2,1,0,0)	{0,1,3,7,8}	5-20	34	(2,1,1,2,3,1)
{0,1,2,3,5}	5-2	11	(3,3,2,1,1,0)	{0,1,4,5,8}	5-21	21	(2,0,2,4,2,0)
{0,1,2,4,5}	5-3	14	(3,2,2,2,1,0)	{0,1,4,7,8}	5-22(12)	8	(2,0,2,3,2,1)
{0,1,2,3,6}	5-4	12	(3,2,2,1,1,1)	{0,2,3,5,7}	5-23	25	(1,3,2,1,3,0)
{0,1,2,3,7}	5-5	13	(3,2,1,1,2,1)	{0,1,3,5,7}	5-24	22	(1,3,1,2,2,1)
{0,1,2,5,6}	5-6	27	(3,1,1,2,2,1)	{0,2,3,5,8}	5-25	24	(1,2,3,1,2,1)
{0,1,2,6,7}	5-7	30	(3,1,0,1,3,2)	{0,2,4,5,8}	5-26	26	(1,2,2,3,1,1)
{0,2,3,4,6}	5-8(12)	2	(2,3,2,2,0,1)	{0,1,3,5,8}	5-27	23	(1,2,2,2,3,0)
{0,1,2,4,6}	5-9	15	(2,3,1,2,1,1)	{0,2,3,6,8}	5-28	36	(1,2,2,2,1,2)
{0,1,3,4,6}	5-10	19	(2,2,3,1,1,1)	{0,1,3,6,8}	5-29	32	(1,2,2,1,3,1)
{0,2,3,4,7}	5-11	18	(2,2,2,2,2,0)	{0,1,4,6,8}	5-30	37	(1,2,1,3,2,1)
{0,1,3,5,6}	5-Z12(12)	5	(2,2,2,1,2,1)	{0,1,3,6,9}	5-31	33	(1,1,4,1,1,2)
{0,1,2,4,8}	5-13	17	(2,2,1,3,1,1)	{0,1,4,6,9}	5-32	38	(1,1,3,2,2,1)
{0,1,2,5,7}	5-14	28	(2,2,1,1,3,1)	{0,2,4,6,8}	5-33(12)	6	(0,4,0,4,0,2)
{0,1,2,6,8}	5-15(12)	4	(2,2,0,2,2,2)	{0,2,4,6,9}	5-34(12)	9	(0,3,2,2,2,1)
{0,1,3,4,7}	5-16	20	(2,1,3,2,1,1)	{0,2,4,7,9}	5-35(12)	7	(0,3,2,1,4,0)
{0,1,3,4,8}	5-Z17(12)	10	(2,1,2,3,2,0)	{0,1,2,4,7}	5-Z36	16	(2,2,2,1,2,1)
{0,1,4,5,7}	5-Z18	35	(2,1,2,2,2,1)	{0,3,4,5,8}	5-Z37(12)	3	(2,1,2,3,2,0)
{0,1,3,6,7}	5-19	31	(2,1,2,1,2,2)	{0,1,2,5,8}	5-Z38	29	(2,1,2,2,2,1)

Finally, the six note pitch class sets, or *hexachords*.

Set	Forte	Carter	Vector	Set	Forte	Carter	Vector
{0,1,2,3,4,5}	6-1(12)	4	(5,4,3,2,1,0)	{0,1,3,5,7,8}	6-Z26(12)	26	(2,3,2,3,4,1)
{0,1,2,3,4,6}	6-2	19	(4,4,3,2,1,1)	{0,1,3,4,6,9}	6-27	14	(2,2,5,2,2,2)
{0,1,2,3,5,6}	6-Z3	49	(4,3,3,2,2,1)	{0,1,3,5,6,9}	6-Z28(12)	21	(2,2,4,3,2,2)
{0,1,2,4,5,6}	6-Z4(12)	24	(4,3,2,3,2,1)	{0,1,3,6,8,9}	6-Z29(12)	32	(2,2,4,2,3,2)
{0,1,2,3,6,7}	6-5	16	(4,2,2,2,3,2)	{0,1,3,6,7,9}	6-30(12)	15	(2,2,4,2,2,3)
{0,1,2,5,6,7}	6-Z6(12)	33	(4,2,1,2,4,2)	{0,1,3,5,8,9}	6-31	8	(2,2,3,4,3,1)
{0,1,2,6,7,8}	6-7(6)	7	(4,2,0,2,4,3)	{0,2,4,5,7,9}	6-32(12)	6	(1,4,3,2,5,0)
{0,2,3,4,5,7}	6-8(12)	5	(3,4,3,2,3,0)	{0,2,3,5,7,9}	6-33	18	(1,4,3,2,4,1)
{0,1,2,3,5,7}	6-9	20	(3,4,2,2,3,1)	{0,1,3,5,7,9}	6-34	9	(1,4,2,4,2,2)
{0,1,3,4,5,7}	6-Z10	42	(3,3,3,3,2,1)	{0,2,4,6,8,10}	6-35(2)	1	(0,6,0,6,0,3)
{0,1,2,4,5,7}	6-Z11	47	(3,3,3,2,3,1)	{0,1,2,3,4,7}	6-Z36	50	(4,3,3,2,2,1)
{0,1,2,4,6,7}	6-Z12	46	(3,3,2,2,3,2)	{0,1,2,3,4,8}	6-Z37(12)	23	(4,3,2,3,2,1)
{0,1,3,4,6,7}	6-Z13(12)	29	(3,2,4,2,2,2)	{0,1,2,3,7,8}	6-Z38(12)	34	(4,2,1,2,4,2)
{0,1,3,4,5,8}	6-14	3	(3,2,3,4,3,0)	{0,2,3,4,5,8}	6-Z39	41	(3,3,3,3,2,1)
{0,1,2,4,5,8}	6-15	13	(3,2,3,4,2,1)	{0,1,2,3,5,8}	6-Z40	48	(3,3,3,2,3,1)
{0,1,4,5,6,8}	6-16	11	(3,2,2,4,3,1)	{0,1,2,3,6,8}	6-Z41	45	(3,3,2,2,3,2)
{0,1,2,4,7,8}	6-Z17	35	(3,2,2,3,3,2)	{0,1,2,3,6,9}	6-Z42(12)	30	(3,2,4,2,2,2)
{0,1,2,5,7,8}	6-18	17	(3,2,2,2,4,2)	{0,1,2,5,6,8}	6-Z43	36	(3,2,2,3,3,2)
{0,1,3,4,7,8}	6-Z19	37	(3,1,3,4,3,1)	{0,1,2,5,6,9}	6-Z44	38	(3,1,3,4,3,1)
{0,1,4,5,8,9}	6-20(4)	2	(3,0,3,6,3,0)	{0,2,3,4,6,9}	6-Z45(12)	28	(2,3,4,2,2,2)
{0,2,3,4,6,8}	6-21	12	(2,4,2,4,1,2)	{0,1,2,4,6,9}	6-Z46	40	(2,3,3,3,3,1)
{0,1,2,4,6,8}	6-22	10	(2,4,1,4,2,2)	{0,1,2,4,7,9}	6-Z47	44	(2,3,3,2,4,1)
{0,2,3,5,6,8}	6-Z23(12)	27	(2,3,4,2,2,2)	{0,1,2,5,7,9}	6-Z48(12)	25	(2,3,2,3,4,1)
{0,1,3,4,6,8}	6-Z24	39	(2,3,3,3,3,1)	{0,1,3,4,7,9}	6-Z49(12)	22	(2,2,4,3,2,2)
{0,1,3,5,6,8}	6-Z25	43	(2,3,3,2,4,1)	{0,1,4,6,7,9}	6-Z50(12)	31	(2,2,4,2,3,2)

Complementation takes some hexachords to equivalent ones and some to inequivalent ones. Inequivalent pairs always share an interval vector, and these turn out to be the only coincidences of interval vectors for hexachords. The inequivalent pairs of complements are as follows:

6-Z3	6-Z36	6-Z12	6-Z41	6-Z24	6-Z46
6-Z4(12)	6-Z37(12)	6-Z13(12)	6-Z42	6-Z25	6-Z47
6-Z6(12)	6-Z38(12)	6-Z17	6-Z43	6-Z26(12)	6-Z48(12)
6-Z10	6-Z39	6-Z19	6-Z44	6-Z28(12)	6-Z49(12)
6-Z11	6-Z40	6-Z23(12)	6-Z45(12)	6-Z29(12)	6-Z50(12)

Further reading

Allen Forte (1973), *The Structural Function of Atonal Music.*

David Schiff, *The Music of Elliott Carter.* Ernst Eulenberg Ltd, 1983. Reprinted by Faber and Faber, 1998.

9.15 Pólya's enumeration theorem

In this section, we show how to vamp up Burnside's lemma 9.13.1 to address some more complicated counting problems. By way of illustration, we shall revisit the problem considered in Section 9.14. Suppose we want to know how many pitch class sets there are, consisting of three of the twelve possible pitch classes. Suppose further that we wish to consider two such sets to be equivalent if one can be obtained from the other by means of an operation $\mathbf{T}^n$ for some n. This is a typical kind of problem which can be solved using Pólya's enumeration theorem.

A lot of physical counting problems involving symmetry are of a similar nature. A typical example would involve counting how many different necklaces can be made from three red beads, two sepia beads and five turquoise beads. The symmetry group in this situation is a dihedral group whose order is twice the number of beads.

In the general form of the problem, the *configurations* being counted are regarded as functions from a set X to a set Y, and the symmetry group G acts on the set X. In the bead problem, the set X would consist of the places in the necklace where we wish to put the beads, and the set Y would consist of the possible colours. A function from X to Y then specifies for each place in the necklace what colour bead to use. The group G acts on configurations by rotating and turning over the necklace.

In the pitch class set counting problem, the set X is the set of twelve pitches, and Y is taken to be the set $\{0, 1\}$. A pitch class set corresponds to a function taking the notes in the set to 1 and the remaining notes to 0. This gives a one-to-one correspondence between pitch class sets and functions from X to Y.

In the general setup, we write Y^X for the set of configurations, or functions from the set X to the set Y. The reason for this notation is that the number of elements of Y^X is equal to the number of elements of Y raised to the power of the number of elements of X ($|Y^X| = |Y|^{|X|}$). The action of G on the set Y^X of configurations is given by the formula

$$g(f)(x) = f(g^{-1}(x)).$$

The reason for the inverse sign is so that composition works right. For a group action, we need $g_1(g_2(f)) = (g_1 g_2)(f)$. To see that this holds, we have

$$(g_1(g_2(f)))(x) = (g_2(f))(g_1^{-1}(x)) = f\left(g_2^{-1}(g_1^{-1}(x))\right) = f\left(\left(g_2^{-1} g_1^{-1}\right)(x)\right)$$
$$= f((g_1 g_2)^{-1}(x)) = ((g_1 g_2)(f))(x),$$

whereas without the inverse sign the order of g_1 and g_2 would be reversed. The general problem is to find the number of orbits of G on configurations.

We begin by defining the *cycle index* of G on X as follows. We introduce variables $t_1, t_2, \ldots$, and then the cycle index of an element g on X is

$$P_g(t_1, t_2, \ldots) = t_1^{j_1(g)} t_2^{j_2(g)} \cdots,$$

where $j_k(g)$ denotes the number of cycles of length k in the action of G on X. We define the cycle index of the group to be the average cycle index of an element, namely

$$P_G(t_1, t_2, \ldots) = \frac{1}{|G|} \sum_{g \in G} P_g(t_1, t_2, \cdots) = \frac{1}{|G|} \sum_{g \in G} t_1^{j_1(g)} t_2^{j_2(g)} \cdots \qquad (9.15.1)$$

For example, if G is a dihedral group of order eight acting on the set X consisting of the four corners of a square, then the cycle indices of the eight elements of G are as follows. The identity element has cycle index t_1^4, the two ninety degree rotations have cycle index t_4, the one hundred and eighty degree rotation and the reflections about the horizontal and vertical axes all have cycle index t_2^2, and the two diagonal reflections have cycle index $t_1^2 t_2$. So

$$P_G = \tfrac{1}{8}\left(t_1^4 + 2t_4 + 3t_2^2 + 2t_1^2 t_2\right).$$

Several standard examples of cycle index are worth writing out explicitly. If $G = \mathbb{Z}/n$, cycling a set X of n objects, we get

$$P_{\mathbb{Z}/n} = \frac{1}{n} \sum_{j \mid n} \phi(j) t_j^{n/j}. \qquad (9.15.2)$$

Here, ϕ is the Euler phi function, described on page 336, and $j \mid n$ means j is a divisor of n. The formula is obvious, because there are $\phi(j)$ elements of $\mathbb{Z}/n$ having order j, and each one has n/j cycles of length j.

The next example generalizes the above dihedral calculation. For the dihedral group D_{2n} acting on the n vertices of a regular n-sided polygon, we have to divide into two cases according to whether n is even or odd. If $n = 2m + 1$ is odd, we get

$$P_{D_{4m+2}} = \tfrac{1}{2} P_{\mathbb{Z}/(2m+1)} + \tfrac{1}{2} t_1 t_2^m, \qquad (9.15.3)$$

because each reflection has exactly one fixed point. If $n = 2m$ is even, we get

$$P_{D_{4m}} = \tfrac{1}{2} P_{\mathbb{Z}/2m} + \tfrac{1}{4}\left(t_2^m + t_1^2 t_2^{m-1}\right), \qquad (9.15.4)$$

because half the reflections have no fixed points and half of them have two.

For the full symmetric group S_n on a set X of n elements, the formula is rather messy. But adding up the cycle indices of all the symmetric groups gives a much cleaner answer.

$$\sum_{n=0}^{\infty} P_{S_n} = \exp\left(\sum_{j=1}^{\infty} \frac{t_j}{j}\right) = \prod_{j=1}^{\infty}\sum_{i=0}^{\infty} \frac{1}{i!}\left(\frac{t_j}{j}\right)^i$$

$$= \left(1 + t_1 + \tfrac{1}{2!}t_1^2 + \tfrac{1}{3!}t_1^3 + \tfrac{1}{4!}t_1^4 + \cdots\right)\left(1 + \tfrac{1}{2}t_2 + \tfrac{1}{2^2 \cdot 2!}t_2^2 + \tfrac{1}{2^3 \cdot 3!}t_2^3 + \cdots\right)$$

$$\left(1 + \tfrac{1}{3}t_3 + \tfrac{1}{3^2 \cdot 2!}t_3^2 + \tfrac{1}{3^3 \cdot 3!}t_3^3 + \cdots\right)\left(1 + \tfrac{1}{4}t_4 + \tfrac{1}{4^2 \cdot 2!}t_4^2 + \tfrac{1}{4^3 \cdot 3!}t_4^3 + \cdots\right) \cdots$$

The cycle index for an individual S_n can be extracted by taking the terms with total size n, where each t_j is regarded as having size j. So, for example,

$$P_{S_4} = \tfrac{1}{24}t_1^4 + \tfrac{1}{4}t_1^2 t_2 + \tfrac{1}{8}t_2^2 + \tfrac{1}{3}t_1 t_3 + \tfrac{1}{4}t_4.$$

The corresponding formula for the alternating group A_n (this is the group of even permutations; exactly half the elements of S_n are even) is

$$2 + 2t_1 + \sum_{n=2}^{\infty} P_{A_n} = \exp\left(\sum_{j=1}^{\infty} \frac{t_j}{j}\right) + \exp\left(\sum_{j=1}^{\infty}(-1)^{j+1}\frac{t_j}{j}\right).$$

Next, we assign a weight $w(y)$ to each of the elements y of Y. The weights can be any sorts of quantities which can be added and multiplied (the formal requirement is that the weights should belong to a *commutative ring*). For example, the weights can be independent formal variables, or one of them can be chosen to be 1 to simplify the algebra. The weight of a configuration is then defined to be the product over $x \in X$ of the weight of $f(x)$,

$$w(f) = \prod_{x \in X} w(f(x)).$$

The weights of two configurations in the same orbit of the action of G are clearly equal.

So, for example, if $Y = \{$red, sepia, turquoise$\}$ then we could assign variables $r = w(\text{red})$, $s = w(\text{sepia})$ and $t = w(\text{turquoise})$ for the weights.

We form a power series called the *configuration counting series* C using these weights. Namely, C is the sum, over all orbits of G on the set Y^X configurations, of the weight of a representative of the orbit. In the necklace example, the coefficient of $r^a s^b t^c$ in $C = C(r, s, t)$ gives the number of necklaces in which a beads are red, b are sepia and c are turquoise. So the coefficient of $r^3 s^2 t^5$ would give the number of necklaces in the original problem. Since $a + b + c$ is fixed, if we wanted to simplify the algebra, it would make sense to put $w(\text{turquoise}) = 1$ instead of t. Then the coefficient of $r^3 s^2$ would be the desired number of necklaces. In other words, once we know the number of red and sepia beads, the number of turquoise beads is also known by subtraction.

In the pitch class set example, where $Y = \{0, 1\}$, it would make sense to introduce just one variable z and set $w(0) = 1$ and $w(1) = z$. Then the coefficient of z^a would tell us about pitch class sets with a notes.

Theorem 9.15.1 (Pólya) *The configuration counting series C is given in terms of the cycle index of G on X by*

$$C = P_G \left(\sum_{y \in Y} w(y), \sum_{y \in Y} w(y)^2, \sum_{y \in Y} w(y)^3, \dots \right)$$

We shall prove this theorem after seeing how to apply it.

Example In the pitch class set example, we consider the cases $G = \mathbb{Z}/12$ and $G = D_{24}$, with X is the set of twelve pitch classes, $Y = \{0, 1\}$, $w(0) = 1$ and $w(1) = y$. Equations (9.15.2) and (9.15.4) give the cycle indices as

$$P_{\mathbb{Z}/12} = \tfrac{1}{12}\left(t_1^{12} + t_2^6 + 2t_3^4 + 2t_4^3 + 2t_6^2 + 4t_{12}\right)$$

$$P_{D_{24}} = \tfrac{1}{2}P_{\mathbb{Z}/12} + \tfrac{1}{4}\left(t_2^6 + t_1^2 t_2^5\right)$$

$$= \tfrac{1}{24}\left(t_1^{12} + 6t_1^2 t_2^5 + 7t_2^6 + 2t_3^4 + 2t_4^3 + 2t_6^2 + 4t_{12}\right).$$

Then Theorem 9.15.1 says that we should substitute $1 + z^n$ for t_n to give the configuration counting series C. This gives the following values.

(i) If $G = \mathbb{Z}/12$ then

$$C = 1 + z + 6z^2 + 19z^3 + 43z^4 + 66z^5 + 80z^6 + 66z^7 + 43z^8 1 + 19z^9 + 6z^{10} + z^{11} + z^{12}.$$

So, for example, there are 19 three note sets up to transposition.

(ii) If $G = D_{24}$ then

$$C = 1 + z + 6z^2 + 12z^3 + 29z^4 + 38z^5 + 50z^6 + 38z^7 + 29z^8 + 12z^9 + 6z^{10} + y^{11} + y^{12}.$$

So, for example, there are 12 three note sets and 50 hexachords, up to transposition and inversion. The reason why the coefficients in these polynomials are symmetric was described in Section 9.14. That is, a set can be replaced by its complement, to give a natural correspondence between j note sets and $12 - j$ note sets.

The advantage of using Pólya's enumeration theorem rather than just resorting to Burnside's lemma 9.13.1 is that we do not have to do an explicit computation of numbers of fixed configurations, as we had to in Section 9.14. The disadvantage is that the machinery is harder to understand and remember.

The proof of Pólya's enumeration theorem depends on a weighted version of Burnside's lemma 9.13.1.

Lemma 9.15.2 *Let G be a finite group acting by permutations on a finite set X. Let w be a function on X which takes constant values on orbits, so that we can regard w as a function on the set of orbits of G on X. Then the sum of the weights* *of the orbits is equal to*

$$\frac{1}{|G|} \sum_{g \in G} \sum_{x=g(x)} w(x).$$

Proof Consider the set of pairs (g, x) where $g(x) = x$, and calculate in two different ways the sum over the elements of this set of the weights $w(x)$. If we sum over the elements of the group first, we obtain $\sum_{g \in G} \sum_{x=g(x)} w(x)$. On the other hand, if we sum over the elements of X first, then by equation (9.11.1), for each x, the number of elements of G is $|G|$ divided by the length of the orbit in which x lies. So the sum over the elements of the orbit in which x lies gives $|G| w(x)$. So the sum over all x gives $|G|$ times the sum of the weights of the orbits. □

Proof of Pólya's enumeration theorem We are going to apply the above version of Burnside's lemma to the action of G on the set Y^X of configurations. It tells us that C is equal to

$$\frac{1}{|G|} \sum_{g \in G} \sum_{f=g(f)} w(f). \tag{9.15.5}$$

So we will be finished if we can prove that for each $g \in G$ we have

$$P_g\left(\sum_{y \in Y} w(y), \sum_{y \in Y} w(y)^2, \sum_{y \in Y} w(y)^3, \dots\right) = \sum_{f=g(f)} w(f),$$

because then, comparing (9.15.1) with (9.15.5), we see that averaging over the elements of G gives the formula in the theorem. Recalling that $j_k(g)$ denotes the number of cycles of length k in the action of g on X, by definition the left side of this equation is

$$\left(\sum_{y \in Y} w(y)\right)^{j_1(g)} \left(\sum_{y \in Y} w(y)^2\right)^{j_2(g)} \dots \tag{9.15.6}$$

The right hand side is

$$\sum_{f=g(f)} \prod_{x \in X} w(f(x)). \tag{9.15.7}$$

Now a configuration f satisfies $f = g(f)$ precisely when it is constant on orbits of g on X. So, to pick such a configuration, we must assign an element of Y to each orbit of g on X. So, when we multiply the weights of the $f(x)$, an orbit of length j with image $y \in Y$ corresponds to a factor of $w(y)^j$ in the product.

We regard (9.15.6) as being obtained by multiplying together a factor of $\sum_{y \in Y} w(y)^i$ for each orbit of g on X, where i is the length of the orbit. When these sums are all multiplied out, there will be one term for each way of assigning an element of Y to each orbit of g on X, and that term will exactly be the corresponding term in (9.15.7). $\square$

Further reading

Harald Fripertinger, Enumeration in music theory, *Séminaire Lotharingien de Combinatoire*, **26** (1991), 29–42; also appeared in *Beiträge zur elektronischen Musik* **1**, 1992.

Harald Fripertinger, Enumeration and construction in music theory, *Diderot Forum on Mathematics and Music Computational and Mathematical Methods in Music, Vienna, Austria, December 2-4, 1999.* H. G. Feichtinger and M. Dörfler, editors. Österreichische Computergesellschaft (1999), 179–204.

Harald Fripertinger, Enumeration of mosaics, *Discrete Math.* **199** (1999), 49–60.

Harald Fripertinger, Enumeration of non-isomorphic canons, *Tatra Mountains Math. Publ.* **23** (2001).

Harald Fripertinger, Classification of motives: a mathematical approach, to appear in *Musikometrika.*

Michael Keith (1991), *From Polychords to Pólya; Adventures in Musical Combinatorics.*

G. Pólya, Kombinatorische Anzahlbestimmungen für Gruppen, Graphen und chemische Verbindungen, *Acta Math.* **68** (1937), 145–254.

R. C. Read, Combinatorial problems in the theory of music, *Discrete Mathematics* **167/168** (1997), 543–551.

D. Reiner, Enumeration in music theory, *Amer. Math. Monthly* **92** (1) (1985), 51–54. Note that there is a typographical error in the formula for the cycle index of the dihedral group in this paper.

9.16 The Mathieu group M_{12}

The combinatorics of twelve tone music has given rise to a curious coincidence, which I find worth mentioning. Messiaen, in his *Ile de Feu 2* for piano, nearly rediscovered the Mathieu group M_{12}. On pages 409–414 of Berry (1976), you can read about how Messiaen uses the permutations

$$\begin{pmatrix} 1 & 2 & 3 & 4 & 5 & 6 & 7 & 8 & 9 & 10 & 11 & 12 \\ 7 & 6 & 8 & 5 & 9 & 4 & 10 & 3 & 11 & 2 & 12 & 1 \end{pmatrix}$$

and

$$\begin{pmatrix} 1 & 2 & 3 & 4 & 5 & 6 & 7 & 8 & 9 & 10 & 11 & 12 \\ 6 & 7 & 5 & 8 & 4 & 9 & 3 & 10 & 2 & 11 & 1 & 12 \end{pmatrix}$$

to generate sequences of tones and sequences of durations. These permutations generate a group M_{12} of order $95\,040$ discovered by Mathieu in the nineteenth century.[7]

---

[7] E. Mathieu, Mémoire sur l'étude des fonctions de plusieurs quantités, *J. Math. Pures Appl.* **6** (1861), 241–243; Sur la fonction cinq fois transitive de 24 quantités, *J. Math. Pures Appl.* **18** (1873), 25–46.

A group is said to be *simple* if it has just two normal subgroups, namely the whole group and the subgroup consisting of just the identity element.[8] One of the outstanding achievements of twentieth century mathematics was the classification of the finite simple groups. Roughly speaking, the classification theorem says that the finite simple groups fall into certain infinite families which can be explicitly described, with the exception of 26 *sporadic* groups. Five of these 26 groups were discovered by Mathieu in the nineteenth century, and the remaining ones were discovered in the 1960s and 70s.

Diaconis, Graham and Kantor discovered that M_{12} was generated by the above two permutations, which they call *Mongean shuffles*. Start with a pack of twelve cards in your left hand, and transfer them to your right hand by placing them alternately under and over the stack you have so far. When you have finished, hand the pack back to your left hand. Since I did not tell you whether to start under or over, this describes two different permutations of the twelve cards. These are the permutations shown above. In cycle notation, these permutations are

$$(1, 7, 10, 2, 6, 4, 5, 9, 11, 12)(3, 8)$$

of order ten, and

$$(1, 6, 9, 2, 7, 3, 5, 4, 8, 10, 11)(12)$$

of order eleven. These permutations can be visualized as follows.

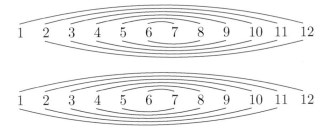

Exercise

(Carl E. Linderholm, 1971) If this book is read backwards (beginning at the last word of the last page), the last thing read is the introduction (reversed, of course). Thus the introduction acts as a sort of extraduction, and is suggested as a simple form of therapy, used in this way, if the reader gets stuck. Read this exercise backwards, and write an extraduction from it.

---

[8] So, for example, the group with only one element is not simple, because it has only one, not two, normal subgroups. Compare this with the definition of a prime number; 1 is not prime.

Further reading

Wallace Berry, *Structural Function in Music*, Prentice-Hall, 1976. Reprinted by Dover, 1987. This book contains a description of the Messiaen example referred to in this section.

J. H. Conway and N. J. A. Sloane, *Sphere Packings, Lattices and Groups*, Grundlehren der mathematischen Wissenschaften 290, Springer-Verlag, 1988. This book contains a huge amount of information about the sporadic groups in general, and Section 11.17 contains more information on Mongean shuffles and the Mathieu group M_{12}.

P. Diaconis, R. L. Graham and W. M. Kantor, The mathematics of perfect shuffles, *Adv. Appl. Math.* **4** (1983), 175–196.

Unlike Mozart's *Requiem* and Bartók's *Third Piano Concerto*, the piece that P. D. Q. Bach was working on when he died has never been finished by anyone else.[9]

---

[9] Professor Peter Schickele, *The Definitive Biography of P. D. Q. Bach (1807–1742)?*, Random House, 1976.

Appendix A

Bessel functions

Table of zeros of Bessel functions

Note: the kth zero of J_n is denoted $j_{n,k}$.

k	J_0	J_1	J_2	J_3	J_4	J_5	J_6	J_7
1	2.40482 55577	3.831706	5.135622	6.380162	7.588342	8.771484	9.936110	11.08637
2	5.52007 81103	7.015587	8.417244	9.761023	11.06471	12.33860	13.58929	14.82127
3	8.65372 79129	10.17347	11.61984	13.01520	14.37254	15.70017	17.00382	18.28758
4	11.79153 44391	13.32369	14.79595	16.22347	17.61597	18.98013	20.32079	21.64154
5	14.93091 77086	16.47063	17.95982	19.40942	20.82693	22.21780	23.58608	24.93493
6	18.07106 39679	19.61586	21.11700	22.58273	24.01902	25.43034	26.82015	28.19119
7	21.21163 66299	22.76008	24.27011	25.74817	27.19909	28.62662	30.03372	31.42279
8	24.35247 15308	25.90367	27.42057	28.90835	30.37101	31.81172	33.23304	34.63709
9	27.49347 91320	29.04683	30.56920	32.06485	33.53714	34.98878	36.42202	37.83872
10	30.63460 64684	32.18968	33.71652	35.21867	36.69900	38.15987	39.60324	41.03077
11	33.77582 02136	35.33231	36.86286	38.37047	39.85763	41.32638	42.77848	44.21541
12	36.91709 83537	38.47477	40.00845	41.52072	43.01374	44.48932	45.94902	47.39417
13	40.05842 57646	41.61709	43.15345	44.66974	46.16785	47.64940	49.11577	50.56818
14	43.19979 17132	44.75932	46.29800	47.81779	49.32036	50.80717	52.27945	53.73833
15	46.34118 83717	47.90146	49.44216	50.96503	52.47155	53.96303	55.44059	56.90525

k	J_8	J_9	J_{10}	J_{11}	J_{12}	J_{13}	J_{14}	J_{15}
1	12.22509	13.35430	14.47550	15.58985	16.69825	17.80144	18.90000	19.99443
2	16.03777	17.24122	18.43346	19.61597	20.78991	21.95624	23.11578	24.26918
3	19.55454	20.80705	22.04699	23.27585	24.49489	25.70510	26.90737	28.10242
4	22.94517	24.23389	25.50945	26.77332	28.02671	29.27063	30.50595	31.73341
5	26.26681	27.58375	28.88738	30.17906	31.45996	32.73105	33.99318	35.24709
6	29.54566	30.88538	32.21186	33.52636	34.82999	36.12366	37.40819	38.68428
7	32.79580	34.15438	35.49991	36.83357	38.15638	39.46921	40.77283	42.06792
8	36.02562	37.40010	38.76181	40.11182	41.45109	42.78044	44.10059	45.41219

Fourier series

$$\sin(z \sin \theta) = 2 \sum_{n=0}^{\infty} J_{2n+1}(z) \sin(2n+1)\theta,$$

$$\cos(z \sin \theta) = J_0(z) + 2 \sum_{n=1}^{\infty} J_{2n}(z) \cos 2n\theta,$$

$$J_n(z) = \frac{1}{\pi} \int_0^{\pi} \cos(n\theta - z \sin \theta) \, d\theta.$$

Differential equation

$$J_n''(z) + \frac{1}{z} J_n'(z) + \left(1 - \frac{n^2}{z^2}\right) J_n(z) = 0.$$

Power series

$$J_n(z) = \sum_{k=0}^{\infty} \frac{(-1)^k \left(\frac{z}{2}\right)^{n+2k}}{k!(n+k)!}.$$

Generating function

$$e^{\frac{1}{2}z\left(t - \frac{1}{t}\right)} = \sum_{n=-\infty}^{\infty} J_n(z) t^n.$$

Limiting values

If n is constant, z is real and $|z| \to \infty$,

$$J_n(z) = \sqrt{\tfrac{2}{\pi z}} \cos\left(z - \tfrac{1}{2}\left(n + \tfrac{1}{2}\right)\pi\right) + O\left(|z|^{-3/2}\right).$$

(Here, $O(|z|^{-3/2})$ represents an error term which is bounded by some constant multiple of $|z|^{-3/2}$.)

If z is constant and $n \to \infty$, $J_n(z) \sim \frac{1}{\sqrt{2n\pi}} \left(\frac{ez}{2n}\right)^n$.

For n fixed, as $k \to \infty$, $j_{n,k} \sim (k + \tfrac{1}{2}n - \tfrac{1}{4})\pi$.

Other formulas

$$J_{-n}(z) = (-1)^n J_n(z),$$

$$J_n'(z) = \tfrac{1}{2}(J_{n-1}(z) - J_{n+1}(z)),$$

$$J_n(z) = \tfrac{z}{2n}(J_{n-1}(z) + J_{n+1}(z)),$$

$$\frac{\mathrm{d}}{\mathrm{d}z}(z^n J_n(z)) = z^n J_{n-1}(z),$$

$$1 = \sum_{n=-\infty}^{\infty} J_n(z) = J_0(z) + 2J_2(z) + 2J_4(z) + 2J_6(z) + \cdots$$

$$1 = \sum_{n=-\infty}^{\infty} J_n(z)^2 = J_0(z)^2 + 2J_1(z)^2 + 2J_2(z)^2 + 2J_3(z)^2 + \cdots$$

In particular, $|J_n(z)| \le 1$ for all n and z, and if $n \ne 0$ then $|J_n(z)| \le \frac{1}{\sqrt{2}}$.

Computation

Although the power series converges very quickly for small values of z, and converges for all values of z, rounding errors tend to accumulate for larger z because a small number is resulting from addition and subtraction of very large numbers.

Instead, a computer program for calculating the Bessel functions can be based on the recurrence relation $J_n(z) = (2(n + 1)/z)J_{n+1}(z) - J_{n+2}(z)$ and normalizing via the relation $J_0(z) + 2J_2(z) + 2J_4(z) + \cdots = 1$. This is called *Miller's backwards recurrence algorithm* (J. C. P. Miller, *The Airy integral*, Cambridge University Press, 1946). Build an array indexed by n and make the last two entries 1 and 0, use the recurrence relation to calculate the remaining entries, and then normalize. An array containing 100 entries gives reasonable accuracy, and does not consume much memory. Here is a simple C++ program which implements this method. I haven't put in any exception checking.

```cpp
/* file bessel.cpp */
#include <iostream.h>
#include <stdio.h>
#define length 100
void main() {
  long double X[length], z, sum;
  int n=0, j=0;
  X[length - 2]=1; X[length - 1]=0;
  while (1)
  {
    printf("\n\nOrder (integer); -1 to exit: ");
    cin>>n;
    if (n<0)
      break;
    printf("Argument (real): ");
    cin>>z;
    if (z==0)            // prevent divide by zero
      {printf("J_0(0)=1; J_n(0)=0 (n>0)");}
    else
      {for(j=length - 3; j>=0; --j)
        {X[j]=(2*(j+1)/z)*X[j+1] - X[j+2];}
      sum=X[0];
      for(j=2; j < length; j=j+2)
```

```
      {sum+=2*X[j];}
     printf("J_%d(%Lf)= %11.10Lf",n,z,X[n]/sum);
   }
 }
}
```

I compiled this program using Borland C++. It prints out the answer to 10 decimal places, and at least for reasonably small values of *n* and *z*, up to about 50, the answers it gives agree with published tables to this accuracy. If you need more accuracy, I recommend the standard Unix multiple precision arithmetic utility bc. If invoked with the option -l (which loads the library mathlib of mathematical functions), it recognizes the syntax j(n,z) and calculates $J_n(z)$ using the algorithm above. The number of digits after the decimal point is set to 50, for example, by using the command scale=50. Windows users can use bc in the free Unix environment Cygwin (www.cygwin.com); there is also a (free) version compiled for MS-DOS in UnxUtils.zip (unxutils.sourceforge.net). Here is a sample session:

```
$ bc -l
j(1,1)
.44005058574493351595
scale=50
for (n=0;n<5;n++) {j(n,1)}
.76519768655796655144971752610266322090927428975532
.44005058574493351595968220371891491312737230199276
.11490348493190048046964688133516660534547031423020
.01956335398266840591890532162175150825450895492805
.00247663896410995504378504839534244418158341533812
quit
$
```

Appendix B

Equal tempered scales

q	p_3	e_3	p_5	e_5	p_7	e_7	e_{35}	e_{357}	$e_5.q^2$	$e_{35}.q^{\frac{3}{2}}$	$e_{357}.q^{\frac{4}{3}}$
2	1	+213.686	1	−101.955	2	+231.174	166.245	190.365	392	470	480
3	1	+13.686	2	+98.045	2	−168.826	70.000	112.993	882	**364**	489
4	1	−86.314	2	−101.955	3	−68.826	94.459	86.760	1631	756	551
5	2	+93.686	3	+18.045	4	−8.826	67.464	55.319	451	754	473
6	2	+13.686	4	+98.045	5	+31.174	70.000	59.922	3530	1029	653
7	2	−43.457	4	−16.241	6	+59.746	32.804	43.672	796	608	585
8	3	+63.686	5	+48.045	6	−68.826	56.410	60.831	3075	1276	973
9	3	+13.686	5	−35.288	7	−35.493	26.764	23.104	2858	723	433
10	3	−26.314	6	+18.045	8	−8.826	22.561	19.113	1804	713	412
11	4	+50.050	6	−47.410	9	+12.992	48.748	40.503	5737	1778	991
12	4	+13.686	7	−1.955	10	+31.174	9.776	19.689	**282**	**406**	541
13	4	−17.083	8	+36.507	10	−45.749	28.500	35.202	6170	1336	1076
14	5	+42.258	8	−16.241	11	−25.969	32.012	30.132	3183	1677	1017
15	5	+13.686	9	+18.045	12	−8.826	16.015	14.034	4060	930	519
16	5	−11.314	9	−26.955	13	+6.174	20.671	17.250	6900	1323	695
17	5	−33.373	10	+3.927	14	+19.409	23.761	22.404	1135	1665	979
18	6	+13.686	11	+31.378	15	+31.174	24.207	26.732	10167	1849	1261
19	6	−7.366	11	−7.218	15	−23.457	7.293	13.745	2606	604	697
20	6	−26.314	12	+18.045	16	−8.826	22.561	19.113	7218	2018	1038
21	7	+13.686	12	−16.241	17	+2.603	15.018	12.354	7162	1445	716
22	7	−4.496	13	+7.136	18	+12.992	5.964	8.943	3454	615	551
31	10	+0.783	18	−5.181	25	−1.084	3.705	3.089	4979	639	**301**
41	13	−5.826	24	+0.484	33	−2.972	4.134	3.786	814	1085	535
53	17	−1.408	31	−0.068	43	+4.759	0.997	2.866	**192**	**385**	570
65	21	+1.379	38	−0.417	52	−8.826	1.018	5.163	1760	534	1349
68	22	+1.922	40	+3.927	55	+1.762	3.092	2.722	18160	1734	755
72	23	−2.980	42	−1.955	58	−2.159	2.520	2.406	10135	1540	721
84	27	−0.599	49	−1.955	68	+2.603	1.446	1.911	13794	1113	703
99	32	+1.565	58	+1.075	80	−0.871	1.343	1.206	10539	1323	552
118	38	+0.127	69	−0.260	95	−2.724	0.205	1.582	3621	**262**	915
130	42	+1.379	76	−0.417	105	+0.405	1.018	0.864	7040	1509	569
140	45	−0.599	82	+0.902	113	−0.254	0.766	0.642	17682	1269	467
171	55	−0.349	100	−0.201	138	−0.405	0.285	0.330	5866	636	**313**
441	142	+0.081	258	+0.086	356	−0.118	0.083	0.096	16689	772	**324**
494	159	−0.079	289	+0.069	399	+0.405	0.074	0.241	16909	815	943
612	197	−0.039	358	+0.006	494	−0.198	0.028	0.117	2166	**424**	607
665	214	−0.148	389	−0.0001	537	+0.197	0.105	0.142	**50**	1798	825

This table shows how well the scales based around equal divisions of the octave approximate the 5:4 major third, the 3:2 perfect fifth and the 7:4 seventh harmonic. The first column (q) gives the number of divisions to the octave. The second column (p_3) shows the scale degree closest to the 5:4 major third (counting from zero for the tonic), and the next column (e_3) shows the error in cents:

$$e_3 = 1200 \left(\frac{p_3}{q} - \log_2 \left(\frac{5}{4} \right) \right).$$

Similarly, the next two columns (p_5 and e_5) show the scale degree closest to the 3:2 perfect fifth and the error in cents:

$$e_5 = 1200 \left(\frac{p_5}{q} - \log_2 \left(\frac{3}{2} \right) \right).$$

The two columns after that (p_7 and e_7) show the scale degree closest to the 7:4 seventh harmonic and the error in cents:

$$e_7 = 1200 \left(\frac{p_7}{q} - \log_2 \left(\frac{7}{4} \right) \right).$$

We write e_{35} for the root mean square (RMS) error of the major third and perfect fifth:

$$e_{35} = \sqrt{(e_3^2 + e_5^2)/2}$$

and e_{357} for the RMS error for the major third, perfect fifth and seventh harmonic:

$$e_{357} = \sqrt{(e_3^2 + e_5^2 + e_7^2)/3}.$$

Theorem 6.2.3 shows that the quantity $e_5.q^2$ is a good measure of how well the perfect fifth is approximated by p_5/q of an octave, with respect to the number of notes in the scale. This theorem shows that there are infinitely many values of q for which $e_5.q^2 < 1200$, while on average we should expect this quantity to grow linearly with q.

Similarly, Theorem 6.2.5 with $k = 2$ shows that the quantity $e_{35}.q^{\frac{3}{2}}$ is a good measure of how well the major third and perfect fifth are simultaneously approximated, and shows that there are infinitely many values of q for which $e_{35}.q^{\frac{3}{2}} < 1200$, while on average we should expect this quantity to grow like the square root of q. Theorem 6.2.5 with $k = 3$ shows that the quantity $e_{357}.q^{\frac{4}{3}}$ is a good measure of how well all three intervals: major third, perfect fifth and seventh harmonic are simultaneously approximated, and shows that there are infinitely many values of q for which $e_{357}.q^{\frac{4}{3}} < 1200$, while on average we should expect this quantity to grow like the cube root of q.

Particularly good values of $e_5.q^2$, $e_{35}.q^{\frac{3}{2}}$ and $e_{357}.q^{\frac{4}{3}}$ are indicated in bold face in the last three columns of the table.

Appendix C

Frequency and MIDI chart

This table shows the frequencies and MIDI numbers of the notes in the standard equal tempered scale, based on the standard A4 = 440 Hz.

	MIDI	Hz	USA	Eur
piano ↑	108	4186.01	C8	c'''''
violin ↑	107	3951.07	B7	
	106	3729.31		
	105	3520.00	A7	
	104	3322.44		
	103	3135.96	G7	
	102	2959.96		
	101	2793.83	F7	
	100	2637.02	E7	
	99	2489.02		
flute ↑	98	2349.32	D7	
	97	2217.46		
	96	2093.00	C7	c''''
	95	1975.53	B6	
	94	1864.66		
	93	1760.00	A6	
	92	1661.22		
—	91	1567.98	G6	
	90	1479.98		
	89	1396.91	F6	
—	88	1318.51	E6	
leger	87	1244.51		
lines	86	1174.66	D6	
	85	1108.73		
—	84	1046.50	C6	c'''
	83	987.767	B5	
	82	932.328		
—	81	880.000	A5	
	80	830.609		
	79	783.991	G5	
	78	739.989		
⌐	77	698.456	F5	
|	76	659.255	E5	
|	75	622.254		
⊢	74	587.330	D5	
	73	554.365		
| treble	72	523.251	C5	c''
⊢ clef	71	493.883	B4	
|	70	466.164		
|	69	440.000	A4	
|	68	415.305		
⊢	67	391.995	G4	
|	66	369.994		
|	65	349.228	F4	
L	64	329.628	E4	
	63	311.127		
	62	293.665	D4	
	61	277.183		
middle c	60	261.626	C4	c'

	MIDI	Hz	USA	Eur
flute ↓	59	246.942	B3	
	58	233.082		
⌐	57	220.000	A3	
|	56	207.652		
violin ↓	55	195.998	G3	
|	54	184.997		
⊢	53	174.614	F3	
|	52	164.814	E3	
| bass	51	155.563		
⊢ clef	50	146.832	D3	
|	49	138.591		
|	48	130.813	C3	c
⊢	47	123.471	B2	
|	46	116.541		
|	45	110.000	A2	
|	44	103.826		
L	43	97.9989	G2	
	42	92.4986		
	41	87.3071	F2	
—	40	82.4069	E2	
	39	77.7817		
	38	73.4162	D2	
	37	69.2957		
—	36	65.4064	C2	C
leger	35	61.7354	B1	
lines	34	58.2705		
—	33	55.0000	A1	
	32	51.9131		
	31	48.9994	G1	
	30	46.2493		
—	29	43.6535	F1	
	28	41.2034	E1	
	27	38.8909		
	26	36.7081	D1	
	25	34.6478		
	24	32.7032	C1	C_1
	23	30.8677	B0	
	22	29.1352		
piano ↓	21	27.5000	A0	
	20	25.9565		
	19	24.4997	G0	
	18	23.1247		
	17	21.8268	F0	
	16	20.6017	E0	
	15	19.4454		
	14	18.3540	D0	
	13	17.3239		
	12	16.3516	C0	C_2
	11	15.4339		

Appendix D

Intervals

This is a table of intervals not exceeding one octave (or a tritave in the case of the Bohlen–Pierce, or BP scale). A much more extensive table may be found in Appendix XX to Helmholtz (1877) (page 453), which was added by the translator, Alexander Ellis. Names of notes in the BP scale are denoted with a subscript BP, to save confusion with notes which may have the same name in the octave based scale.

The first column is equal to 1200 times the logarithm to base two of the ratio given in the second column. Logarithms to base two can be calculated by taking the natural logarithm and dividing by $\ln 2$. So the first column is equal to

$$\frac{1200}{\ln 2} \approx 1731.234$$

times the natural logarithm of the second column.

We have given all intervals to three decimal places for theoretical purposes. While intervals of less than a few cents are imperceptible to the human ear in a melodic context, in harmony very small changes can cause large changes in beats and roughness of chords. Three decimal places gives great enough accuracy that errors accumulated over several calculations should not give rise to perceptible discrepancies.

If more accuracy is needed, I recommend using the multiple precision package bc (see page 364) with the -l option. The following lines can be made into a file to define some standard intervals in cents. For example, if the file is called music.bc then the command 'bc -l music.bc' will load them at startup.

```
scale=50 /* fifty decimal places - seems like plenty but you
   never know */
octave=1200
savart=1.2*l(10)/l(2)
syntoniccomma=octave*l(81/80)/l(2)
pythagoreancomma=octave*l(3^12/2^19)/l(2)
septimalcomma=octave*l(64/63)/l(2)
schisma=pythagoreancomma-syntoniccomma
diaschisma=syntoniccomma-schisma
perfectfifth=octave*l(3/2)/l(2)
equalfifth=700
meantonefifth=octave*l(5)/(4*l(2))
perfectfourth=octave*l(4/3)/l(2)
```

```
justmajorthird=octave*l(5/4)/l(2)
justminorthird=octave*l(6/5)/l(2)
justmajortone=octave*l(9/8)/l(2)
justminortone=octave*l(10/9)/l(2)
```

Cents	Interval ratio	Eitz	Name, etc.	Ref
0.000	1:1	C^0, C^0_{BP}	Fundamental	§4.1
1.000	$2^{\frac{1}{1200}}:1$		Cent	§5.4
1.805	$2^{\frac{1}{665}}:1$		Degree of 665 tone scale	§6.4
1.953	32805:32768	$B\sharp^{-1}$	Schisma	§5.8
3.986	$10^{\frac{1}{1000}}:1$		Savart	§5.4
14.191	245:243	C^{+1}_{BP}	BP-minor diesis	§6.7
19.553	2048:2025	$D\flat\flat^{+2}$	Diaschisma	§5.8
21.506	81:80	C^{+1}	Syntonic, or ordinary comma	§5.5
22.642	$2^{\frac{1}{53}}:1$		Degree of 53 tone scale	§6.3
23.460	$3^{12}:2^{19}$	$B\sharp^0$	Pythagorean comma	§5.2
27.264	64:63		Septimal comma	§5.8
35.099			Carlos' γ scale degree	§6.6
41.059	128:125	$D\flat\flat^{+3}$	Great diesis	§5.12
49.772	$7^{13}:3^{23}$	$D\flat\flat^0_{BP}$	BP 7/3 comma	§6.7
63.833			Carlos' β scale degree	§6.6
70.672	25:24	$C\sharp^{-2}$	Small (just) semitone	§5.5
77.965			Carlos' α scale degree	§6.6
90.225	256:243	$D\flat^0$	Diesis or limma	§5.2
100.000	$2^{\frac{1}{12}}:1$	$\approx C\sharp^{-\frac{7}{11}}$	Equal semitone	§5.14
111.731	16:15	$D\flat^{+1}$	Just minor semitone (ti–do, mi–fa)	§5.5
113.685	2187:2048	$C\sharp^0$	Pythagorean apotomē	§5.2
133.238	27:25	$D\flat^{-2}_{BP}$		§6.7
146.304	$3^{\frac{1}{13}}:1$		BP-equal semitone	§6.7
182.404	10:9	D^{-1}	Just minor tone (re–mi, so–la)	§5.5
193.157	$\sqrt{5}:2$	$D^{-\frac{1}{2}}$	Meantone whole tone	§5.12
200.000	$2^{\frac{1}{6}}:1$	$\approx D^{-\frac{2}{11}}$	Equal whole tone	§5.14

Cents	Interval ratio	Eitz	Name, etc.	Ref
203.910	9:8	D^0	Just major tone (do–re, fa–so, la–ti);	§5.5
			Pythagorean major tone;	§5.2
			Ninth harmonic	§4.1
294.135	32:27	$E\flat^0$	Pythagorean minor third	§5.2
300.000	$2^{\frac{1}{4}}{:}1$	$\approx E\flat^{+\frac{3}{11}}$	Equal minor third	§5.14
315.641	6:5	$E\flat^{+1}$	Just minor third (mi–so, la–do, ti–re)	§5.5
386.314	5:4	E^{-1}	Just major third (do–mi, fa–la, so–ti);	§5.5
			Meantone major third;	§5.12
			Fifth harmonic	§4.1
400.000	$2^{\frac{1}{3}}{:}1$	$\approx E^{-\frac{4}{11}}$	Equal major third	§5.14
407.820	81:64	E^0	Pythagorean major third	§5.2
498.045	4:3	F^0	Perfect fourth	§5.2
500.000	$2^{\frac{5}{12}}{:}1$	$\approx F^{+\frac{1}{11}}$	Equal fourth	§5.14
503.422	$2{:}5^{\frac{1}{4}}$	$F^{+\frac{1}{4}}$	Meantone fourth	§5.12
551.318	11:8		Eleventh harmonic	§4.1
600.000	$\sqrt{2}{:}1$	$\approx F\sharp^{-\frac{6}{11}}$	Equal tritone	§5.14
611.731	729:512	$F\sharp^0$	Pythagorean tritone	§5.2
696.579	$5^{\frac{1}{4}}{:}1$	$G^{-\frac{1}{4}}$	Meantone fifth	§5.12
700.000	$2^{\frac{7}{12}}{:}1$	$\approx G^{-\frac{1}{11}}$	Equal fifth	§5.14
701.955	3:2	G^0	Just and Pythagorean (perfect) fifth;	§5.2
			Third harmonic	§4.1
792.180	128:81	$A\flat^0$	Pythagorean minor sixth	§5.2
800.000	$2^{\frac{2}{3}}{:}1$	$\approx A\flat^{+\frac{4}{11}}$	Equal minor sixth	§5.14
813.687	8:5	$A\flat^{+1}$	Just minor sixth	§5.5
840.528	13:8		Thirteenth harmonic	§4.1
884.359	5:3	A^{-1}	Just major sixth	§5.5
889.735	$5^{\frac{3}{4}}{:}2$	$A^{-\frac{3}{4}}$	Meantone major sixth	§5.12
900.000	$2^{\frac{3}{4}}{:}1$	$\approx A^{-\frac{3}{11}}$	Equal major sixth	§5.14
905.865	27:16	A^0	Pythagorean major sixth	§5.2
968.826	7:4		Seventh harmonic	§4.1

Cents	Interval ratio	Eitz	Name, etc.	Ref
996.091	16:9	$B\flat^{0}$	Pythagorean minor seventh	§5.2
1000.000	$2^{\frac{5}{6}}{:}1$	$\approx B\flat^{+\frac{2}{11}}$	Equal minor seventh	§5.14
1082.892	$5^{\frac{5}{4}}{:}4$	$B^{-\frac{5}{4}}$	Meantone major seventh	§5.12
1088.269	15:8	B^{-1}	Just major seventh;	§5.5
			Fifteenth harmonic	§4.1
1100.000	$2^{\frac{11}{12}}{:}1$	$\approx B^{-\frac{5}{11}}$	Equal major seventh	§5.14
1109.775	243:128	B^{0}	Pythagorean major seventh	§5.2
1200.000	2:1	C^{0}	Octave; second harmonic	§4.1
1466.871	7:3	A^{0}_{BP}	BP-tenth	§6.7
1901.955	3:1	C^{0}_{BP}	BP-Tritave	§6.7

Appendix E

Just, equal and meantone scales compared

Figure E.1 has its horizontal axis measured in multiples of the (syntonic) comma, and the vertical axis measured in cents. Each vertical line represents a regular scale, generated by its fifth. The size of the fifth in the scale is equal to the Pythagorean fifth (ratio of 3:2, or 701.955 cents) minus the multiple of the comma given by the position along the horzontal axis. The three sloping lines show how far from the just values the fifth, major third and minor third are in these scales. This figure is relevant to Exercise **2** in Section 6.4.

It is worth noting that if $\frac{1}{11}$ comma meantone were drawn on this diagram, it would be indistinguishable from 12 tone equal temperament; see Section 5.14.

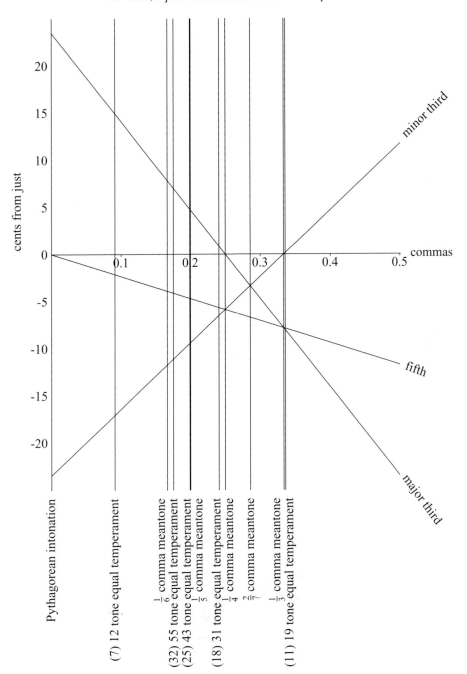

Figure E.1 Regular scales and their deviations from just intonation.

Appendix F

Music theory

This appendix consists of the background in elementary music theory needed to understand the main text. The emphasis is slightly different from that of a standard music text. We begin with the piano keyboard, as a convenient way to represent the modern scale (see Figure F.1 and also Appendix C).

Both the black and the white keys represent notes. This keyboard is periodic in the horizontal direction, in the sense that it repeats after seven white notes and five black notes. The period is one *octave*, which represents doubling the frequency corresponding to the note. The principle of *octave equivalence* says that notes differing by a whole number of octaves are regarded as playing equivalent roles in harmony. In practice, this is not completely true.

On a modern keyboard, each of the twelve intervals making up an octave represents the same frequency ratio, called a *semitone*. The name comes from the fact that two semitones make a *tone*. The twelfth power of the semitone's frequency ratio is a factor of 2:1, so a semitone represents a frequency ratio of $2^{\frac{1}{12}}:1$. The arrangement where all the semitones are equal in this way is called *equal temperament*. Frequency is an exponential function of position on the keyboard, and so the keyboard is really a *logarithmic* representation of frequency.

Because of this logarithmic scale, we talk about *adding* intervals when we want to *multiply* the frequency ratios. So, when we add a semitone to another semitone, for example, we get a tone with a frequency ratio of $2^{\frac{1}{12}} \times 2^{\frac{1}{12}}:1$ or $2^{\frac{1}{6}}:1$. This transition between additive and multiplicative notation can be a source of great confusion.

Staff notation works in a similar way, except that the logarithmic frequency is represented vertically, and the horizontal direction represents time. So music notation paper can be regarded as graph paper with a linear horizontal time axis and a logarithmic vertical frequency axis.

In Figure F.2, each note is twice the frequency of the previous one, so they are equally spaced on the logarithmic frequency scale (except for the break between the bass and treble clefs). The gap between adjacent notes is one octave, so the gap between the lowest and highest note is described *additively* as five octaves, representing a *multiplicative* frequency ratio of $2^5:1$.

There are two clefs on this diagram. The upper one is called the *treble clef*, with lines representing the notes E, G, B, D, F, beginning with the E two white notes above middle C and working up the lines. The spaces between them represent the notes F, A, C, E between them, so that this takes care of all the white notes between the E above middle C and the F

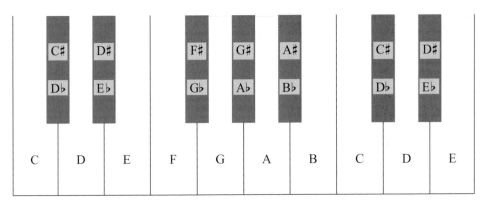

Figure F.1 Piano keyboard.

log (frequency)   time   ⟶

Figure F.2 Octave Cs.

an octave and a semitone above that. The black notes are represented by using the line or space with the likewise lettered white note with a sharp (♯) or flat (♭) sign in front.

The lower clef is called the bass clef, with lines representing the notes G, B, D, F, A, with the last note representing the A two white notes below middle C and the first note representing the G an octave and a tone below that.

Middle C itself is represented using a *leger line*, either below the treble clef or above the bass clef.

The frequency ratio represented by seven semitones, for example the interval from C to the G above it, is called a *perfect fifth*. Well, actually, this isn't quite true. A perfect fifth is supposed to be a frequency ratio of 3:2, or 1.5:1, whereas seven semitones on our modern equal tempered scale produce a frequency ratio of $2^{\frac{7}{12}}$:1 or roughly 1.4983:1. The perfect fifth is a consonant interval, just as the octave is, for reasons described in Chapter 4. So seven semitones is very close to a consonant interval. It is very difficult to discern the difference between a perfect fifth and an equal tempered fifth except by listening for beats; the difference is about one fiftieth of a semitone.

The perfect fourth represents the interval of 4:3, which is also consonant. The difference between a perfect fourth and the equal tempered fourth of five semitones is exactly the same as the difference between the perfect fifth and the equal tempered fifth, because they are obtained from the corresponding versions of a fifth by subtracting from an octave.

F G A B C D E F G A B C   =   C D E F G A B C D E F G

Figure F.3 The treble and bass clefs.

Figure F.4 C major chord.

Figure F.5 Inverted C major chord.

The frequency ratio represented by four semitones, for example the interval from C to the E above it, is called a *major third*. This represents a frequency ratio of $2^{\frac{4}{12}}:1$ or $\sqrt[3]{2}:1$, or roughly 1.25992:1. The *just major third* is defined to be the frequency ratio of 5:4 or 1.25:1. Again it is the just major third which represents the consonant interval, and the major third on our modern equal tempered scale is an approximation to it. The approximation is quite a bit worse than it was for the perfect fifth. The difference between a just major third and an equal tempered major third is quite audible; the difference is about one seventh of a semitone.

The frequency ratio represented by three semitones, for example the interval from E to the G above it, is called a *minor third*. This represents a frequency ratio of $2^{\frac{3}{12}}:1$ or $\sqrt[4]{2}:1$, or roughly 1.1892:1. The consonant *just minor third* is defined to be the frequency ratio of 6:5 or 1.2:1. The equal tempered minor third again differs from it by about a seventh of a semitone.

A major third plus a minor third makes up a fifth, either in the just/perfect versions or the equal tempered versions. So the intervals C to E (major third) plus E to G (minor third) make C to G (fifth). In the just/perfect versions, this gives ratios 4:5:6 for a *just major triad* C–E–G. We refer to C as the *root* of this chord. The chord is named after its root, so that is a C major chord.

If we used the frequency ratios 3:4:5, it would just give an *inversion* of this chord, which is regarded as a variant form of the C major chord, because of the principle of octave equivalence, while the frequency ratios 2:3:4 give a much simpler chord with a fifth and an octave, as shown in Figure F.6.

So the just major triad 4:5:6 is the chord that is basic to the western system of musical harmony. On an equal tempered keyboard, this is approximated with the chord $1:2^{\frac{4}{12}}:2^{\frac{7}{12}}$, which is a good approximation except for the somewhat sharp major third.

The *major scale* is formed by taking three major triads on three notes separated by intervals of a fifth. So, for example, the scale of C major is formed from the notes of the F

Figure F.6 Fifth and octave.

C	D	E	F	G	A	B	C	D
$\frac{1}{1}$	$\frac{9}{8}$	$\frac{5}{4}$	$\frac{4}{3}$	$\frac{3}{2}$	$\frac{5}{3}$	$\frac{15}{8}$	$\frac{2}{1}$	$\frac{9}{4}$

4 : 5 : 6 : (8)

4 : 5 : 6

(3) : 4 : 5 : 6

Figure F.7 Frequency ratios for C major in just intonation.

major, C major and G major triads. Between them, these account for the white notes on the keyboard, which make up the scale of C major. So, in just intonation, the C major scale would have the frequency ratios given in Figure F.7.

Here, we have made use of 2:1 octaves to transfer ratios between the right and left end of the diagram.

The basic problem with this scale is that the interval from D to A is almost, but not quite, equal to a perfect fifth. It is just close enough that it sounds like a nasty, out of tune fifth. It is short of a perfect fifth by a ratio of 81:80. This interval is called a *syntonic comma*. In this text, when we use the word comma without further qualification, it will always mean the syntonic comma. This and other commas are investigated in Section 5.8.

The *meantone* scale addresses this problem by distributing the syntonic comma equally between the four fifths C–G–D–A–E. So, in the meantone scale, the fifths are one quarter of a comma smaller than the perfect fifth, and the major thirds are just. In the meantone scale, a number of different keys work well, but the more remote keys do not. For further details, see Section 5.12.

To make all keys work well, the meantone scale must be bent to meet around the back. A number of different versions of this compromise have been used historically, the first ones being due to Werckmeister. Some of these well tempered scales are described in Section 5.13. Meantone and well tempered scales were in common use for about four centuries before equal temperament became widespread in the late nineteenth and early twentieth centuries.

A *minor triad* is obtained by inverting the order of the intervals in a major triad. So, for example, the minor triad on the note C consists of C, E♭ and G. In just intonation, the frequency ratios are 5:6 for C–E♭ and 4:5 for E♭–G, so that C–G still makes a perfect fifth. So the ratios are 10:12:15. See Section 5.6 for a discussion of the role of the minor triad. A *minor scale* can be built out of three minor triads in the same way as we did for the major scale, to give the frequency ratios in Figure F.8.

This is called the *natural minor* scale. Other forms of the minor scale occur because the sixth and seventh notes can be varied by moving one or both of them up a semitone to their major equivalents.

C	D	E♭	F	G	A♭	B♭	C	D
$\frac{1}{1}$	$\frac{9}{8}$	$\frac{6}{5}$	$\frac{4}{3}$	$\frac{3}{2}$	$\frac{8}{5}$	$\frac{9}{5}$	$\frac{2}{1}$	$\frac{9}{4}$

10 : 12 : 15

              10 : 12 : 15

                            10 : 12 : 15

Figure F.8 Frequency ratios for C minor.

   G♭   D♭   A♭   E♭   B♭   F   C   G   D   A   E   B   F♯

Figure F.9 Key signatures.

The concept of *key signature* arises from the following observation. If we look at major scales which start on notes separated by the interval of a fifth, then the two scales have all but one of the notes in common. For example, in C major, the notes are C–D–E–F–G–A–B–C, while in G major, the notes are G–A–B–C–D–E–F♯–G. The only difference, apart from a cyclic rearrangement of the notes, is that F♯ appears instead of F. So to indicate that we are in G major rather than C major, we write a sharp sign on the F at the beginning of each stave.

Similarly, the key of F major uses the notes F–G–A–B♭–C–D–E–F, which only differs from C major in the use of B♭ instead of B.

This means that key signatures are regarded as 'adjacent' if they begin on notes separated by a fifth. So the key signatures form a 'circle of fifths', as shown in Figure F.9.

In the above sequence of key signatures, the first and last are *enharmonic* versions of the same key. This means that in equal temperament, they are just different ways of writing the same keys, but in other systems such as meantone, the actual pitches may differ.

There is an easy way to memorize the correspondence between key signatures and the names of the major keys. For key signatures with sharps, the last sharp in the signature is the leading note of the key (i.e., a semitone below the note describing the key signature). So, for example, with four sharps, the last sharp is D♯ and so the key is E major. For key signatures with flats, the second to last flat gives the key signature. So, for example, with four flats, the second to last flat is A♭, so the key is A♭ major. The only case where this fails is if there is only one flat, and this is such a familiar key signature that most people find it easy to remember that it's F major.

The notes which occur in a natural minor scale are the same as the notes which occur in the major scale starting three semitones higher. For example, the notes of A minor are A–B–C–D–E–F–G–A. So the same key signature is used for A minor as for C major, and we say that A minor is the *relative minor* of C major.

The note on which a scale starts is called the *tonic*. The word *dominant* refers to the fifth above the tonic. The *roman numeral notation* is a device for naming triads relative to

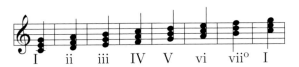

Figure F.10 Roman numeral notation.

the tonic. So, for example, the major triad on the dominant is written V. Upper case roman numerals refer to major triads and lower case to minor. So, for example, in C major, the chords are as shown in Figure F.10.

In D major, each chord would be a whole tone higher; so V would refer to the chord of A major instead of G major. So the roman numeral refers to the harmonic function of the chord within the key signature, rather than giving the absolute pitches.

The only triad here which is neither major nor minor is the *diminished* triad on the seventh note of the scale. This is denoted vii°, and consists of two intervals of a minor third with no major thirds.

Mode

The word *mode* refers to an arrangement of tones and semitones, with the tones approximately twice the size of the semitones (exact size depending on choice of scale), to form an octave. The naming of the modes can be a source of considerable confusion. The problem is that the names of the medieval church modes conflict with the names of the ancient Greek *tonoi*, because of a misreading of the ancient literature by some tenth century authors. The two definitions of *Hypodorian* agree, but then the medieval church modes go the wrong way around the circle.

Each mode can be considered to be the set of white keys on the piano, for a given choice of starting point. So, for example, *Hypodorian* goes from A to A, so that the arrangement of tones and semitones, from bottom to top, is TSTTSTT, like the minor scale. Of course, it should be realized that the pitches in a mode are not absolute, so the entire discussion can be transposed into any other key signature. For convenience, we stick to the 'white note' key signature of C.

The medieval church modes also come with a choice of *finalis* or *final note*, which would normally be used as the last note of the melody. The *authentic* modes start and end with the *finalis*, while the *plagal* mode has its *finalis* on the fourth note of the scale. The four choices of *finalis* were D, E, F, G, corresponding to the authentic modes *Dorian*, *Phrygian*, *Lydian* and *Mixolydian*. The prefix *Hypo-* then turns it into the plagal mode with the same *finalis*.

To add to the confusion, the sixteenth century Swiss theorist Glareanus added four more modes with *finalis* A and C, whose authentic forms he called *Aeolian* and *Ionian*. He did not consider B to be a valid choice of *finalis*, because the fifth above it has the wrong size. More information can be found in the excellent discussion of mode in Grout and Palisca, *A History of Western Music*, fifth edn, Norton, 1996.

We summarize with a table. The first column gives the pattern of semitones and tones, from the bottom to the top of the scale. The *finalis* column only refers to the medieval church modes, not to the Greek *tonoi*. The numbers 1 to 8 are used in most medieval treatises rather than the names, and 9 to 12 are from Glareanus. Modern books on music theory often use the names for numbers 1, 3, 5, 7, 9, 4 and 11 in the following table as their names of the modes.

obtained from Blackwood Enterprises, 5300 South Shore Drive, Chicago, IL 60615, USA for a nominal cost.

Dietrich Buxtehude, *Orgelwerke*, Volumes 1–7, recorded by Harald Vogel, published by Dabringhaus and Grimm. These works are recorded on a variety of European organs in different temperaments. Extensive details are given in the liner notes.

CD1 Tracks 1–8: Norden – St. Jakobi/Kleine organ in Werckmeister III;

Tracks 9–15: Norden – St. Ludgeri organ in modified $\frac{1}{5}$ Pythagorean comma meantone with $C\sharp^{-\frac{6}{5}p}$, $G\sharp^{-\frac{6}{5}p}$, $B\flat^{+\frac{1}{5}p}$ and $E\flat^{0}$;

CD2 Tracks 1–6: Stade – St. Cosmae organ in modified quarter comma meantone with[1] $C\sharp^{-\frac{3}{2}}$, $G\sharp^{-\frac{3}{2}}$, F^{0}, $B\flat^{0}$, $E\flat^{-\frac{1}{5}}$;

Tracks 7–15: Weener – Georgskirche organ in Werckmeister III;

CD3 Tracks 1–10: Grasberg organ in Neidhardt No. 3;

Tracks 11–14: Damp – Herrenhaus organ in modified meantone with pitches taken from original pipe lengths;

CD4 Tracks 1–8: Noordbroeck organ in Werckmeister III;

Tracks 9–15: Groningen – Aa-Kerk organ in (almost) equal temperament;

CD5 Tracks 1–5: Pilsum organ in modified $\frac{1}{5}$ Pythagorean comma meantone (the same as the Norden – St. Ludgeri organ described above);

Tracks 6–7: Buttforde organ;

Tracks 8–10: Langwarden organ in modified quarter comma meantone with $G\sharp^{-\frac{7}{4}}$, $B\flat^{-\frac{1}{4}}$, $E\flat^{-\frac{1}{4}}$;

Tracks 11–13: Basedow organ in quarter comma meantone;

Tracks 14–15: Groß Eichsen organ in quarter comma meantone;

CD6 Tracks 1–10: Roskilde organ in Neidhardt (no. 3?);

Track 11: Helsingør organ (unspecified temperament);

Tracks 12–15: Torrlösa organ (unspecified temperament);

CD7 Tracks 1–10 modified $\frac{1}{5}$ comma meantone with[2] $C\sharp^{-\frac{6}{5}}$, $G\sharp^{-\frac{6}{5}}$, $B\flat^{+\frac{1}{5}}$ and $E\flat^{\frac{1}{5}-\frac{1}{10}p}$.

William Byrd, *Cantones Sacrae 1575, The Cardinall's Music*, conducted by David Skinner. Track 12, *Diliges Dominum*, exhibits temporal reflectional symmetry, so that it is a perfect palindrome (see Section 9.1).

Wendy Carlos, *Beauty in the Beast*, Audion, 1986, Passport Records, Inc., SYNCD 200. Tracks 4 and 5 make use of Carlos' just scales described in Section 6.1.

Wendy Carlos, *Switched-On Bach 2000*, 1992. Telarc CD-80323. Carlos' original *Switched-On Bach* recording was performed on a Moog analog synthesizer, back in the late 1960s. The Moog is only capable of playing in equal temperament. Improvements in technology inspired her to release this new recording, using a variety of temperaments and modern methods of digital synthesis. The temperaments used are $\frac{1}{5}$ and $\frac{1}{4}$ comma meantone, and various circular (irregular) temperaments.

Wendy Carlos, *Tales of Heaven and Hell*, 1998. East Side Digital, ESD 81352. The third track, *Clockwork Black*, uses $\frac{1}{5}$th comma meantone temperament. The sixth track, *Afterlife*, uses 15 tone equal temperament, alternating with another more *ad hoc* scale. The seventh and final track uses a variation of Werckmeister III.

---

[1] The liner notes are written as though $G\sharp^{-\frac{3}{2}}$ were equal to $A\flat^{-\frac{2}{5}}$, which is not quite true. But the discrepancy is only about 0.2 cents.

[2] The liner notes identify $A\flat^{-\frac{1}{10}p}$ with $G\sharp^{-\frac{6}{5}}$, in accordance with the approximation of Kirnberger and Farey described in Section 5.14.

Charles Carpenter has two CDs, titled *Frog à la Pêche* (Caterwaul Records, CAT8221, 1994) and *Splat* (Caterwaul Records, CAT4969, 1996), composed using the Bohlen–Pierce scale, and played in a progressive rock/jazz style. Although Carpenter does not restrict himself to sounds composed mainly of odd harmonics, his compositions are nonetheless compelling.

Jacques Champion de Chambonnières, *Pièces pour Clavecin*, Françoise Lengellé, Clavecin. Lyrinx, LYR CD066, France. These pieces were recorded on copies of original harpsichords, tuned in quarter comma meantone, with A at 415 Hz.

Jane Chapman, *Beau Génie: Pièces de Clavecin from the Bauyn Manuscript, Vol. I*, Collins Classics 14202, 1994. These pieces were recorded on a 1614 Ruckers harpsichord, tuned in quarter comma meantone with A at 415 Hz.

Marc Chemillier and E. de Dampierre, *Central African Republic. Music of the Former Bandia Courts*, CNRS/Musée de l'Homme, Le Chant du Monde, CNR 2741009, Paris, 1996.

Perry Cook (ed.) (1999), *Music, Cognition and Computerized Sound. An Introduction to Psychoacoustics* comes with an accompanying CD full of sound examples.

Jean-Henry d'Anglebert, *Harpsichord Suites and Transcriptions*, Byron Schenkman, Harpsichord. Centaur CRC 2435, 1999. These pieces were recorded on a copy of an original 1638 harpsichord, tuned in quarter comma meantone.

Johann Jakob Froberger, *The Complete Keyboard Works*, Richard Egarr, Harpsichord and Organ. Globe GLO 6022–5, 1994. The organ works in this collection were recorded on the organ at St. Martin's Church in Cuijk, tuned in 1/5 comma meantone with A at 413 Hz. The suites for harpsichord were recorded in 'the tuning described by Marin Mersenne in his *Harmonie Universelle* of 1636 (generally known as '*Ordinaire*')'. The remaining harpsichord works were recorded in quarter comma meantone. The harpsichords were tuned with A at 415 Hz.

Lou Harrison, *Complete Harpsichord Works; Music for Tack Piano and Fortepiano; in Historic and Experimental Tunings*, New Albion Records (2002). Linda Burman-Hall, solo keyboards. The pieces on this recording are: *A Sonata for Harpsichord* (Kirnberger II with A at 415 Hz), *Village Music* (a well temperament with A at 415 Hz), *Six Sonatas for Cembalo* (Werckmeister III with A at 440 Hz), *Instrumental Music for Corneille's 'Cinna'* (7 limit just intonation), *A Summerfield Set* (Werckmeister III), *Triphony* (modified well temperament based on Charles, Earl of Stanhope), *A Twelve-tone Morning After to Amuse Henry*, and *Largo Ostinato* (both in the same unspecified temperament based on tuning its core sonorities in just intonation).

Michael Harrison, *From Ancient Worlds*, for Harmonic Piano, New Albion Records, Inc., 1992. NA 042 CD. The pieces on this recording all make use of his 24 tone just scale, described in Section 6.1.

Michael Harrison has also released another CD using his Harmonic Piano, *Revelation*, recorded live in the Lincoln Center in October 2001 and issued in January 2002. In this recording, the harmonic piano is tuned to a just scale using only the primes 2, 3 and 7 (not 5). The 12 notes in the octave have ratios

$$1:1, 63:64, 9:8, 567:512, 81:64, 21:16, 729:512, 3:2,$$
$$189:128, 27:16, 7:4, 243:128, (2:1).$$

The scale begins on F, and has the peculiarity that ♯ lowers a note by a septimal comma.

Jonathan Harvey, *Mead: Ritual Melodies*, Sargasso CD #28029, 1999. Track two on this CD, *Mortuos Plango, Vivos Voco*, makes use of a scale derived from a spectral analysis of the Great Bell of Winchester Cathedral.

Neil Haverstick, *Acoustic Stick*, Hapi Skratch, 1998. The pieces on this CD are played on custom made guitars using 19 and 34 tone equal temperament.

In Joseph Haydn's *Sonata 41* in A (Hob. XVI:26), the movement *Menuetto al Rovescio* is a perfect palindrome (see Section 9.1). This piece can be found as track 16 on the Naxos CD number 8.553127, Haydn, *Piano Sonatas, Vol. 4*, with Jenõ Jandó at the piano.

A. J. M. Houtsma, T. D. Rossing and W. M. Wagenaars, *Auditory Demonstrations*, Audio CD and accompanying booklet, Philips, 1987. This classic collection of sound examples illustrates a number of acoustic and psychoacoustic phenomena. It can be obtained from the Acoustical Society of America at asa.aip.org/discs.html for $26 + shipping.

Ben Johnson, *Music for Piano*, played by Phillip Bush, Koch International Classics CD #7369. Pieces for piano in a microtonal just scale.

Enid Katahn, *Beethoven in the Temperaments* (Gasparo GSCD-332, 1997). Katahn plays Beethoven's Sonatas Op. 13, *Pathétique* and Op. 14 Nr. 1 using the Prinz temperament, and Sonatas Op. 27 Nr. 2, *Moonlight* and Op. 53 *Waldstein* in Thomas Young's temperament. The instrument is a modern Steinway concert grand rather than a period instrument. The tuning and liner notes are by Edward Foote.

Enid Katahn and Edward Foote have also brought out a recording, *Six Degrees of Tonality* (Gasparo GSCD-344, 2000). This begins with Scarlatti's *Sonata* K. 96 in quarter comma meantone, followed by Mozart's *Fantasie* K. 397 in Prelleur temperament, a Haydn sonata in Kirnberger III, a Beethoven sonata in Young temperament, Chopin's *Fantaisie-Impromptu* in DeMorgan temperament, and Grieg's *Glochengeläute* in Coleman 11 temperament. Finally, and in many ways the most interesting part of this recording, the Mozart *Fantasie* is played in quarter comma meantone, Prelleur temperament and equal temperament in succession, which allows a very direct comparison to be made. Unfortunately, the tempi are slightly different, which makes this recording not very useful for a blind test.

Bernard Lagacé has recorded a CD of music of various composers on the C. B. Fisk organ at Wellesley College, Massachusetts, USA, tuned in quarter comma meantone temperament. This recording is available from Titanic Records Ti-207, 1991.

Guillaume de Machaut (1300–1377), *Messe de Notre Dame* and other works. The Hilliard Ensemble, Hyperíon, 1989, CDA66358. This recording is sung in Pythagorean intonation throughout. The mass alternates polyphonic with monophonic sections. The double leading-note cadences at the end of each polyphonic section are particularly striking in Pythagorean intonation. Track 19 of this recording is *Ma fin est mon commencement* (My end is my beginning). This is an example of retrograde canon, meaning that it exhibits temporal reflectional symmetry (see Section 9.1).

Mathews and Pierce (1989), *Current Directions in Computer Music Research* comes with a companion CD containing numerous examples; note that track 76 is erroneous, cf. Pierce (1983), page 257 of 2nd edn.

Microtonal Works, Mode CD #18, contains microtonal works of Joan la Barbara, John Cage, Dean Drummond and Harry Partch.

Edward Parmentier, *Seventeenth Century French Harpsichord Music*, Wildboar, 1985, WLBR 8502. This collection contains pieces by Johann Jakob Froberger, Louis Couperin, Jacques Champion de Chambonnières, and Jean-Henri d'Anglebert. The recording was made using a Keith Hill copy of a 1640 harpsichord by Joannes Couchet, tuned in $\frac{1}{3}$ comma meantone temperament.

Many of Harry Partch's compositions have been rereleased on CD by Composers Recordings Inc., 73 Spring Street, Suite 506, New York, NY 10012-5800. As a starting point, I would recommend *The Bewitched*, CRI CD 7001, originally released on Partch's own label, Gate 5. This piece makes extensive use of his 43 tone just scale, described in Section 6.1.

A number of Robert Rich's recordings are in some form of just scale. His basic scale is mostly 5-limit with a 7:5 tritone:

$$1:1, 16:15, 9:8, 6:5, 5:4, 4:3, 7:5, 3:2, 8:5, 5:3, 9:5, 15:8.$$

This appears throughout the CDs *Numena*, *Geometry*, *Rainforest*, and others. One of the nicest examples of this tuning is *The Raining Room* on the CD *Rainforest*, Hearts of Space HS11014-2. He also uses the 7-limit scale

$$1:1, 15:14, 9:8, 7:6, 5:4, 4:3, 7:5, 3:2, 14:9, 5:3, 7:4, 15:8.$$

This appears on *Sagrada Familia* on the CD *Gaudi*, Hearts of Space HS11028-2.

William Sethares, *Xentonality*, Music in 10-, 13-, 17- and 19-tone equal temperament using spectrally adjusted instruments. Frog Peak Music www.frogpeak.org, 1997.

William Sethares (1998), *Tuning, Timbre, Spectrum, Scale* comes with a CD full of examples.

Isao Tomita, *Pictures at an Exhibition* (Mussorgsky), BMG 60576-2-RG. This recording was made on analogue synthesizers in 1974, and is remarkably sophisticated for that era.

Johann Gottfried Walther, *Organ Works*, Volume 1, played by Craig Cramer on the organ of St. Bonifacius, Tröchtelborn, Germany. Naxos CD number 8.554316. This organ was restored in Kellner's reconstruction of Bach's temperament, see Section 5.13. For more information about the organ (details are not given in the CD liner notes), see www.gdo.de/neurest/troechtelborn.html.

Aldert Winkelman, *Works by Mattheson, Couperin, and Others*. Clavigram VRS 1735-2. This recording is hard to obtain. The pieces by Johann Mattheson, François Couperin, Johann Jakob Froberger, Joannes de Gruytters and Jacques Duphly are played on a harpsichord tuned to Werckmeister III. The pieces by Louis Couperin and Gottlieb Muffat are played on a spinet tuned in quarter comma meantone.

References

G. Assayag, H. G. Feichtinger and J. F. Rodrigues (eds.), 2002. *Mathematics and Music, a Diderot Mathematical Forum*, Springer-Verlag.

Pierre-Yves Asselin, 1997. *Musique et tempérament*, Éditions Costallat, 1985; reprinted by Jobert.

J. Murray Barbour, 1951. *Tuning and Temperament, a Historical Survey*, Michigan State College Press. Reprinted by Dover, 2004.

James Beament, 1997. *The Violin Explained: Components, Mechanism, and Sound*, Oxford University Press.

Georg von Békésy, 1960. *Experiments in Hearing*, McGraw-Hill.

Richard Charles Boulanger (ed.), 2000. *The CSound Book: Perspectives in Software Synthesis, Sound Design, Signal Processing, and Programming*, MIT Press.

Murray Campbell and Clive Greated, 1986. *The Musician's Guide to Acoustics*, Oxford University Press, reprinted 1998.

John Chowning and David Bristow, 1986. *FM Theory and Applications*, Yamaha Music Foundation.

Thomas Christensen (ed.), 2002. *The Cambridge History of Western Music Theory*, Cambridge University Press.

Perry R. Cook (ed.), 1999. *Music, Cognition, and Computerized Sound. An Introduction to Psychoacoustics*, MIT Press.

Lothar Cremer, 1984. *The Physics of the Violin*, MIT Press.

Diana Deutsch (ed.), 1982. *The Psychology of Music*, Academic Press; 2nd edn, 1999.

B. Chaitanya Deva, 1981. *The Music of India: a Scientific Study*, Munshiram Manoharlal Publishers Pvt. Ltd.

Dominique Devie, 1990. *Le tempérament musical: philosophie, histoire, théorie et practique*, Société de musicologie du Languedoc Béziers.

William C. Elmore and Mark A. Heald, 1969. *Physics of Waves*, McGraw-Hill. Reprinted by Dover, 1985.

Neville H. Fletcher and Thomas D. Rossing, 1991. *The Physics of Musical Instruments*, Springer-Verlag.

Allen Forte, 1973. *The Structure of Atonal Music*, Yale University Press.

Steve De Furia and Joe Scacciaferro, 1989. *MIDI Programmer's Handbook*, M & T Publishing, Inc.

H. Genevois and Y. Orlarey, 1997. *Musique & Mathématiques*, Aléas–Grame.

Karl F. Graff, 1975. *Wave Motion in Elastic Solids*, Oxford University Press. Reprinted by Dover, 1991.

R. W. Hamming, 1989. *Digital Filters*, Prentice Hall. Reprinted by Dover Publications, 1998.

G. H. Hardy and E. M. Wright, 1980. *An Introduction to the Theory of Numbers*, Oxford University Press, fifth edn.

Hermann Helmholtz, 1877. *Die Lehre von den Tonempfindungen*, Longmans & Co., fourth German edn. Translated by Alexander Ellis as *On the Sensations of Tone*, Dover, 1954 (and reprinted many times).

Hua, 1982. *Introduction to Number Theory*, Springer-Verlag.

Ian Johnston, 1989. *Measured Tones: the Interplay of Physics and Music*, Institute of Physics Publishing.

Owen H. Jorgensen, 1991. *Tuning*, Michigan State University Press.

Michael Keith, 1991. *From Polychords to Pólya; Adventures in Musical Combinatorics*, Vinculum Press.

T. W. Körner, 1988. *Fourier Analysis*, Cambridge University Press, reprinted 1990.

Patricia Kruth and Henry Stobart (eds.), 2000. *Sound*, Cambridge University Press.

Marc Leman, 1997. *Music, Gestalt, and Computing; Studies in Cognitive and Systematic Musicology*, Lecture Notes in Computer Science, vol. 1317, Springer-Verlag.

Ernő Lendvai, 1993. *Symmetries of Music*, Kodály Institute, Kecskemét.

Carl E. Linderholm, 1971. *Mathematics Made Difficult*, Wolfe Publishing, Ltd.

Mark Lindley and Ronald Turner-Smith, 1993. *Mathematical Models of Musical Scales*, Verlag für systematische Musikwissenschaft GmbH.

Max V. Mathews and John R. Pierce, 1989. *Current Directions in Computer Music Research*, MIT Press. Reprinted 1991.

Brian C. J. Moore, 1997. *Psychology of Hearing*, Academic Press.

F. Richard Moore, 1990. *Elements of Computer Music*, Prentice Hall.

Philip M. Morse and K. Uno Ingard, 1968. *Theoretical Acoustics*, McGraw Hill. Reprinted with corrections by Princeton University Press, 1986.

Bernard Mulgrew, Peter Grant and John Thompson, 1999. *Digital Signal Processing*, Macmillan Press.

Cornelius Johannes Nederveen, 1998. *Acoustical Aspects of Woodwind Instruments*, Northern Illinois Press.

Harry Partch, 1974. *Genesis of a Music*, Second edn, enlarged. Da Capo Press, 1979 (pbk).

George Perle, 1977. *Twelve-tone Tonality*, University of California Press. Second edn, 1996.

James O. Pickles, 1988. *An Introduction to the Physiology of Hearing*, Academic Press, second edn.

John Robinson Pierce, 1983. *The Science of Musical Sound*, Scientific American Books; 2nd edn, W. H. Freeman & Co, 1992.

Jean-Philippe Rameau, 1722. *Traité de l'harmonie*, Ballard. Reprinted as *Treatise on Harmony* in English translation by Dover, 1971.

J. W. S. Rayleigh, 1896. *The Theory of Sound* (2 vols.), Second edn, Macmillan. Dover, 1945.

Curtis Roads, 1989. *The Music Machine. Selected Readings from Computer Music Journal*, MIT Press.

Curtis Roads, 1996. *The Computer Music Tutorial*, MIT Press.

Curtis Roads, 2001. *Microsound*, MIT Press.

Curtis Roads, Stephen Travis Pope, Aldo Piccialli and Giovanni De Poli (eds.), 1997. *Musical Signal Processing*, Swets & Zeitlinger Publishers.

Curtis Roads and John Strawn (eds.), 1985. *Foundations of Computer Music. Selected Readings from Computer Music Journal*, MIT Press.

Thomas D. Rossing (ed.), 1988. *Musical Acoustics, Selected Reprints*, American Association of Physics Teachers.

Thomas D. Rossing, 1990. *The Science of Sound*, Addison-Wesley, Reading, Mass., Second edn.

Thomas D. Rossing, 2000. *Science of Percussion Instruments*, World Scientific.

Joseph Rothstein, 1992. *MIDI, a Comprehensive Introduction*, Oxford University Press.

William A. Sethares, 1998. *Tuning, Timbre, Spectrum, Scale*, Springer-Verlag.

Stan Tempelaars, 1996. *Signal Processing, Speech and Music*, Swets & Zeitlinger Publishers.

Martin Vogel, 1991. *Die Naturseptime*, Verlag für systematische Musikwissenschaft.

G. N. Watson, 1922. *A Treatise on the Theory of Bessel Functions*, Cambridge University Press. Reprinted, 1996.

Joseph Yasser, 1932. *A Theory of Evolving Tonality*, American Library of Musicology, Inc.

William A. Yost, 1977. *Fundamentals of Hearing. an Introduction*, Academic Press.

Eberhard Zwicker and H. Fastl, 1999. *Psychoacoustics: Facts and Models*, Springer-Verlag, second edn.

Bibliography

John Backus, 1969. *The Acoustical Foundations of Music*, W. W. Norton & Co. Reprinted 1977.

Patrice Bailhache, 2001. *Une histoire de l'acoustique musicale*, CNRS Éditions.

Scott Beall, 2000. *Functional Melodies: Finding Mathematical Relationships in Music*, Key Curriculum Press.

James Beament, 2001. *How We Hear Music: the Relationship Between Music and the Hearing Mechanism*, The Boydell Press.

Arthur H. Benade, 1990. *Fundamentals of Musical Acoustics*, Oxford University Press. Reprinted by Dover, 1990.

Richard E. Berg and David G. Stork, 1982. *The Physics of Sound*, Prentice-Hall. Second edn, 1995.

Easley Blackwood, 1985. *The Structure of Recognizable Diatonic Tunings*, Princeton University Press.

Pierre Buser and Michel Imbert, 1992. *Audition*, MIT Press.

Peter Castine, 1994. *Set Theory Objects: Abstractions for Computer-aided Analysis and Composition of Serial and Atonal Music*, European University Studies, vol. 36, Peter Lang Publishing.

David Colton, 1988. *Partial Differential Equations, an Introduction*, Random House.

Deryck Cooke, 1959. *The Language of Music*, Oxford University Press, reprinted in paperback, 1990.

David H. Cope, 1989. *New Directions in Music*, Wm. C. Brown Publishers, fifth edn. Sixth edn, Waveland Press, 1998.

David H. Cope, 1991. *Computers and Musical Style*, Oxford University Press.

David H. Cope, 1996. *Experiments in Musical Intelligence*, Computer Music and Digital Audio, vol. 12, A-R Editions.

David H. Cope, 2001. *Virtual Music*, MIT press, 2001.

Malcolm J. Crocker (ed.), 1998. *Handbook of Acoustics*, Wiley Interscience.

Alain Daniélou, 1967. *Sémantique musicale. Essai de psycho-physiologie auditive*, Hermann, Reprinted 1978.

Alain Daniélou, 1995. *Music and the Power of Sound*, Inner Traditions, revised from a 1943 publication.

Peter Desain and Henkjan Honig, 1992. *Music, Mind and Machine: Studies in Computer Music, Music Cognition, and Artificial Intelligence (Kennistechnologie)*, Thesis Publishers.

Charles Dodge and Thomas A. Jerse, 1997. *Computer Music: Synthesis, Composition, and Performance*, Simon & Schuster, second edn.

W. Jay Dowling and Dane L. Harwood, 1986. *Music Cognition*, Academic Press Series in Cognition and Perception.

Laurent Fichet, 1996. *Les théories scientifiques de la musique aux XIXe et XXe siècles*, Librairie J. Vrin.

Trudi Hammel Garland and Charity Vaughan Kahn, 1995. *Math and Music: Harmonious Connections*, Dale Seymore Publications.

Ben Gold and Nelson Morgan, 2000. *Speech and Audio Signal Processing: Processing and Perception of Speech and Music*, Wiley & Sons.

Heinz Götze and Rudolf Wille (eds.), 1995. *Musik und Mathematik. Salzburger Musikgespräch 1984 unter Vorsitz von Herbert von Karajan*, Springer-Verlag.

Penelope Gouk, 1999. *Music, Science and Natural Magic in Seventeenth-century England*, Yale University Press.

Niall Griffith and Peter M. Todd (eds.), 1999. *Musical Networks: Parallel Distributed Perception and Performance*, MIT Press.

Donald E. Hall, 1980. *Musical Acoustics*, Wadsworth Publishing Company.

W. M. Hartmann, 1998. *Signals, Sound and Sensation*, Springer-Verlag.

Michael Hewitt, 2000. *The Tonal Phoenix; a Study of Tonal Progression Through the Prime Numbers Three, Five and Seven*, Verlag für systematische Musikwissenschaft GmbH.

Douglas R. Hofstadter, 1979. *Gödel, Escher, Bach*, Harvester Press. Reprinted by Basic Books, 1999.

David M. Howard and James Angus, 1996. *Acoustics and Psychoacoustics*, Focal Press.

Stuart M. Isacoff, 2001. *Temperament: the Idea that Solved Music's Greatest Riddle*, Knopf; paperback 2003.

Sir James Jeans, 1937. *Science & Music*, Cambridge University Press. Reprinted by Dover, 1968.

Franck Jedrzejewski, 2002. *Mathématiques des systèmes acoustiques: Tempéraments et modèles contemporains*, L'Harmattan.

Franck Jedrzejewski, 2004. *Dictionnaire des musiques microtonales*, L'Harmattan.

Jeffrey Johnson, 1997. *Graph Theoretical Methods of Abstract Musical Transformation*, Greenwood Publishing Group.

Tom Johnson, 1996. *Self-similar Melodies*, Editions 75.

Lawrence E. Kinsler, Austin R. Frey, Alan B. Coppens and James V. Sanders, 2000. *Fundamentals of Acoustics*, John Wiley & Sons, fourth edn.

Albino Lanciani, 2001. *Mathématiques et musique. Les Labyrinthes de la phénoménologie*, Éditions Jérôme Millon.

J. Lattard, 1988. *Gammes et tempéraments musicaux*, Masson.

J. Lattard, 2003. *Intervalle, échelles, tempéraments et accordages musicaux: De Pythagore à la simulation informatique*, L'Harmattan.

Marc Leman, 1995. *Music and Schema Theory: Cognitive Foundations of Systematic Musicology*, Springer Series on Information Science, vol. 31, Springer-Verlag.

David Lewin, 1987. *Generalized Musical Intervals and Transformations*, Yale University Press.

Llewelyn S. Lloyd and Hugh Boyle, 1963. *Intervals, Scales and Temperaments*, Macdonald.

R. Duncan Luce, 1993. *Sound and Hearing, a Conceptual Introduction*, Lawrence Erlbaum Associates, Inc.

Charles Madden, 1999. *Fractals in Music – Introductory Mathematics for Musical Analysis*, High Art Press.

Max V. Mathews, 1969. *The Technology of Computer Music*, MIT Press.

W. A. Mathieu, 1997. *Harmonic Experience*, Inner Traditions International.

Guerino Mazzola, 1985. *Gruppen und Kategorien in der Musik*, Heldermann-Verlag.

Guerino Mazzola, 1990. *Geometrie der Töne: Elemente der Mathematischen Musiktheorie*, Birkhäuser.

Guerino Mazzola, with contributions by Stefan Göller, Stefan Müller and Karin Ireland, 2002. *The Topos of Music: Geometric Logic of Concepts, Theory, and Performance*, Birkhäuser.

Ernest G. McClain, 1976. *The Myth of Invariance: the Origin of the Gods, Mathematics and Music from the R̩g Veda to Plato*, Nicolas-Hays, Inc. Paperback edn, 1984.

Joseph Morgan, 1980. *The Physical Basis of Musical Sounds*, Robert E. Krieger Publishing Company.

Erich Neuwirth, 1997. *Musical Temperaments*, Springer-Verlag.

Harry F. Olson, 1952. *Musical Engineering*, McGraw Hill. Revised and enlarged version, Dover, 1967, with new title: *Music, Physics and Engineering*.

Jack Orbach, 1999. *Sound and Music*, University Press of America.

Charles A. Padgham, 1986. *The Well-tempered Organ*, Positif Press.

Hermann Pfrogner, 1976. *Lebendige Tonwelt*, Langen Müller.

Dave Phillips, 2000. *Linux Music and Sound*, Linux Journal Press.

Ken C. Pohlmann, 2000. *Principals of Digital Audio*, McGraw-Hill, fourth edn.

Giovanni De Poli, Aldo Piccialli and Curtis Roads (eds.), 1991. *Representations of Musical Signals*, MIT Press.

Stephen Travis Pope (ed.), 1991. *The Well-tempered Object: Musical Applications of Object-oriented Software Technology*, MIT Press.

Daniel R. Raichel, 2000. *The Science and Applications of Acoustics*, American Institute of Physics.

Joan Reinthaler, 1990. *Mathematics and Music: Some Intersections*, Mu Alpha Theta.

Geza Révész, 1946. *Einführung in die Musikpsychologie*, Amsterdam. Translated by G. I. C. de Courcy as *Introduction to the Psychology of Music*, University of Oklahoma Press, 1954, and reprinted by Dover, 2001.

John S. Rigden, 1977. *Physics and the Sound of Music*, Wiley & Sons. Second edn, 1985.

Juan G. Roederer, 1995. *The Physics and Psychophysics of Music*, Springer-Verlag.

Thomas D. Rossing (ed.), 1984. *Acoustics of Bells*, Van Nostrand Reinhold.

Thomas D. Rossing and Neville H. Fletcher (contributor), 1995. *Principles of Vibration and Sound*, Springer-Verlag.

Heiner Ruland, 1992. *Expanding Tonal Awareness*, Rudolf Steiner Press.

Joseph Schillinger, 1941. *The Schillinger System of Musical Composition,* Carl Fischer, Inc. Reprinted by Da Capo Press, 1978.

Albrecht Schneider, 1997. *Tonhöhe, Skala, Klang: akustiche, tonometrische und psychoakustische Studien auf Vergleichender Grundlage*, Verlag für systematische Musikwissenschaft.

Günter Schnitzler, 1976. *Musik und Zahl*, Verlag für systematische Musikwissenschaft.

Ken Steiglitz, 1996. *A Digital Signal Processing Primer: with Applications to Digital Audio and Computer Music*, Addison-Wesley.

Reinhard Steinberg (ed.), 1995. *Music and the Mind Machine. the Psychophysiology and Psychopathology of the Sense of Music*, Springer-Verlag.

Charles Taylor, 1992. *Exploring Music: the Science and Technology of Tones and Tunes*, Institute of Physics Publishing. Reprinted 1994.

David Temperley, 2001. *The Cognition of Basic Musical Structures*, MIT Press.

Martin Vogel, 1975. *Die Lehre von den Tonbeziehungen*, Verlag für systematische Musikwissenschaft.

Martin Vogel, 1984. *Anleitung zur harmonischen Analyse und zu reiner Intonation*, Verlag für systematische Musikwissenschaft.

Scott R. Wilkinson, 1988. *Tuning in: Microtonality in Electronic Music*, Hal Leonard Books.

Fritz Winckel, 1967. *Music, Sound and Sensation, a Modern Exposition*, Dover.

Iannis Xenakis, 1971. *Formalized Music: Thought and Mathematics in Composition*, Indiana University Press. Revised edn (four new chapters and a new appendix), Pendragon Press, 1992. Paperback edn, 2001.

Index

Bold page numbers indicate a main entry.

393

Menuetto al rovescio (Haydn), 321
Mercer, Johnny (1909–1976), 184
Mercury, rotation of, 222
Mersenne, Marin (1588–1648), 37
 improved meantone temperament, 192
 law of stretched strings, 99
 picture, 99
 pitch and frequency, 149
 spinet/lute tunings, 176
message, system exclusive, 251
Messiaen, Olivier (1908–1992), 358
Metamagical Themas (Hofstadter), 223
MetaPost, xiii
Meyer, Alfred, 12
middle ear, 7
MIDI, 251, 291, 367
 baud rate, 251
 files, 188
 to CSound, 291
MIDI2CS, 291
Miller, James Charles Percy
 algorithm, 363
minor
 scale, 163
 semitone, 163, 226
  just, 369
 seventh, 163
 sixth, 163, 227
  just, 168
 third, 163, 227
  just, 152, 168
 tone, just, 369
 triad, 169, 377
  just, 173
Mirror Duet (attr. Mozart), 315
missing fundamental, 6, 117, 157
MIT Media Lab, 291
Mixolydian, 379
mode, 379
 Dorian, 202
 vibrational, 20
modelling, physical, 270
Modern Major General, 36
modification, 170
modiolus, 9
modulation
 amplitude, 276
 frequency, 276
 index of, 281
 pulse width, 71, 268
 ring, 277
modulus
 bulk, 108, 142
 Young's, 125, 133
moment
 bending, 124
 sectional, 126, 133
Mongean shuffle, 359
monochord
 Agricola's, 175
 BP, 236

de Caus's, 175
Erlangen, 174
Euler's, 178, 240
Fogliano's, 175
Gibelius', 186
Kepler's, 175
Malcolm's, 178
Marpurg's, 177
Montvallon's, 178
Ramis', 174
Romieu's, 178
Rousseau's, 179
monomorphism, 332
Monteverdi, Claudio (1567–1643), 204
Montvallon, André Barrigue de
 monochord, 178
Moog, Robert A. (1934–)
 synthesizer, 207, 382
Moon, 222
Moonlight Sonata (Beethoven), 312
motion
 Brownian, 47, 81
 circular, 26
 damped harmonic, 17, 28
 harmonic, 18
 planetary, 68, 147
 simple harmonic, 17
Mozart, Wolfgang Amadeus (1756–1791)
 Fantasie (K. 397), 184, 189, 198, 384
 pitch, 21
 Sinfonia Concertante, 190
 Sonata (K. 333), 183
 Spiegel, 313
MP3 sound file, 248
MPEG, 249
MPEG 4 Audio, 250
MS-DOS, 291
Muffat, Gottlieb (1690–1770), 189
multiplication table, 325
Musæ Sioniæ (M. Praetorius), 16
music
 atonal, 205, 348
 baroque, 181
 digital, 245
 folk, 181
 Greek, 202
 of the spheres, 147
 polyphonic, 203
 rock, 181
 romantic, 181
 theory, 374
 twelve tone, 205
Musical Offering (J. S. Bach), 314
Musical World, 161
MusicTeX, xiii
Mussorgsky, Modest (1839–1881), 266, 385

N (sample rate), 251
nabla squared (∇^2), 118
Nachbaur, Fred, 313